# Advanced Systems Thinking, Engineering, and Management

For a listing of recent titles in the *Artech House Technology Management and Professional Development Library*, turn to the back of this book.

# Advanced Systems Thinking, Engineering, and Management

Derek K. Hitchins

Artech House
Boston • London
www.artechhouse.com

Library of Congress Cataloging-in-Publication Data
Hitchins, Derek K.
   Advanced systems thinking, engineering, and management / Derek K. Hitchins.
       p. cm. — (Artech House technology management and professional development
   library)
   Includes bibliographical references and index.
   ISBN 1-58053-619-0 (alk. paper)
   1. System analysis. 2. Industrial management. 3. Technological complexity. I. Title
II. Series.

T57.6.H58 2003
004.2'1—dc22

                                                                    2003057714


British Library Cataloguing in Publication Data
Hitchins, Derek K.
   Advanced systems thinking, engineering, and management. —
   (Artech House technology management and professional development library)
   1. Systems engineering
   I. Title
   620'.001171

   ISBN 1-58053-619-0

Cover design by Gary Ragaglia

© 2003 ARTECH HOUSE, INC.
685 Canton Street
Norwood, MA 02062

International Standard Book Number: 1-58053-619-0
Library of Congress Catalog Card Number: 2003057714

10 9 8 7 6 5 4 3 2 1

To my beloved wife,
without whom…
very little

# Contents

# Preface

"Systems" brings different things to mind: to some it means computers, to others it is the nonmechanical parts of an aircraft or ship, while to yet others it invokes softer ideas of organization and method.

Systems thinking is relatively recent, at least as an identifiable practice. There are books on the subject, but different authors view the subject differently, so it is an as-yet unconstrained discipline.

Systems engineering is not new, so you might imagine that by now it is an established discipline. Not so. To some, it is a multidiscipline; an amalgam intended to cover the cracks (interfaces?) between electrical, electronic, mechanical, instrument, aeronautical, civil, and many other engineering disciplines. To others, it is a metadiscipline, and sits somehow "above" those engineering disciplines, drawing them together under a single management umbrella. Oh! And by the way, it is just common sense....

Most would agree that systems engineering is a practice, a way of approaching things, but is not scientific as such—so, is it neither a distinct discipline, nor scientific?

In writing this book, my objective is to bring systems theory, systems thinking, systems engineering, and systems management together into a single framework, and to integrate them in such a way that systems engineering is seen as a distinct discipline, founded on a system-scientific basis.

That systems engineering exists without a scientific underpinning is no surprise. Engineering was practiced for centuries before any mathematically based science came along. When it did, engineering performance improved immeasurably.

It is my fervent wish and belief that the establishment of a scientific basis will similarly boost systems engineering, and that it will, in consequence, be able to expand its horizons to encompass all kinds and types of system: business, industry, socioeconomic, governmental, ecological, and even global. The book is filled with examples of such systems, so that you, the reader, may judge for yourself.

# PART I

# SYSTEMS

# PHILOSOPHY,

# SYSTEMS SCIENCE

# Chapter 1

# The Need for, and Value of, Systems

*Observe how system into system runs,*
*what other planets circle other suns*
*Alexander Pope, 1688 - 1744*

## THE BOOK AS A SYSTEM

This book is about systems science and systems engineering. The goal of the book is to create and present a rational, scientific basis for systems engineering. That goal will be pursued through a sequence of connected parts by following the thread from early developments of systems ideas into system science and systems thinking and from there into a scientifically based systems engineering process, that is, one that stands up to refutation [1].[1]

## ORIGINS OF SYSTEMS SCIENCE

### A Mechanistic World

The Western world as we see it today has been greatly influenced by the Industrial Revolution. The European Renaissance inspired that revolution, partly through the work of René Descartes, the French philosopher of cogito ergo sum fame. His name is enshrined in Cartesian reductionism; his philosophy is that of breaking down big things into ever-smaller, and hence more understandable, pieces before assembly or reassembly into something larger. Cartesian reductionism is alive and well today in almost every walk of life. Every time we list, prioritize, disassemble, disaggregate, decompose, and so on, we pay implicit homage to Descartes.

---

[1] A scientific theory that lends itself to prediction, Karl Popper declared, can be proven false if that prediction proves false. Falsifiability is the touchstone that distinguishes science from nonscience.

## Perceived Limitations in the Traditional Sciences

In the 1930s and before, it was becoming apparent that reductionism and mechanistic views of the world were limited. Initially, the main issue was Life. Life seemed to confound the traditional sciences: physics and chemistry. Organisms could clearly be alive, decaying, or dead. Detailed physical or chemical examination, however, revealed no difference between those states. DNA, for instance, is precisely the same whether the organism, for which it acts as template, is alive or has been dead for millions of years. Life also represented organization, yet physicists believed implicitly that order should decrease in any closed system over time.

Things other than living organisms were also displaying characteristics that were inconsistent with a simple mechanistic viewpoint. Stability in physics is associated with a rest state of low energy. A bus, for instance, is stable when forces restore its center of gravity to its lowest point, minimizing potential energy.

Like organisms, mass production systems stabilize at high energy with parts entering, being assembled, and finally ejected, to be sold in the marketplace. As long as the mass of parts entering is roughly equal to the mass of things leaving, and so long as energy is being expended to assemble and move the parts internally, the whole is evidently stable, yet at a high, not low, energy state.

Analogies were presenting themselves between organisms on the one hand and organizations and civilizations on the other. This is the so-called "organismic analogy [2]," which upset some historians by highlighting similarities in behavior and composition between organisms and large-scale human activity systems. This was not to suggest that civilizations were organisms, but that each constituted a reduction in entropy, each had a life cycle, with growth, stability, and finally death–often sudden death–and each appeared to have an essential internal structure with organizational parts contributing to the viability of the whole.

It had been the practice to compare the various parts of the human body to supposed mechanical equivalents. The heart was a pump, made of muscle and tissue to be sure, but essentially just like a mechanical pump. The kidneys were filters. The eyes were cameras. The brain was a computer, and so on. Looking at the human body in this way, on a piecemeal basis, it seemed apparent to the observer that there was nothing really inexplicable about life. Man was little different from the robot of science fiction, or the automaton of the ballet. Earlier, it was Descartes himself who introduced the notion of the animal as machine. The animal was complex clockwork.

Later, the idea emerged that man was a heat engine, then a cybernetic machine, then a molecular machine. Such ideas face problems. Ludwig von Bertalanffy [3] identified three:

1. *The origin of the machine.* Descartes relied on creation by a blind Watchmaker, but how do machines come about in a universe of undirected physico-chemical events? Clocks do not create themselves in nature.

2. *The problem of regulation.* Machines can regulate themselves, of course. The problem concerns regulation and repair after arbitrary disturbance. Can an embryo or a brain be programmed for regulation after disturbances of an indefinite, possibly immense, number?

3. *The continuous exchange of components.* Metabolism is a basic characteristic of living things. Life is a machine composed of fuel spending itself continually, yet maintaining itself. This creates a paradox. A machinelike structure of the organism cannot be the ultimate reason for the *order* of life processes because the machine itself is maintained in an ordered flow of processes. The order, then, must lie in the overall process itself.

## Life and the Second Law

The Second Law of Thermodynamics dictates that entropy will increase with time in a closed system. Life appeared to confound the Second Law, the physicists' touchstone. Toward the end of life, when decay set in, it seemed that the Second Law regained lost ground.

At one level, the solution to this issue was evident: organisms could ingest food, and use the ordered substances to maintain and build their structures. In other words, an organism could feed on negative entropy, that is, food, and so decrease or hold steady its overall entropy. This did not confound the Second Law which, after all, referred to closed systems, since an organism that could ingest was clearly an open system. It was becoming apparent that all systems were open, and that–in consequence–the Second Law may prove difficult to apply rigorously on any scale smaller than that of the universe itself.

## Information and Entropy

Entropy is a measure of disorder. Information theory shows that information reduces uncertainty in the receiver, and hence reduces entropy. Information cannot be readily related to energy, but it can to entropy. If we are trying to choose between eight options, and we receive information that identifies the best choice, that information has reduced our uncertainty eightfold, or the amount of information is $\log_2 8 = 3$ bits (*bi*nary digi*ts*).

Another form of information is feedback. The primary regulation in organisms is derived from dynamic interactions, generally between two substances, muscles, and so on. Increasing blood sugar level is regulated by the generation of insulin. Embryonic bud growth is directed by the creation of chemical gradients, for instance. Feedback mechanisms within the body are secondary forms of regulation, such as maintenance of body temperature. Dynamics are at the heart of regulation in organisms, rather than control by feedback.

## Causality and Teleology

In the mechanistic world view, the goal of science was analysis, the breaking down of phenomena into ever-smaller parts, and the isolation of individual causal factors. Organisms were split into cells, processes into activities, behavior into reflexes, mass into atoms, and so on. Causality was unidirectional. One gene corresponded to one deficiency in the organism. One bacterium caused one disease.

This idea of individual units acting on their own in one-way causality proved insufficient to explain observed phenomena. It became necessary to consider groups of parts mutually interacting. The need for a holistic viewpoint, to consider wholes, to be organismic, for *gestalt*, emerged. It was proving necessary in many fields of scientific endeavor to consider systems of elements in mutual interaction.

In a similar vein, the mechanistic world view had difficulties with ideas of directed behavior, or teleology. Analysis down to individual isolated components erases all trace of directed, adaptive, or goal-seeking behavior, which was therefore viewed as mysterious and beyond the realm of scientific research. Yet the evidence of teleology, of purpose, and of goal-seeking behavior in organisms, was unmistakable.

Organization was also inaccessible to mechanistic science. Organization of people or of organisms is concerned with growth, hierarchy, structure, dominance and submission, control, and so on, none of which appeared in traditional physics.

## Concept of Open Systems

The concept of the open system emerged in response to many of the shortcomings of the mechanistic viewpoint. In modern biology, the open system is fundamental. The human body is an open system, as are many of its internal organic subsystems. Human activity systems, organizations, and many technological systems are open systems. Some systems may be considered more open than others, but all systems must be open in some degree; otherwise we would not be aware of their existence.

The basis of the open system model is the dynamic interactions of its components. In this it is differentiated from the cybernetic model, which is based on feedback. The open system ingests, and removes waste. The open system responds to stimuli. The open system can exhibit growth, can be stable at high energy levels, and can collapse and disintegrate. The open system can maintain and reproduce itself. The theory of open systems is part of general systems theory (GST).

## General Systems Theory

The Society for General Systems Research was organized in 1954 with the following principal contributors: Ludwig von Bertalanffy, a biologist; Kenneth

Boulding, an economist; Ralph Gerard, a physiologist; and A. Rapoport, a biomathematician. Their general systems theory had four aims:

1. To investigate the isomorphs of concepts, laws, and models in various fields, and to help in useful transfers from one field to another.
2. To encourage the development of adequate theoretical models in areas that lack them.
3. To eliminate the duplication of theoretical efforts in different fields.
4. To promote the unity of science through improving the communication between specialists.

They used a variety of means, methods, and models to achieve their ends, including:

- Classical mathematics;
- Set theory;
- Graph theory;
- Information theory;

- Game theory;
- Decision theory;
- Computing and simulation; and
- Theory of automata.

GST was postulated to be a science of wholeness. In elaborate form it was a logico-mathematical discipline applicable to the empirical sciences. For those sciences concerned with organized complex wholes, it is a formal mathematical discipline with application in diverse applied sciences such as genetics, life insurance, management, behavioral science, politics, economics, and so on.

Typically, von Bertalanffy developed general transport equations for open systems as follows:

$$\frac{\partial Q_i}{\partial t} = T_i + P_i \qquad (1.1)$$

where:

$Q_i$ = is a measure of the *ith* element of a system
$T_i$ = the velocity of transport of $Q_i$ at that point in space
$P_i$ = the rate of production or destruction of $Q_i$ at a certain point in space.

A system defined by equation 1.1 may have three types of solution: first, there may be an unlimited growth in the system, $Q$; second, a time-independent state may be reached; and third, there may be periodic solutions.

In the case where a time-independent solution is reached:

$$T_i + P_i = 0 \qquad (1.2)$$

In these two simple equations can be seen both the conservation laws of physics and the open systems stability of organisms. However, it was perhaps more the general ideas and models emanating from GST, particularly of the organismic analogy and the open system, that most influenced the fledgling disciplines of systems science and engineering.

## IDENTIFYING AND DEFINING A SYSTEM

It might seem simple to define the concept "system." It should be simple, but there are so many aspects and viewpoints that agreement, even between experts, is elusive. As we have seen, it is not even straightforward to decide whether a system is abstract or real. A process made from many activities may be a system. A product made from many parts may be a system. Before defining a system, it may be useful to examine some of the many aspects, to see if there are common threads that we may pick up on, to develop a useful definition of a system.

### Basic System Models

Entropy Models

Entropy "maps" might be used to present patterns of organizational order as it developed and decayed in societies, or the growth and collapse of civilizations over time. The surface of a pool, with water pumped up from below, exemplifies turbulent entropy maps.

As pumped water wells up from underneath, whirlpools and eddies form, move across the surface, and fade, to be replaced by others. Bubbles form and burst. Waves spread from eddies, and interfere with each other to produce fleeting standing waves. With increased energy from the pump, the rate of the activity increases, with more, deeper eddies, and more bubbling. Since the whirlpools, eddies, waves, and bubbles are features of order, no matter how transitory, it seems that the energy causing the water to well up from the bottom is causing order as well as disorder, and the greater the inflowing energy, the greater the rate of entropy generation and collapse as the energy is dissipated.

The spread of humans out of Africa[2] and across the globe might be presented as another real-world entropy-time map. Meteorologists' weather maps are also presented as virtual entropy maps, with cyclones, anticyclones, ridges, waves, and fronts; similar to the pond surface.

---

[2] Curiously, if the "out of Africa" theory is correct, then all Americans are African-Americans, just as all Europeans are African Europeans, and all Asians are African Asians.

This macroscopic view of systems as local, transient reductions in entropy is valid, aids understanding, and helps to maintain objectivity and perspective. Energy increases disorder. For open, interacting systems, energy also increases the rate at which order forms from disorder, and at which that new order decays in its turn. These observations are entirely consistent with the Second Law, but bring a new perspective to its implications for open, interacting systems.

A different kind of entropy map is named for Jules Henri Poincaré, the eminent French mathematician who is famous for addressing the three-body problem. Newtonian physics allows sensible analysis and prediction of the paths of two bodies orbiting around each other in space. For three bodies, the problem is much more complex.

Poincaré won a prize for solving the mathematical three-body puzzle, only to realize afterwards that his solution was incorrect, since it did not allow for chaotic interaction. Poincaré maps reduce the three-dimensional celestial dynamics to a two-dimensional picture by recording the passage through a vertical plane of each orbiting star as a dot. Chaotic orbits make a pattern of dots characteristic of the complex pattern of interwoven orbits.

Poincaré maps can be used in thought experiments. Consider two opposing soccer teams. Imagine a vertical plane, say, 10 meters from one end, set up so that a spot is marked on a Poincaré map every time the ball passes through the plane. As the game progresses, and the ball passes back and forth through the vertical plane, a pattern of dots will develop, recording the progress of the game. The pattern will record more than this, however. The distribution of the dots will indicate the offensive and defensive strategies of the opposing teams. If the attackers elected, as a strategy, to go down the right wing and cross to the center, then there would develop a predominance of dots at the right lower part of the map. So a team's strategy would start to emerge in the patterns on the map.

The notion encourages further questions. Could the pattern of dots ever repeat? Would the same pattern emerge if the two teams played each other again, or would it change as the defenders adapted to the attackers' strategy? Would the same pattern appear if the successful attacking team played a quite different team, or is the pattern only the result of interactions between two teams, and not exclusive to any one team?

The two phase plane[3] graphs in Figure 1.1 were formed by simulating three closely coupled systems, mutually exchanging energy. At left, the three-system simulation is started from rest; represented by the center of the pattern. As the energy exchanges increase, the helical pattern forms from the center outwards, describing the onset and growth of oscillation. The system stabilizes on the outer rim of the pattern. At right, the same system is simulated; this time with slightly greater coupling. This counterintuitive pattern of interchanges is chaotic, but is

---

[3] Phase plane maps were formed by plotting the contents of two of the three systems on a scatter diagram. The result is similar to a classic Poincaré map.

also stable and bounded; there is no continuum. The structure is convoluted, revealing unexpected structure in interchange behavior.

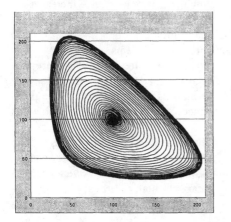 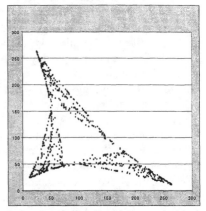

**Figure 1.1** *Phase plane maps showing the effects of coupling on three interacting bodies.*

The Poincaré map is interesting for two reasons. First, it is specifically about the entropy of interactions. Second, it leads directly to systems thinking, as evidenced by the observations of unexpected structure above, and by the questions that they prompt.

Poached Egg (Hierarchy) Model

A more prosaic view of systems is presented in the so-called poached egg[4] model, Figure 1.2. It is, nonetheless, a most useful and insightful model. The universe of discourse is seen as a world made up of systems. A system of interest (SOI) is seen in context as interacting with other, sibling systems within a mutual environment, the whole existing within a containing system. This containing system is similarly connected to other containing systems, not shown, and they also exist and interact within some mutual environment. The SOI also contains subsystems that are intraconnected.

The model shows three levels of hierarchy, and implies a further two: one, higher level, that contains the containing system; and another, lower level, within each subsystem. The model can be applied to any subject, problem, issue, or situation. The act of ascribing problem components to the poached egg model is one of organization and simplification, or reduction in perceived complexity.

So, words form sentences. Several sentences form a paragraph. Several paragraphs constitute a topic. Complementary topics comprise a chapter. And

---

[4] A colleague conferred the title "poached egg model," and it stuck.

several related chapters form a book. These words describe five levels of hierarchy.

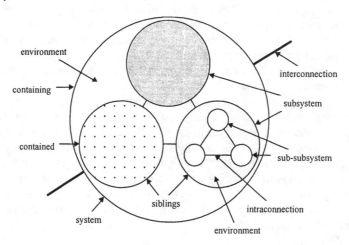

**Figure 1.2** *The "poached egg" model of systems, containment, and hierarchy.*

Several men with different skills come together to form a maintenance team. That team combines with an operations team and a logistics team to create a self-sufficient operational unit. The operational unit comes together with a helicopter unit and a communications unit to form a rapid reaction service. These words describe four levels of hierarchy.

In each instance, and at each level in the ascending hierarchy, the various siblings complement each other in some way. Differences between siblings[5] are essential to the idea of systems. Differences that complement each other allow a group of interacting systems to create new properties; properties of the whole that cannot be ascribed exclusively to any one member of the group. Self sufficiency is a property of the operational service above, but not of any of the three teams (maintenance, logistics, and operations) from which it is formed.

Differences between members of a maintenance team enable the team to maintain a full spread of facilities that no one member could accommodate on its own. Differences between modules allow assemblies to perform functions that no one module could achieve on its own. And, conversely, unrelated, or identical topics and chapters do not create a coherent book.

The poached egg model may be functional or physical. Members of the maintenance team may be physically collocated in some maintenance facility, but they might equally be distributed among the various operating units. The same model could be applied in both cases.

---

[5] Siblings may be identical in some cases to provide redundancy, for example, to accommodate failures in an airliner auto-landing system. However, the other siblings will still exist to create the system.

Note that interconnections in the model are not shown specifically connecting through the container to one or other of the contained systems. Such connections are not always easy to define. For the modules within subassemblies, the connections will be apparent. For topics within chapters the connection is more likely to be some relationship between different topics in different chapters; perhaps an earlier topic informs a later topic, for instance. Without such coherence, the book would be difficult to follow and understand; it would not form a system, that is, it would be disordered.

The various systems are shown with line boundaries. This is a convenience for presenting the model; in practice, the system boundary may be fuzzy or not evident. It is often invalid to correlate some physical boundary with a functional system boundary. In the case of the maintenance team, distributed throughout the operating units, a maintenance system is at work, but there is no clear physical boundary to the maintenance system. Nonetheless, developing the model as comprising three interacting systems—maintenance, logistics and operations—reduces the perceived complexity that the distribution of the individuals might create.

Recursion Model

The recursion model, Figure 1.3, presents several interacting systems. Each system is receiving energy and dissipating energy. Each system is receiving resources and discarding residues. Each system is receiving information and sending information. The systems are interconnected such that the dissipation of some systems provides the energy input to others, and the residues from some systems form the resources of others. This is an organismic view of systems. For both energy and resources there is a sense of conservation. For each system (sphere in the model), energy supplied will equal work done, plus energy stored

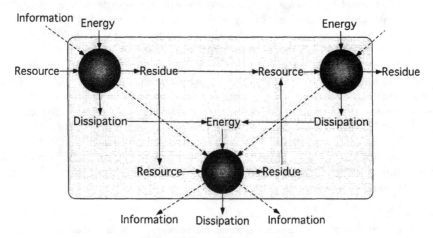

Figure 1.3 *Systems recursion model.*

plus energy dissipated. Similarly, resources supplied will equal resources utilized plus resources stored plus residues. Information is different, as indicated by the dotted line. Thus, information is not a conserved item; information can be given away without being lost.

In the figure, the three systems (spheres) each receive information, energy, and resource, and each system also gives out residue, dissipation, and information. The gray envelope is also a system, but at one hierarchy level higher. Note that it also receives information, energy, and resource, and gives out residue, dissipation, and information. This is the basis of recursion, and also of encapsulation.

If enough of the right subsystems were brought together, such that each received what it needed from the others while each gave out what the others needed, then there would be a complementary set of systems, which might form a stable, higher level system. If such a system could be described/labeled adequately, then in some contexts its contents would not be of interest, the label alone being sufficient for many purposes.

An orbiting satellite relay might be one such system. The relay has many subsystems, it receives energy–perhaps, from the Sun–it may have stored resources (such as hydrazine) necessary to maintain its station, the dissipation from various components is used to maintain internal temperature, and it receives and transmits information; perhaps providing a transparent multiplexing or protocol exchange service. In practice, many systems–manmade, natural, and organizational–can be usefully viewed using this model. The organismic view becomes particularly appropriate when there is some overall constraint on a system, such as mass, volume, or cost. For the satellite it might be form/shape, mass, and mass distribution.

Open System Model

In principle, systems may be open or closed, meaning that they receive and emit energy, matter, and/or information, or they do not. If a system were truly closed, we would be unaware of its existence–even a black hole reveals its presence by virtue of the effect its gravity has on other stellar objects. Pragmatically, then, all systems may be considered open, at least in some degree.

Open systems may be stable over a period of interest, yet at the same time receive inflows and emit outflows. Figure 1.4 shows a notional open system containing several subsystems. The system with its contained systems can be dynamically stable in spite of, or because of, the inflows and outflows. For instance, a baby is an open system and can remain stable, provided it receives energy and other resources, and provided it is able to dispose of residues and dissipate excess heat energy. Unlike simple physical artifacts, open systems stabilize at high energy levels.

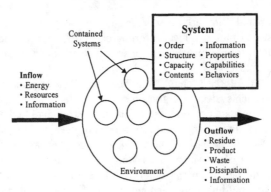

**Figure 1.4** *Dynamic open system.*

Open systems comply with physical conservation laws. A baby takes in food with nutritional value. That source of energy is either converted and used as work done by the infant in moving its limbs, maintaining its immune system, and so on, or is stored as energy in body fat, or is converted into growth, or is ejected as waste matter or is dissipated as heat, perspiration, gases, and so on What goes in must equal what is used, plus what is stored plus what is emitted. This is a transport model viewpoint. Conservation of energy is as true for the satellite relay, the operations unit, the radar; indeed, for all the systems we are likely to come across. The only exception to the conservation principle is information that, as has been noted, is not conserved, in that it can be given away yet still retained.

Open systems may exhibit a variety of properties (see the panel in Figure 1.4) including behavior. Behavior is a response to stimulus, which could be part of either inflow or outflow, or both.

Cybernetic Model

Cybernetics offers a control view of the world. Figure 1.5 shows the classic view of a control system. The objective of the system is to cause an output to correspond to some desired input, in the presence of some environmental disturbance that militates against such input-output correspondence.

In a manmade system, the components would generally be technological. For an aircraft control system, the input might be, say, a desired aileron setting. The environmental disturbance might be the airflow, opposing the aileron displacement. Differences between the desired aileron setting and the actual aileron displacement are used to make the two measures correspond, using power in the forward loop to overcome air resistance.

The model simplifies the detailed process, which may be required to respond very rapidly without oscillation or overshoot. The information system may, for instance, pass information not only about the aileron's position, but also about the rate of change of that position. Since the aileron may be subject to static friction, it may also be subjected to continuous "flutter" to overcome the threshold effect that

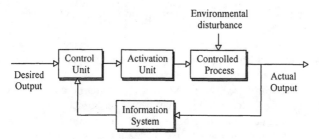

Figure 1.5 *The cybernetic model.*

• CU compares actual with desired output.
• AU receives signals from CU and responds by making changes in CP.
• CP is that which is being controlled.
• IS measures actual output and relays information to CU.

would otherwise make the aileron judder and jump when a control signal was applied.

The cybernetic model has been applied to management. A manager sets a task or objective for his staff. The manager receives information about their progress, which he compares with his planned expectation, and he then applies corrective measures to restore actual progress to planned progress. There are difficulties with this application of the model, since humans have a propensity for behaving unexpectedly and for being reluctant to operate like robots. Managers nonetheless continue to use the model.[6]

Open Loop Control Model

In practice, control between humans tends to follow a process like that illustrated in Figure 1.6. There are generally two parties: the director, and the directed or actor. They share a common mental model about their environment and the nature of the task to be achieved. The director tells the actor to take some action; one that perhaps the director would be unable to perform himself because he might be unskilled, infirm, or whatever. Because of their shared mental model, the actor is able to "fill in the gaps" and to undertake the task, bringing specialist skills to bear.

So, for instance, the foreman tells a plumber to fit a new sink in a house. The foreman does not need to detail how the job is to be done, nor precisely when. He does not need to tell the plumber where to get the sink, nor does he need to say that the sink must match the other items in the bathroom, and that it must be connected to hot and cold water supplies and to waste pipes. All of these factors

---

[6] The cybernetic model is not a true systems model. It presumes all regulation to be through feedback, whereas in systems the primary means of regulation is dynamic interaction between subsystems. Nonetheless, the cybernetic model can be a useful aid to understanding.

are the stock in trade of the plumber, and the foreman knows it. Hence control is light and brief.

As the figure shows, the actor executes the instructions as he or she understands them, and the foreman finds out about the task at some later time. Control in the cybernetic sense is not evident.

This type of control is more common and acceptable among humans and is employed by military forces, where it may be called "directive control," as well as by commercial managers. One reason is that it allows the doer to use his or her initiative and skill, it prevents the doer from feeling oppressed and controlled, and it obviates the need for the director to know as much detail as the actor, and to convey all the relevant detail in every instruction.

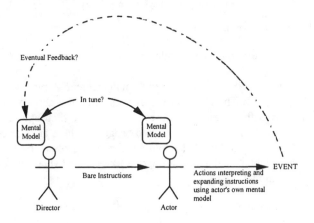

**Figure 1.6** *Control in management.*

Process Model

Processes can be systems, too. All of the basic models above refer equally to processes and groups of processes. A process is a set of interrelated activities, usually with some purpose or outcome. See Figure 1.7.

The process qualifies as a system if the activities fully complement each other to produce the product or outcome, and if activities are able to interact. So, an upstream activity that is accomplished in a particular way (e.g., incompletely) should be able to affect the nature and perhaps the duration of later activities.

For example, if an upstream activity that is intended to detect and eradicate all errors is ineffective, then later, downstream activities may face the prospect of working with erroneous inputs that require correction (or workaround). Such correction may necessitate additional time and effort; perhaps even rework.

A systems process model will allow and enable such interactions. Thus, in the example, a systems process model would relate a change in upstream error detection to consequent downstream activity duration and resource demand. The activities may be grouped into sets and subsets, or they may be subdivided into phases, with intermediate objectives or perhaps part products, or build standards.

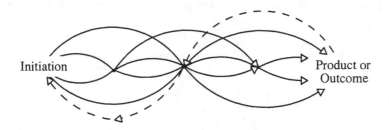

**Figure 1.7** *Interacting processes.*

A real-world model would have some means of labeling, describing each activity. It may also have some indication of resources and costs. Figure 1.8, shows the environment surrounding a typical process. A process model that shows activities being undertaken in a particular manner using particular resources supposes those skills, facilities, and resources to be available. As the figure shows, resources of all kinds may be the subject of competition and uncertainty.

A full systems model of a process would include complementary interacting sibling systems and their containing systems; nothing less is meaningful. However, this need is rarely met in practice, where dynamical modeling of the full process, together with its enabling support systems, is unusual.

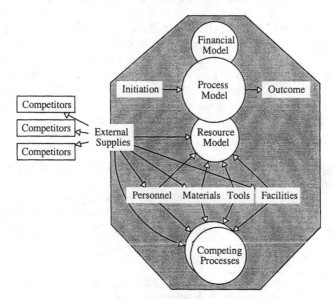

**Figure 1.8** *The process model in context.*

Pipeline Models

The pipeline model is an extension of the process model, in which the flow through of entities is subjected to a series of processes, with outputs from one being inputs to the next; see Figure 1.9.

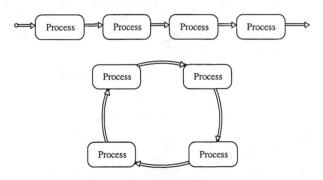

**Figure 1.9** *Pipeline and circle models.*

A classic example of pipelining is that of society dealing with criminals; see Figure 1.10. The police first catch a suspect. He or she is then taken to court and tried. The convicted criminal is then sentenced and imprisoned. After serving the sentence, the criminal is returned to society. In this example, the pipeline may well be circular, as the released criminal may offend again, and continue to go around the pipeline, becoming more familiar with each process as he or she passes through.

Pipeline models are ubiquitous.[7] In an analog radio receiver, sequential processes include: radio frequency pre-amplification, mixing, frequency changing, intermediate frequency amplification, detection, audio amplification, and conversion to sound waves. End-to-end processes are controlled by manual controls (e.g., volume and tuning) and by automatic controls (e.g., automatic gain control and automatic frequency control).

Transport Models

In physics, transport phenomena generally refer to molecular diffusion, thermal conduction, viscosity, and electrical conduction. In the systems context, the mathematics associated with these phenomena are applicable either directly because the systems are physically appropriate, or by analogy, where the systems can be seen to offer clear parallels. Such analogies have to be approached with care.

---

[7] A washing machine offers an example of a series of processes that are *not* pipelined. The clothes in the machine stay put in the drum, while different processes are applied to them in sequence, using stored materials (soap, softener) and external inflows/outflows (hot and cold water, and waste).

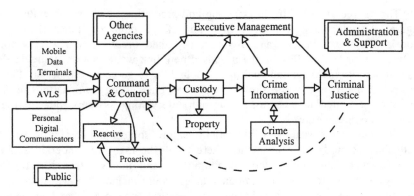

**Figure 1.10** *Police and criminal justice as a pipeline system. (AVLS stands for automatic vehicle location system.)*

For example, road transport experts have, in the past, used a so-called gravitational model to predict traffic flows on new freeways. The idea was to use an analogy from the mathematical formula relating the force of gravitational attraction between two bodies:

$$f = G \cdot M_1 \cdot M2/d^2 \tag{1.3}$$

where:   $f$ is the force of gravitational attraction;
         G is the universal gravitational constant;
         $M_1$ and $M_2$ are the masses of the two bodies; and
         $d$ is the distance between the two bodies.

To use this inverse square law, the idea was to substitute the two masses with the populations of the two cities at each end of the new freeway, and to estimate the analogous value of G by observing the actual volume of transport flowing between two towns of known population.

The results were a gross underestimate of the eventual volume of traffic. With hindsight, the reasons are not hard to find. The new freeway could be accessed at various points along its route, so traffic was attracted to use it from many other towns and cities besides those at either end. The convenience of the new freeway encouraged new towns and villages to spring up along its path, creating so-called "ribbon" developments. And people were encouraged to commute greater distances to and from their daily work by the lure of reduced journey times using the faster freeway.[8]

In retrospect, the use of a simple gravitational model turned out to be inappropriate. Criticism is easy in retrospect; the prediction of traffic for new freeways and town bypasses is still problematic.

---

[8] Sadly, so many commuters were attracted by the lure of reduced journey times that congestion and lines on the freeway made journey times longer, not shorter.

The degree of difficulty in assessing and predicting flow rates is evident in Figure 1.11, which suggests only a small part of the complexity inherent in such models. Suppose Systems A and D were to be linked *directly* by a new Transport 2 link. The amount of flow in the proposed new link would not be immediately evident, and it would also be far from obvious how to go about estimation. Flow between A and B is possible already by several routes using combinations of the two transport media and passage through intervening systems, but there is presently no direct route.

To resolve the issue sensibly would require some kind of simulation or mathematical model, and even then there would be many imponderables. For instance, were the four systems communities, and the two transport media railways and private cars, then imponderables would include: the willingness or not of motorists to switch to rail travel; the ease and time to traverse intervening systems (which would change once the new link was introduced); the influx of outsiders seeking to benefit from the new link; the acceptability of, and reaction to, overcrowding on the various links, including the proposed new link; and many more.

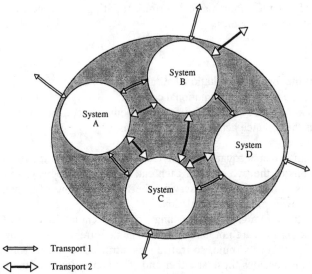

**Figure 1.11** *System transport model.*

Queuing Models

Queuing models are well known, and there are many books [4] on the subject. Queues form in everyday situations, at bus stops, in check-outs, at check-ins, and so on. Queues also form in digital communication networks, in traffic flows, while waiting to cross the road, when repairing faulty devices using automatic test equipment, in aircraft stacking up to await landing at an airport, when ancient

Egyptians waited to raise stones up the sides of pyramids, and so on. Wherever there are bottlenecks, there will be queues.

As Figure 1.12 shows, a service channel may have limited capacity and a limited entry capability. Thus, supermarket checkout can service only one customer at a time, and by design, only one trolley can enter the service channel; an escalator may hold scores of people at once.

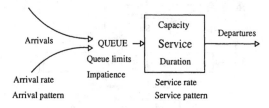

**Figure 1.12** *Classic queuing model.*

A service channel may take different amounts of time to service entities passing through, either because the different entities have different needs (more or less items in the check-out basket?) or because the service channel may have different rates of working (is the check-out staff tiring?).

Similarly, the queue may have a limited capacity; for example, the queue of people waiting for a bus under a small roadside shelter during a thunderstorm. People in a line may exhibit impatience at the length of the line, or at the time they have had to wait, and may either jump to other lines, or elect not to check out at all. The pattern of arrivals in a line may by regular or distributed in many different ways.

For simple queues, with exponential interarrival and service times, the mathematics gives seductively simple results, as follows:

Channel utilization $\rho = \lambda / \mu$        (1.4)
where:

$\rho$ is the proportion of time for which the channel is in use;
$\lambda$ is the mean arrival rate;
$\mu$ is the mean service rate.

The number of items, L, queuing and in the service channel, is:

$L = \rho / (1 - \rho)$        (1.5)

The number of items, $L_q$ in the queue, is given by:

$L_q = \rho^2 / (1 - \rho)$        (1.6)

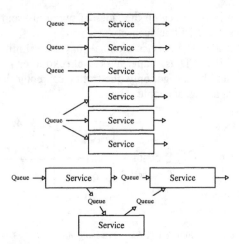

**Figure 1.13** *Different queuing arrangements.*

L and $L_q$ go to infinity as $\rho$ approaches unity, that is, as the mean arrival rate approaches the mean service rate.[9] Even seasoned engineers have been known to overlook this factor.

For instance, the capacity of a digital communications network such as Mil-Std 1553B or Ethernet[10] might be stated as, say, 1 Mbps. If the facilities using the network are putting data on to the network and taking data off randomly, then the working capacity is only about some 600 Kbps, not the 1 Mbps that might be expected. Going above c.600 Kbps starts to form a queue of messages on the network, increasing the chance that messages will be lost in transmission. Using the figures quoted, $\rho = 600/1000$, so:

$$Lq = 0.6^2 / (1 - 0.6) = 0.9 \qquad\qquad (1.7)$$

Ideally, the system will accommodate only one message "in the queue," so 0.9 messages, or 90% of one message, is about right.[11]

Figure 1.13 shows some of the many ways in which service channels and queues may be formed. Factories may apply many processes to items during volume production. Queues may form ahead of some or all processes, as the rate of throughput increases. Such queues can be a cause of concern: not only do the queues represent delay, but also the items in the queue may be expensive. Filling

---

[9] These simple formulas have proven so useful over the years that I have committed them to memory.

[10] MilStd 1553B is a serial digital highway used by the military. Ethernet is also a serial digital highway, used commercially to link computers and their peripherals.

[11] It is possible to approach the maximum capacity of such networks by synchronizing inputs and outputs. The formulas and figures quoted apply only to exponential interarrival and service times, which apply for many practical systems.

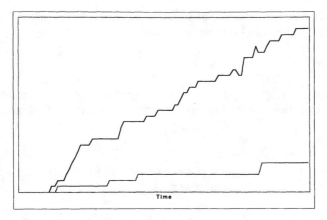

**Figure 1.14** *First in /first out queue build-up.*

the factory floor with expensive, part-manufactured items may cost so much that it raises the unit production cost to uneconomic levels and can damage a business.

Figure 1.14 shows a typical simulation of a first in/first out (FIFO) queue with exponential interarrival and service patterns. In the lower diagram on Figure 1.13, the lower line shows items exiting the system. The upper line shows items queuing to enter the system. The difference between the lines represents the numbers of items in the system; either queuing or being serviced. Using simulation methods allows the dynamics of parallel queues and of complex queuing configurations and rules, to be explored more easily than is practicable with mathematical methods.

**Fundamental System Constructs**

Without being specific about any particular system, it is possible to identify a set of constructs, which are both fundamental and common to all systems, of whatever kind, real or abstract.

Interaction

All the elements of any system are interconnected–directly or indirectly–and they interact. Connectivity and relationship are fundamentals that enable interactions. Conservation laws apply, except for information flows.

Configuration

Because the elements of any system are interrelated, the parts and their interrelationships form patterns. The pattern is referred to as the configuration, and it may be more or less disordered. Configuration entropy is an extensive property of the pattern, and may be used as a measure of "systemness," that is, the less the entropy, the more the order, or "systemness."

## Architecture

The architecture, related to configuration, is fundamentally the structure created by grouping and linking parts to form interacting subsystems and systems.

## Containment

Containment is the demarcation of a number of complementary parts within a set boundary, to form a system or subsystem.

## Complementation

Complementation is the ability of parts of a system to complement each other—to "fit" with the others—such that all mutually contribute through interaction to the whole system. A full set of complementary parts makes a complete whole.

## Hierarchy

Hierarchy is vertical structure formed when complementary parts form a complete whole that can then be considered as a unit with its own properties, capabilities, and behaviors. If that unit complements other units then a higher level of hierarchy emerges. Hierarchy is indicated by emergence, and vice versa.

## Emergence

Emergence is the phenomenon of properties, capabilities, and behaviors evident in the whole system that are not exclusively ascribable to any of its parts. Classic examples of emergence include: self awareness from the human brain; the pungent smell of ammonia emerging from two colorless, odorless gases–nitrogen and hydrogen; and so on. More mundane examples of emergence include: all-up weight emerging from the parts of a floating ship, reliability emerging from the redundant configuration of many unreliable parts, and so on.

The relationship between emergence and hierarchy is illustrated in Figure 1.15. Emergent properties generally refer to attributes observable from "outside" of the system. Emergent capabilities generally refer to limits of functional ability. Emergent behaviors generally refer to responses to stimuli.

In the past, something mysterious, unexpected, or counterintuitive has been associated with the notion of emergence.[12] An early example of emergence was

---

[12] Cartesian reductionism denies emergence. In the process of analyzing the parts of any system separately to understand the whole, interactions between the parts are invariably overlooked.

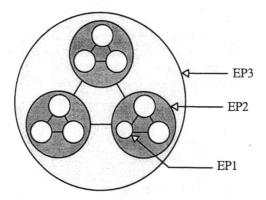

**Figure 1.15** *Emergence and hierarchy. EP1 (emergent property 1) emerges from a sub-subsystem. EP2, subsystem emergent properties, emerge from a group of three open interacting, sub-subsystems. EP3, system emergent properties, emerge from the three open, interacting subsystems.*

the observation that flipping through a number of still photographs of a galloping horse, taken at short intervals, gave the appearance of movement in the eye of the observer.

Emergence is not really mysterious, although it may be complex. Emergence is brought about by the interactions between the parts of a system. The galloping horse illusion depends upon the persistence of the human retina/brain combination, for instance. Elemental gases bond in combination by sharing outer electrons, thereby altering the appearance and behavior of the combination. In every case of emergence, the source is interaction between the parts—sometimes, as with the brain, very many parts—so that the phenomenon defies simple explanation.

## Definitions and Justification

### Definitions

A system can be defined only in the context of its environment; of those other systems within which it is contained and with which it interacts. Many definitions of systems are introspective, and may miss the role of interactions, fundamental to systems. Some dictionary definitions are better than others:

> *"Anything formed of parts placed together or adjusted into a regular and connected whole; a set of things considered as a connected whole."*

> *Chambers Dictionary*

The Chambers definition is good, although the essential idea of emergence is missing, and the perception conveyed is of something static, like a jigsaw puzzle–the parts are seen as connected, but not necessarily interacting.

There are so many definitions that it can be confusing. The following seems to capture the idea:

> A system is *an open set of complementary, interacting parts*
> *with properties, capabilities, and behaviors emerging both*
> *from the parts and from their interactions.*

This definition contains the essential ingredients:

- Open–for interchanges with other systems, which implies hierarchy and architecture;
- Set–whereby the parts have something in common, or are related;
- Complementary–which implies order, structure, mutual contribution and completeness of the whole;
- Interacting–which is the basis of emergence, implying a dynamic within every system;
- Parts–which are generally subsystems; that is, systems in their own right;
- Properties, capabilities, and behaviors, explained above;
- Emerging–which indicates the roots of emergence; and
- Parts and interactions–which, together, show that emergence derives both from parts and from interactions.

**Systems Principles**

The definition of a system leads on to a simple, but valuable system principle, the first principle of systems:

> *The properties, capabilities, and behaviors of a system*
> *derive from its parts, from interactions between those*
> *parts, and from interactions with other systems.*

The principle emerges directly from the definition of a system. One noteworthy aspect concerns the observation that properties, capabilities, and behaviors emerge, at least in part, from "interactions with other systems." This is an inevitable consequence of systems being open.

For instance, the road surface, traffic conditions, crosswinds, availability of appropriate fuel and oil, and the nerve of the rider limit the top speed of a motorcycle. Similarly, human creativity is stimulated by interactions with other creative humans, and an amplifier's performance is limited by the characteristics of the source signal, the power supplies, and local electromagnetic interference. The idea that systems exhibit properties, capabilities, and behaviors in isolation is untenable.

A corollary to the first principle is:

> *Altering the properties, capabilities, or behavior of any of the parts, or any of their interactions, affects other parts, the whole system, and interacting systems.*

The corollary may be obvious, but its implications are widely overlooked. Closing down railroad tracks for maintenance during a public holiday might reasonably be expected to exacerbate traffic jams on the roads, as holiday makers travel to their destinations. Declaring a multipurpose "cocktail" inoculation against several infantile ailments to be dangerous might reasonably be expected to increase demand for separate inoculations for each ailment. Increasing income tax on company cars might reasonably be expected to reduce the numbers of company cars, and increase salaries to compensate. These and many, many more instances came as surprises to politicians, suggesting either that they do not understand systems, or that they choose to overlook the implications.

**Classifying Systems**

Boulding's Classification

Kenneth Boulding, one of the founding fathers of general systems theory, developed a system classification in 1956. As Table 1.1 shows, Boulding's Classification was a mixture of ideas. Increasing levels of complexity, autonomy, and political sensitivity seem to characterize the classification.

Level 1 identifies static structures: bridges, roads, railways, arteries, telephone lines, and so on. Level 2 introduces the idea of predictable motion, such as clockwork, the heart, hydraulic pumps, orbiting satellites and planets, and so on. Level 3 introduces feedback, or closed loop control, such as thermostats, automatic volume and frequency control, Watt governors, and so on. Level 4 makes a big jump to address open, self-maintaining systems such as biological cells, but presumably excludes cells in groups, since these would be included in the next level. Level 4 seems to address very little other than cells, which receive inputs, conduct internal processes, and dispense outputs. Many cells do not grow significantly, which may have attracted Boulding's attention. Would a human being in a coma, being drip fed, constitute a level 4 system?

**Table 1.1**
Boulding's Classification of Systems

| Level | Characteristic | Examples |
|---|---|---|
| 1. Structures | Static | Bridges |
| 2. Clockworks | Predetermined motion | Solar System |
| 3. Controls | Closed loop control | Thermostats |
| 4. Open | Self maintaining | Biological Cells |
| 5. Lower organisms | Growth, reproduction | Plants |
| 6. Animals | Brain, learning | Birds |
| 7. Man | Knowledge, symbolism | Humans |
| 8. Social | Communication, values | Families |
| 9. Transcendental | Unknowable | God |

The characteristic of level 5 systems is that they exhibit growth and reproduction, as with plants and animals, but also bacteria. A bonfire may exhibit the same traits, with sparks starting new fires that may be much bigger than the original source. A key differentiator of life from fires is that biological reproduction does *not* create identical offspring.

Level 6, animals, is characterized by having brains and learning. There are creatures at the junction between plant and animal, such as anemones, for which this characterization is not clearly applicable.

Level 7 separates man from the other animals. This separation, based in ideas of symbolism and knowledge, is debatable, and suggests that Boulding was carefully skirting around political and religious issues. Homo sapiens are primates; to separate them from the other primates, who appear in Level 6, may be seen as a little strange today. Cetaceans such as whales and dolphins also appear to be intelligent mammals, and the octopus, not a mammal, is evidently intelligent, too. As we learn more about other intelligent species, the intellectual boundary between man and the others seems to become fuzzier. Primates make and use tools; so do some birds. Primates are capable of conducting conversations and forming sentences. One thing that does seem to separate man from the other animals is brain size in relation to body mass, which has enabled imagination, and which has lead to reproductive success such that our species has spread across the land surface of the planet, some would say like an infestation.

Level 8, social, characterized by communications and values, appears to be a hierarchy shift from man as an individual to humans in social groups, including the basic human social unit, the family. There are, of course, many social animals, including herds of elephants, flocks of birds, and shoals of fish. Honeybees have a complex social structure; they communicate with each other, they expel idle workers from their communities, they remove the dead from the hive, they defend against intruders, they are willing to die for the hive, and so on. An objective view of honeybees, ants, termites, elephants, cetaceans, and primates might suggest that they, too, communicate and have values.

Level 9, transcendental, is another level intended, it might be thought, to placate politico-religious critics. There is always a risk when classifying on the grand scale of ending up with a few things left over; many classifications include a ragbag of leftover bits and pieces. Boulding chose to address this problem by classifying some systems as unknowable. While many things are unknown, and some may be unknowable in the sense that we may never be able to find out about them, it seems unscientific to include the supernatural in such a classification.

## Classifying Systems by Complexity

It is possible in principle to measure the configuration entropy of a system, and hence to classify it on some scale of entropy or complexity. Boulding appears to have used that approach qualitatively. One difficulty with this is that complexity is a perception, and that what seems complex at first, may become less so. Similarly, what seems complex to one person, need not seem so to another.

## Classifying Systems by Morphology

### Basic System Shapes

Complex or "tangled" systems, shown at the top left in Figure 1.16, may conceal distinctive system shapes. It is by the characteristic shapes that many systems may be identified, and for good reason.

The architecture formed by the linkages within and between systems lends itself to classifying systems by their structure or architecture. Untangling, as at the top right of the figure, may reveal a product view, so called because there may be an implication that the elements are physical elements. From the tangle, three interconnected groups are revealed; these may be subsystems.

A useful way of identifying a subsystem is to observe that the intra-relationships, that is, those between the elements in the subset, are greater than the interrelationships; those between the subset and other subsets. A subset with many intrarelationships is tightly functionally bound, because all the members of the set mutually interact, closely affect each other, and tend therefore to act as one–a subsystem. Where a subsystem has few connections to other subsystems, it is loosely coupled, and the impact of such coupling will be relatively minor.

Figure 1.16 also shows a process view. The elements could have been related activities or processes; untangling then reveals subsets and subsystems that are subsets of intrarelated and interrelated activities. In this case, there is an implication of sequence and of simultaneity, rather than of static physical elements.

### Layered Architectures

Many systems are layered, in that substance or information has to pass through a number of layers; each perhaps processing the throughput in turn. Notable

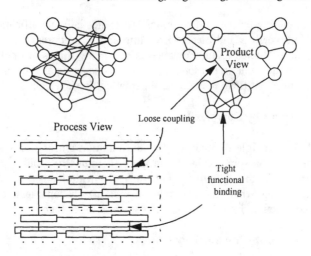

**Figure 1.16** *Untangling to reveal system morphology.*

examples include animal architectures, security systems, medieval motte-and-bailey castles, internet firewalls, chemical processing plants, chlorophyll formation in plants, ICBM defenses, and many, many more. Because all of these different systems share the same architecture, dynamic models of each of them would look much the same, although they may have different numbers of layers.

A vitally important capability of all layered systems is the ability to dispose of waste. One of the weaknesses of medieval castles was the inability to dispose of rubbish and human detritus. The result was susceptibility to disease when besieged. Some early artificial intelligence (AI) systems had the same problem: the inability to discard information, which consequently choked them.

Linked Closed Loops

Many systems comprise two loops; often operating antagonistically. Perhaps the classic example is that of population, where there is a system for creating new members–birth, and a system for discarding old members–death. See Figure 1.17.

If the birth rate and the death rate are equal, then, in principle, the population should remain constant. If the birth rate increases, then the population will increase, and if the death rate increases, the population will decrease. Such models represent a minimalist viewpoint; eliminating detail and cutting through complex system structures.

For example, births cause an instantaneous population increase, but the new babies have to wait several decades before they can, in their turn, contribute to births. There is a delay, which would need to be represented in a more detailed model. Nonetheless, such interacting loop systems are common, and their representation as causal loop models can be insightful.

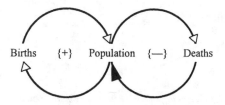

**Figure 1.17** *Causal loop model of population dynamics.*

Other instances of linked closed loop models include the heart, with its two closed loops (one delivering arterial blood to tissues, and the other re-oxygenating blood via the lungs), as well as growth models, feedback models, and social models.

Figure 1.18 shows a double causal loop model which suggests/explains why attempts to innovate within an organization may prove to be less than successful. The two loops in the figure oppose each other, suggesting that there might be a system for innovation and a system for opposing innovation. There are, however, many systems implied by the figure: research and development; sales; the market-as-a-system; accounting, financial, and investment management; training; manufacturing; general management; and working staff/personnel. The causal links in the model cut through these organizational systems, showing loops of causality that feed back and feed forward.

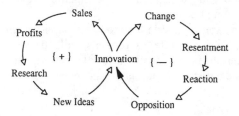

**Figure 1.18** *Causal loop model showing the dynamics of resistance to innovation.*

The causal loop model shows how the overall combination of systems *behaves*, while at the same time ignoring the organizational structure within the company. On the other hand, the causal loop model suggests how the issue of opposition to innovation might be tackled; supposing that to be the desired outcome. Since the antagonistic loop at right is opposing innovation, reducing the effect of the right-hand loop would allow innovation to flourish. This might be achieved by involving the organization in the innovative process, by training, by financial inducement, or even by confining innovation to an offshoot start-up organization.

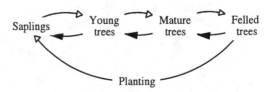

Figure 1.19 *Causal loop model showing tree maturation dynamics. [5]*

So perhaps, after all, there is a system for innovating and a system for opposing innovation, but these two do not map on to the organizational structure within the company. In general, where it is possible within an overall set of structural interacting systems to trace out closed and interacting loops, it is likely that these loops, and the implicit[13] systems they identify, will dominate the behavior of the set, since the positive feedback loops will reinforce themselves and build up, while the negative feedback loops will regulate and sustain themselves. This is the beginning of systems thinking.

Multiloop models are used to represent chains or sequences. Pipeline architectures may sometimes be represented in this manner, and so too may sequences. For instance, in a forest there may be saplings, young trees, mature trees, and felled trees. An individual tree can be in one category only at any one time, so the growth of trees may be represented as follows.

In the causal loop model of Figure 1.19, no tree is moving through a pipeline, since there is none. Instead, trees are categorized by age or condition, and it is their categorization that changes with time. When a young tree is recategorized as a mature tree, a new entry appears under the mature tree heading, and a tree is removed from the young tree heading.

Several systems are implied in the model: an environment conducive to tree growth (with heat, light, moisture, saprobionts, soil, space, and so on); a system for harvesting mature trees; and a system for planting saplings as trees are felled, which might indicate a system for sustaining the forest.

## Anthropomorphism

It is perhaps inevitable that many manmade systems are reminiscent of the humans that designed and created them. Nowhere is this truer than with defensive systems, where platforms (ships, planes, and tanks) seem to take on the shapes of humans.

One only has to look at the plan view of a modern fighter aircraft to see the resemblance to the human frame; see Figure 1.20. Considering the fighter as being an upright figure, note that the sensors are at the top (representing eyes and ears), the weapons and/or spare fuel are carried on the wings (arms), fuel (consumable energy) is carried (mainly) in the fuselage (stomach), and motive power and

---

[13] An implicit system is one that exists and is real, but that is not evident as a discrete entity within an organizational structure.

agility are generated via the engines, jet pipes, and tailerons (legs, feet, and toes). The brain is invested in the pilot, of course, although modern fighter aircraft can be quite intelligent in their own right. Like the dinosaur, perhaps we could view the human/fighter combination as having more than one brain.

Of course such ideas are fanciful. No one would consciously design an aircraft to look like a human standing on his or her legs. But then, need such ideas be conscious? Is it perhaps impossible for people to divorce themselves from their humanity when designing? Should we accept the inevitable, and exploit the excellence of our own, human design to the benefit of manmade systems?

The more we learn about nature, the more we find that there is little new under the sun. Nature's environments and designs are so many and various that most situations have been tested and exploited. Japanese research into artificial intelligence found that their initial attempts at creating autonomous robots were not too successful. An autonomous robot needs to find its way about, say, in a house, and it needs to plug itself into a wall socket when it is running low on power. Initially, the problem of finding one's way about was tried by causing the robot to wander about finding objects and spaces, and in so doing forming a detailed map of where everything in the house was.

The idea was then to enable the robot to preplan any move from one location to another by planning out the route in the optimum manner, prior to moving. This may seem logical, but it is not what humans do. If we are going, say, to the store, we set off in the general direction of the shops, but without any detailed plan. We then make decisions at junctions which may be influenced by such factors as available time, whether or not it is going to rain, a wish to look in passing into some new shop, and so on. Having made a choice, we travel on the new route to

**Figure 1.20** *Anthropomorphic view of a fighter aircraft.*

the next junction, repeat the exercise of choosing, and so on. At no time do we create a detailed mental map of the whole journey; it simply isn't necessary and would require unreasonable effort.

Researchers found that the robot's processor was overloaded by having to make a full detailed plan in advance of any move, and it ran out of battery power so often that it remained virtually tied to an electrical power socket. By changing to the method used by humans, the problem was substantially eased.

A similar issue arose over visual perception. Humans form binocular, three-dimensional images–at least that is what we think. In practice, our binocular vision works out to about only four yards; thereafter we use a technique that engineers call optic flow analysis. If you sit in a moving train and look sideways out the window, you get a strong sense of three-dimensionality (3-D) out to the horizon. This is because items that are close to the track move across the line of sight faster than those that are further off. Our brains can use this differential crossing rate to create a 3-D map of our environment–without binocular vision. Robots can do the same, with single sensors.

Sometimes a trick is needed: watch a cat hunting in grass, and see it moving its head from side to side. This is so that it can cancel out the closer blades of grass and concentrate on the prey, which, being further off, does not move laterally with the cat's head movement. Car drivers do the same when their windshield wipers fail, to cancel out the drops of rain on the windshield while concentrating on the road ahead. Robots can also use this behavior.

People Systems

Another, common way to classify systems is by their principal content. Many of the systems that interest us most might be best described as people systems. These include organizations, social systems, and cultural, religious, educational, and similar systems.

Systems such as legal or governmental systems are frameworks of rules created by humans to regulate human behavior. It would be incorrect to describe a legal system as a people system, in that it is made up of rules, regulations, and penalties; however, the legal system is strongly people related, since it is created and imposed by people, and acts upon them. People systems also include groups of people who design and implement manmade technological systems, such as teams of systems engineers.

Technological Systems

Technological systems perform various functions, usually in support of humans. Guns, knives, radios, toasters, washing machines, computers, houses, communication satellites, and many, many more manmade artifacts come under this category. Technological systems are an outcome of their time and culture. Flint knives, arrowheads, and the pyramids were the technological output of

Neolithic man, but they disappeared during the Bronze and Iron Ages. Medieval architects built gothic structures, like Salisbury Cathedral in England; their working systems were made up from, and filled with, sacred symbols. In Victorian times, the equivalent of software was made up of cams and profiles to make machines behave in particular ways; today's software engineer may not recognize such things as software. Forty years ago, magnetic amplifiers were used in rugged applications; today, they are hardly ever used in this way.

From a systems' viewpoint, we may be able to overlook the technology in favor of the purpose, and effect of such systems. If a flint arrowhead did the job as well as an iron arrowhead, then for some purposes we may not need to be interested in the composition of the arrow, just its effect. If magnetic amplifiers operated well in hostile environments, and today's potted solid-state amplifier performs as well, we need not be concerned; each would be able to make equivalent contributions to its respective systems.

Looked at in this way, the world has not changed quite so much over the last 5,000 years. We may have made outstanding technological advances in telecommunications, but we could already communicate over great distances millennia ago, using messengers, smoke signals, flags, drums, semaphores, heliographs, and so on. If you watch an old movie from the 1930s you may be surprised at how little we have advanced at a personal and domestic level since then; it seems that the only thing they lacked was television.

Technological systems are useful, and they help us to do more, faster and with less effort, but they do not have quite the effect on the human condition that might be imagined.

## Composite–IDA Systems

Some (most?) manmade systems are composites–part technology, part man. Such composite systems may be dubbed sociotechnical. A classic example is air traffic management (ATM). An ATM system controls aircraft around airports and along flight corridors. A typical system is comprised of controllers who assess situations and issue instructions to the pilots of aircraft. Controllers employ radar, radio, telecommunications, weather predictors, and a host of other technologies to ensure that they issue sound instructions. On its own, the technology would be pointless. On their own, the controllers would be blind, deaf, speechless, and powerless.

ATM is one of a class of systems I call information, decision, action (IDA) systems. Others in the class include: command and control ($C^2$) systems, and command, control, communications, and intelligence ($C^3I$) systems, ship control systems, flight control systems, fire brigades, police systems, and many, many more. The essential ingredients of systems in the class are that they receive information, and present it to human operators, who make decisions and act, all in real, or near real, time. Such systems operate at high stress levels, making great demands on operators and technologies not only for performance, but also for integrity and reliability in decision making.

Although IDA systems come in many forms, they all exhibit the same range of capabilities. See Table 1.2.

**Table 1.2**
*Information Decision Action Systems Capabilities*

| Sensing | Intelligence | Information Handling |
|---------|--------------|----------------------|
| Situation presentation | Threat/opportunity identification | Resource information |
| Resource allocation (Decision) | Results assessment | Communications |

An IDA system without all of these interacting capabilities would be incomplete. It is possible to identify each of the functions with those of a human actively engaged on some real-time task, and, since IDA systems generally have human operators at their core, the idea that IDA system designs not only are, but should be, anthropomorphic may seem less unreasonable.

Many animals may be classified as IDA systems. Consider lions. A pride of lions has hunting down to a fine art, but only with years of practice. The lionesses in particular look out for prey, watch a herd for some time, and then set a trap designed to minimize the predator's loss of energy in the chase. Lionesses are distributed at several spots before one reveals herself, causing the herd to panic in the direction of other concealed lionesses. Even then, every outing does not result in a kill. It takes time and patience to kill, and the acceptance of failure.

But, is it the individual lioness, or the pride as a whole, that constitutes the IDA system? The lionesses conform to a strategy that coordinates their behavior, and that requires each of them to adopt a distinct role. While each lioness is a finely honed IDA system in her own right, then, it is difficult to avoid the conclusion that there is an overarching IDA system that regulates the pride as a whole. Put another way, the pride exhibits emergent behavior, that is, behavior that cannot be attributed exclusively to any one member of that pride. This overarching IDA system may be abstract, but its effects are tangible.

(The notion that animals such as lions and the great apes are not conscious seems prima facie to be unreasonable. Human consciousness seems possibly to be an awareness of self and situation brought about by a perceived order. While we humans, with our unique cortex development, are particularly good at perceiving order, we are not the exclusive owners of this capability. Perhaps consciousness exists on a sliding scale, so that some animals are conscious but at a lower level than others. After all, some humans seem to be more or less conscious than others, too! And we have all experienced periods of heightened awareness.)

Composites–Information Systems

Information systems are generally thought of as being computer based, although this is by no means a necessary condition. Library, telephone directory, and news services provided information for years without computers; many still do. So, too, did so-called Hollerith recording systems, where information was stored on punched cards, any of which could be manually extracted from the mass of cards according to specified selection criteria. Fiche and microfiche systems employ film, and may–or may not–employ processors to store and retrieve masses of data. The introduction of computers has made information management faster and easier; in consequence, it is now practicable to tackle monumental information correlation and extraction tasks that would previously have been impossible.

The human brain is an excellent information system in most respects apart, perhaps, from its ability to recall data with precise faithfulness and clarity. On the other hand, the brain has the ability to make connections between items of information, which sets it apart from the logical procedural processor, with its humdrum, if thorough, software instructions. The human brain also has the ability to see similarities in pattern between disparate sets of information, which presumably emanates from its drive to reduce perceived entropy. Some people are gifted in this respect, and we tend to see this as the mark of high intelligence; even of genius.

Information durability/volatility is an issue with processor-based storage and retrieval systems. The technology changes so quickly that records archived as little as 10 years previously may no longer be accessible. This is proving to be a problem for epidemiologists who look over durations of 50 or more years to see the trends in the development of certain diseases, for example. Old paper records may still exist, and can be examined, but more recent computer-based records may be either volatile (if the record has disappeared from the machine) or the technology has so changed that the record is effectively lost[14].

Information management systems (IMSs) come in different forms, some manual using paper and card files, some computer assisted, and some simply in our brains. They all tend to address the following functions:

| | | |
|---|---|---|
| Gathering | Categorizing | Storing |
| Safeguarding | Searching | Retrieving |
| Presenting | Archiving | Destroying |

It would not be unreasonable to suggest that all IMSs are functionally isomorphic. There is also something about these various functions that suggests parallels with how we suppose our brains to work. We gather information willy-

---

[14] I wrote a program in 1987 that is no longer accessible to me (without major effort, and not a little luck) because it was stored on a 5.25-inch floppy disk using a disk drive, processor, and operating system that are no longer available.

nilly every day, transferring it in sleep to longer term memory. We categorize information–as those who try to remember a name that is just on the tip of their tongue will know. (*"I'm sure it starts with a 'B'."*) However, we can visualize information recalled from memory, seeing a loved one's face in the "mind's eye," apparently without going through a mental index; suggesting that visual and language information are stored and recalled differently.

It is possible to identify the workings of the human brain with each of the IMS categories above, even that of destroying, which can occur (in effect, if not in fact) when a traumatized person excludes/buries/forgets a memory that is too painful or shocking to recall. So, when people design and create information systems, are they replicating their perceptions of how their own minds work? And, if they are, as seems not unlikely, then is the act conscious or unconscious?

An issue recurs in the design of systems. Should a system that humans will use be designed to be "human like" in its structure, architecture, and ways of working; or should systems be mathematically and physically optimized to perform their intended functions more efficiently or faster perhaps, even though that makes them look and work in a way quite unlike the humans with whom they will interact?

## Classifying Systems by Behavior

Systems are quite naturally classified by their behavior. Classically and clinically, behavior is "response to stimulus."

### Linear Behavior

Linear systems vary their response to stimulus by producing output that varies in sympathy with the stimulus. A mechanical lever is a linear device. The load lifted is proportional to the effort applied. The distance the load moves is directly proportional to the distance the effort moves. In much the same way, a pulley, or block and tackle, is a linear device. A gear train is also a linear device, with a fixed ratio between the input gear and the output or driving gear. Double the input rotational rate, and the output rotational rate will also double. This is behavior in a highly predictable sense.

Many systems are quasilinear. A violinist bowing on a string expects to increase volume by drawing the string faster. Doubling the number of postal workers sorting mail would be expected to double the sorting rate.[15] Depressing the accelerator on a vehicle is expected to increase vehicle speed in a corresponding, if not precisely linear, way.

---

[15] This probably will not double the sorting rate, since workers will get in each others' way, there is an organizational overhead in managing more workers, and introducing more workers requires the latecomers to "come up to speed." It is also not unlikely that the workforce will "ease off," knowing that help has arrived, and so reduce individual sorting rates. Humans are not mechanical machines.

Doubling the number of fishing boats putting out to sea might be expected to double the catch. This expectation may not be fulfilled, as the boats may go over the same area, because the fish are not uniformly distributed, and because over fishing reduces stocks. Nonetheless, planners tend to assume just such a linear relationship when they reduce numbers of authorized boats to conserve fish stocks. Thus, many systems are expected, or presumed, to be linear. Few live up to the expectation.

Project managers planning a project tend to presume that work will proceed at a rate determined by the number of workers put on each task. Increase the numbers, and the task duration will fall proportionately. The fallacy of this simple notion is evident using reductio ad absurdum. An infinite number of workers will not reduce the time to zero.

The project plan is an interesting example of a supposed linear system, to which we will return; in fact, the project plan is neither linear nor a system. One reason for the nonlinearity is that the execution of projects invariably incurs errors, and these inevitably necessitate rework. Rework is feedback, and feedback tends to make systems nonlinear. (The classic project plan does not represent a process-as-a-system because, although lines showing sequence may relate various activities to each other, changes in the performance of any activity have no effect on the performance of other activities. That is, there is no behavior. Hence, the parts of the plan, although connected, are unable to interact with each other, a basic requirement for system status.)

Nonlinear Dynamic

Many real-world systems are nonlinear. Thus the various systems in animals are generally nonlinear: pulmonary, cardiovascular, central nervous, immune, and other systems. Positive, reinforcing feedback is a common cause of nonlinear dynamic behavior [6]. Cities and thunderclouds both grow because of positive feedback. Populations grow if the birth rate exceeds the death rate; this is positive feedback, in that people are creating more people, who grow up to create even more people. If that condition persists, the population will grow exponentially–at least according to some theories. In practice, populations do not grow exponentially. Disease and natural disasters take their toll, and crowding increases turbulent behavior and the spread of disease. Dwindling resources may cause famine or water shortage, and war may break out—often over scarce resources—although the excuse for war is rarely honest in that respect. So, population growth appears to be nonlinear, and the demographic profile may be dynamic.

Ecologies and economies [7] are nonlinear dynamic. Complex interactions result in multiple feedback loops; some positive, some negative, and some even alternating. In a classic system with several subsystems, a shock within one system may send waves through other systems that absorb some of the energy, passing some on to yet further systems. Eventually, some of the energy may return to the originating system. The whole reverberates like a spring mattress, gradually dying away as energy is dissipated.

Chaotic

Chaos has a mathematically definable meaning to scientists, but the term is also used in everyday parlance to indicate a situation out of control. It is remarkably difficult to classify a real-world system as chaotic in the mathematical sense [8]. There are many systems that appear to be chaotic in a more general sense, however, and some of these are very familiar to us.

For example, large earthquakes, those scoring high on the Richter scale, are less common than medium-sized earthquakes; these in turn are rarer than smaller earthquakes and tremors. Plotting the sizes of earthquakes against the frequency of their occurrence on a logarithmic graph results in a straight line, indicating that earthquake magnitudes conform to a so-called power law, of the form:

$$y = a \cdot x^{b} \qquad\qquad (1.8)$$

where a and b are arbitrary constants, appropriate to the subject being analyzed.

Adherence to a power law appears to be common in natural phenomena, but is not simply explained. Bak and Chen [9] referred to it as "weak chaos." The number of meteors entering the earth's atmosphere obeys the same general law, although the values of a and b are different.

Other examples include:

• Distances between cars on a busy road;
• Stock market price jumps;
• Increases and decreases in annual crime figures;
• Deaths in war;
• "Noise" in an electrical conductor.

The factor that appears to be common to these weakly chaotic systems is the idea of a build-up of energy caused by resistance to movement, followed by a sudden release. In the last instance, the conductor, electrons flow through and around a stable grid of metallic ions. Electrons "pile up" upstream of an ion, causing queues, which then collapse. The size of each queue and the time it takes to build vary, giving the characteristic noise.

For earthquakes, one mechanism might be that tectonic plates, which normally rub past each other, snag and build up both stress and strain before releasing suddenly. A firmer snag would result in a greater build-up of energy, and a greater release; that is, a larger earthquake. Most snags are minor, only a few are major.

Knowing that earthquakes and stock market prices are weakly chaotic does not, unfortunately, allow prediction of when the next big event is going to occur. However, the perception of some underlying order to such events is tantalizing, and offers some hope that we may yet find value in such mathematics and statistics.

## SUMMARY

Systems philosophy has emerged as a reaction to the limitations evident in the philosophy of Cartesian reductionism in a mechanistic world. The most evident limitations include the inability to distinguish living from the nonliving parts, and to address the ability of life to decrease entropy in apparent contradiction of the Second Law of Thermodynamics.

Observing the behavior of whole systems, and seeing the parallels between different systems, enabled system philosophers to perceive a reduction in complexity. A "systems view" made complex things and situations simpler to understand, organize, and manage.

An "organismic analogy" was observed, in which organizations behaved like organisms, and civilizations were observed to have life cycles analogous to those of organisms. This is not to say that organizations were considered to be organisms, but that they behaved in many ways as organisms behaved, and, in particular, that they were made of many complementary parts mutually co-operating to create the whole system.

The notion of the open system was developed; a system in which there were inflows and outflows, and in which stability occurred at high energy, not low as with physical entities. Many real-world systems were recognized as open systems, including sociotechnical systems, social systems, industrial process systems, and so on.

General systems theory described open systems and their behavior mathematically, and using models. These models in particular influenced the fledgling disciplines of operational analysis, systems thinking, and systems engineering. It became practicable to define and identify open systems, and to categorize different open systems according to their shape, structure, purpose, and so on; without necessarily identifying their content in any detail.

Instead of looking inside a system, it was possible to identify its emergent properties, capabilities, and behaviors. These emerged from the system parts, but also from the interactions between those parts. The notion of emergence became central to systems philosophy, systems science, and systems engineering, together with three tenets: holistic, or concerned with wholes; synthetic, or built up from parts; and organismic or built up from complementary parts that interact to support each other and to create emergent properties, capabilities, and behaviors.

## ASSIGNMENTS

1. Represent either a complete railway system, or a general hospital, using the "poached egg" model to show functional subsystems and sub-subsystems, together with environments. Show the principal interactions between subsystems and with the outside world as annotated, directed lines.

2. Define and explain "emergence." Give examples of emergent properties, emergent capabilities, and emergent behavior, justifying why each case exemplifies emergence.

3. State the corollary to the first principal of systems. It is normal practice when designing a development project plan to create phases. Is this practice consistent with the corollary? If you believe it is, justify your belief. If you believe it is not, identify what effect the inconsistency may have on the development.

4. Classify the following systems: a beehive, a football team in action, a parliament or senate, a damage repair crew on a naval destroyer, a systems engineering company, a national internal tax/revenue system, a hospital, and a technical library. What does each of these systems have in common with the others?

5. A car driving along a busy street is an instance of an IDA system. Draw a diagram showing the sources of information, the information sensing mechanisms, the decision making process, and the action subsystems. Your diagram should illustrate the IDA constituents shown in Table 1.2 above. Do you consider this a helpful way to illustrate the way in which a car is driven? How else might you illustrate or explain the process?

6. "Systems, like societies, do not exist in reality. They are simply collections or groups of things or people, designed to comfort and reduce complexity in the eye of the beholder." Discuss this statement.

---

## REFERENCES

[1] Popper, K., *Conjectures and Refutations: the Growth of Scientific Knowledge*, London, England: Routledge and Kegan Paul, 1972.

[2] von Bertalanffy, Ludwig, "General Systems Theory – A Critical Review," *General Systems* Vol. VII, 1962.

[3] von Bertalanffy, Ludwig, *General Systems Theory*, New York, NY: George Braziller Inc., 1968, pp.140-141.

[4] Ruiz-Palá, E., *Waiting-Line Models*, New York, NY: Reinhold Publishing Corp, 1967.

[5] Richmond, Barry, *An Introduction to Systems Thinking*, Hanover NH: HPS-Inc., 2001.

[6] Forrester, J. W., *Industrial Dynamics*, Cambridge, MA: MIT Press, 1961.

[7] Arthur, Brian W., "Positive Feedbacks in the Economy," *Scientific American*, Vol. 262, No. 2, 1990.

[8] Gutzwiller, M. C., "Quantum Chaos," *Scientific American*, Vol. 266, No.1, 1992.

[9] Bak, P. and Chen, K., "Self-Organized Criticality," *Scientific American*, Vol. 264, No.1, 1991.

# Chapter 2

## Measure for Measure

*He gave man speech, and speech created thought,*
*Which is the measure of the Universe.*
                    *Percy Bysshe Shelley, 1792 - 1822*

### MEASURING VALUE

It may sometimes be useful to be able to measure a system. There is, however, no universally accepted measure of the degree of "systemness." Instead, various distinct aspects of a system may be measured: efficiency, cost effectiveness, maximum throughput, cost per mile, and many others.

In the absence of any comprehensive measure of a system, some analysts seize upon a particular aspect of a system, and use that measure as though it somehow represented the whole system. It would be silly to measure a man's value to an organization by the bumps on his head (phrenology), yet some interviewers judge a man's worthiness by his handwriting (graphology). Both smell more of witchcraft than of science.

Pundits may judge a hospital by its waiting times, or a school by its rates of truancy. It does not take a rocket scientist to see that such measures, useful perhaps in narrow contexts, hardly constitute a measure of the whole hospital or school. This raises an issue. Can a whole hospital or school be measured and, if so, how?

One approach to finding a universal measure is to cost out everything. For a hospital it is possible to put a financial cost on the fabric of the building, the costs of the staff, the facilities, the drugs and surgery, the administration, and so on. Summing these figures provides an indication of capital and running costs, but something is obviously missing. The hospital performs a service that has value–it heals people. If we could place a financial figure on value, the argument goes, then we could see if the hospital provides value for money.

Putting a value on such intangibles is difficult; analysts will try, but the results are unconvincing. Insurers put financial value on the lives of individuals, or on parts of individuals such as an eye, or an arm. The results are arguable, to say the least.

The essential meaning of value takes some thought:

> The word VALUE, it is to be observed, has two different meanings, and
> sometimes expresses the utility of some particular object, and sometimes
> the power of purchasing other goods which the possession of the object
> conveys. The one may be called value in use, while the other may be called
> value in exchange.

*Adam Smith, 18th century economist.*

Using Adam Smith's viewpoint, water would be valuable in use, but have little value in exchange,[1] while diamonds have restricted value in use, but great value in exchange. Value in use may be called utility, while value in exchange may be called market price.

In ancient times, before money was invented, trade involved barter. So, a weaver who needed some bread might offer, say, some cloth in exchange. Suppose the baker did not want any cloth; perhaps he had just gained sufficient cloth for his needs in a previous barter. Then the man who needed bread could either offer something else, or might offer so much cloth that the baker would think that it was a worthwhile deal, hoping to offload the excess cloth at a later time.

We can learn about value from barter. Value is variable. The value someone places on an object or service depends upon how much of it he or she already has. Suppose a man has no bread, and his family is hungry. He will value a loaf highly, and will offer much in exchange. Suppose the same man now has nine loaves, and is offered a tenth. He will not value the tenth loaf nearly as much; if he is unable to carry it, he may not value it at all.

This idea of variable value is expressed in Fechner's Law [1]:

$$U = \text{K} \operatorname{Log} S \qquad\qquad (2.1)$$

or, more generally as:

$$U = \text{K} S^{\text{a}} \qquad\qquad (2.2)$$

where:

    $U$ is the utility;
    $S$ is the intensity of the stimulus;
    K and a are constants and a < 1.

For example, 10 miles per hour increase in the speed of a train from 20 to 30 mph has a higher utility than a 10 mph increase from 80 to 90 mph. Fechner's

---

[1] Except, perhaps, in the desert. Situations alter values.

Law is evidently about value in the sense of usefulness, or utility, as economists prefer to state. Economists aver that a rational man seeks to maximize his utility.

Today, we find captains of industry who amass large fortunes, but do not retire to enjoy their fortune; instead, they may work on until they drop in harness. Why? Is this Fechner's Law again, with the wealthy already having so much that even more is of little additional value; or is the money of little interest compared with the thrill of the hunt, the exercise of power, the satisfaction of command, or the fear of being "past it?"

Oscar Wilde defined a cynic as "a man who knows the price of everything and the value of nothing." Since value is in the eye, or mind, of the beholder, it is also true to say that the most important aspect of many systems, their value, cannot be measured.[2]

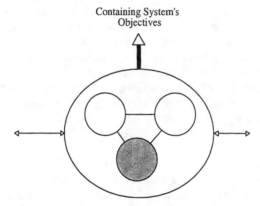

**Figure 2.1** *Determining value.*

This recalls the difficulty with measuring the value of a hospital. We can certainly put a price on running a hospital, but valuing a hospital may not be feasible, at least not in any numerical sense. This puts bureaucrats in a powerful position, since they can say that the cost of some project, say a new football stadium, is too high in relation to its value, and no one can argue rationally against them. Bureaucrats and politicians may also demand "value for money," knowing that they are in the exclusive position of judging value.

Valuing a system, at least qualitatively, may not be altogether impossible. In Figure 2.1, several subsystems mutually interact within their containing system; complementing each other to create emergent properties, capabilities, and behaviors. If the containing system has purpose and objectives, then the value of

---

[2] The most important human values, both in the workplace and in life, must surely include wisdom, judgment, integrity, honesty, and loyalty, yet these are the very values that are not amenable to measurement.

any of the subsystems may be considered as the degree with which it contributes to its containing system's purpose and objectives.

For instance, the value of a particular engine in an F1 racing car is the degree to which it contributes to the team objective of winning the race. Note that many other parts, including the driver, also contribute to the winning formula, and that changing the engine may affect other subsystems; transmission, suspension, center of mass, fuel supply, and so on.

Again, the value of an intensive care unit in a hospital is in the degree to which it can contribute to the hospital's objectives of saving lives and restoring quality of life.

The examples indicate that value is not absolute, but is relative to other subsystems or contributory parts. Fechner's Law also seems to be at work in the examples. Having too powerful an F1 engine may prove an embarrassment, since it will weigh more, use more fuel, encourage skidding on bends, unbalance the suspension, and so on; having too many beds in an intensive care unit would not enhance its value, since they would lie empty and achieve no purpose; and seeking to instill more and more knowledge from a particular department into students would overload them and unbalance their education.

So, to establish the value of something, first establish it in a systems context, interacting with other, complementary systems in a containing system. Second, establish the purpose of the containing system. Last, establish the degree to which the something contributes, along with its siblings, to the containing system's purpose.

For a hospital, the containing system is the area or region served by the hospital, with its fabric of management. The containing system's objective might be to maintain, enhance, and restore the quality of life of the residents. Sibling systems are those that interact with each other to contribute to that objective, and they will include environmental systems, waste disposal systems, education systems, transport and communication systems, sport and leisure systems, policing, and so on.

The value of the hospital can now be seen in context; both in terms of its tangible impact on the quality of life compared with other sibling systems, and in terms of its psychological impact on the quality of life by assuring the society, including those with no immediate need of its services, that it is there, ready, and able to perform should unforeseen difficulties arise.

Note, too, that the various siblings interact. The risk from poor waste disposal is disease. The lack of sport and leisure facilities may lead to heart disease, drug taking, drunkenness, crime, unrest, and disturbance, with injuries to abusers, criminals, rioters and police. This increases the hospital's customers, not to mention the need for more police, and there is a consequent reduction in the amounts of money available for other activities.

To assess the value of a hospital sensibly would require a dynamic model of the society with its various interacting systems, showing how they individually and together contribute to maintaining, enhancing, and restoring the quality of life, measured perhaps by some index. Changing the subsystems in the model would change their interactions and the overall quality of life index. In this way it would be possible to establish not only the degree of contribution for each and every subsystem, but also how they might be changed so as to increase the index overall.[3]

Evidently, value is a transient measure, changing as systems interact. And, as Fechner's Law reminds us, people value things most when they do not have them, and least when they have them in plenty.

## MEASURING PROPERTIES, CAPABILITIES, AND BEHAVIORS

Properties, capabilities, and behaviors emerge from systems; they can be observed and measured. To avoid reductionist bias and imbalance, complete sets, rather than individual, emergent characteristics would be the sensible measures of choice. Extensive, or systemic measures may be used that apply to the whole system; not to parts. Examples of extensive measures include configuration entropy, total mass, moment of inertia, capacity, and symmetry.

### Trade-offs

Quantitative measurement may not always be possible. Suppose for instance, one wished to choose between alternatives; each of which included imponderables. This is a situation arising frequently in systems engineering, where there are options about how to design, what to build, what to buy in, and so on. One rational approach is to conduct a trade-off in which the relative merits of each alternative are compared. For putative alternatives it is impracticable to measure attributes for things that have yet to be built, or processes that have yet to be realized.

In a trade-off, the many and various aspects of the options may be ranked; it is often relatively easy to rate one thing, A, as better than another thing, B. By summing the ranks for many attributes, or using ranking statistical analyses, it is possible to form a view or come to a judgment that, overall, A is to be preferred to B, or vice versa. Such resolutions cannot be perfect, but they are nonetheless rational perceptions within the sphere of knowledge of those seeking to make a choice. Importantly, that choice can–and has to be–made using qualitative criteria, at a point where numerical measurement is impractical.

Ranking methods also have an advantage in giving statistical indications of how good the choice is; if all of A's attributes were ranked first, and all of B's were ranked second, then clearly A would be 100% agreed to as the preferred choice, and B would also be 100% agreed to as the second choice. Where agreement is less than unanimous, degrees of concurrence can be calculated. It is

---

[3] Of course, this procedure would be deemed too difficult, so in practice social valuing reverts to politics, guesswork, and prejudice.

also possible to show, in appropriate cases, that there is no sensible choice between alternatives [2].

Trade-offs are also conducted using weighting and scoring methods, where each option's attribute is allocated a score, often from 1 to 10, and the relative importance of each attribute is numerically weighted, perhaps using numbers 1 to 5. Multiply attribute scores by attribute weights to give the weighted score for each attribute. Practitioners then allocate a score to each attribute for every option, weight the respective scores, and sum the weighted scores, so that the option with the highest number wins.

Weighting and scoring are popular; people feel comfortable and in control of the process of deciding weights and allocating scores. However, it can be highly subjective, tending therefore to produce results that concur with the prejudice of the practitioners.[4] There is no statistical back-up to the results, so generally the option with the highest weighted score wins, even if the amount by which it exceeds other options is so marginal that a small change in, say, weighting, would favor another option.

In practice, groups can get caught up in the mechanics of weighting and scoring, sometimes developing large spreadsheets of calculations. The calculation seduces them into believing that what they are doing is, somehow correct, and that no other outcome could be possible than that resulting from the calculation. Ranking and, particularly, weighting and scoring should be used for mind clearing, and little attention should be paid to the numerical outputs. After all, those seeking to use them for objective decisions created the numbers subjectively in the first place.

**Limitations**

It is not always obvious what we should be measuring. Commonly included candidates include:

- Life cycle costs–the cost of a system from conception to disposal, or from "lust to dust" as it is sometimes called;
- Efficiency–an internal measure of wastage in a system;
- Effectiveness–an external measure, showing the effect that one system might have on others;
- Performance–how far, how fast, how high, and so on;
- Availability (of performance)–the proportion of time for which Performance will be available, often expressed as a percentage probability,

---

[4] My informal experiments have shown that a group of managers and another group of technicians undertaking the same trade-off will, in general, produce scores that are not dissimilar, but the pattern of weights will be radically different, leading to radically different outcomes from the same process.

where the cause of loss of performance is within the system; that is, some element has failed;

- Survivability (of performance)–the proportion of time for which Performance will be available, often expressed as a percentage probability, where the cause of loss of performance is external to the system, for example, hostile action, adverse environment, and so on;

- Process–in terms of duration, cost, degree of concurrency, resource consumption, simplicity, and so on;

- Product–in terms of utility, quality, and fitness for purpose, salability, cost, recyclability, and many, many more; and

- Entropy–the degree of disorder in the system. Entropy is rarely included by name, of course, but may be implied. Entropy is the one candidate that stands to measure the degree of "systemness" in a system.

In the defense industry, systems engineering employs life cycle or through life costs as a basis for calculating monetary costs. Cost drivers include:

- System performance requirements;
- Technology;
- Production quantities;
- Reliability/maintainability requirements;
- System complexity;
- Production learning curves; and
- Maintenance and logistic support plans.

So challenging is the issue of measuring systems that it has burgeoned into a major business. Table 2.1 [3] is extracted from Mil-Hdbk-338, a former handbook for the U.S. defense industry on life cycle costing, and shows a need to measure very many aspects of acquisition, and of operation and support. In this example, the various measures are made in terms of money, so adding them together gives an overall cost. Value does not appear in this context.

Breaking down characteristics and features of a system in such a way is reductionist. The hope is that by addressing every facet of a program or project, the whole will somehow be addressed in the process. Comprehensive reduction is a challenging task however. How can you tell when the exhaustive/exhausting list[5] is complete? Worse, the resulting welter of information can be such that it is difficult, even impossible, to see what is fundamentally important.

While it is not possible to say that any of the items measured in the table below is unimportant, it *is* possible to question whether they have relevance to creating systems. That they appear within standards and books dedicated to systems and systems engineering suggests that Cartesian reductionism is alive and flourishing in the heart of systems creation–a curious and heavy irony, since the fundamental reason for introducing systems engineering in defense was to

---

[5] For example, should system trials appear in the lists? They are important and cost money. Reductionist lists can seldom be shown to be complete.

overcome problems being experienced with piecemeal developments and over attention to detail at the expense of the whole.

**Table 2.1**
Extract from Mil-Hdbk-338 Illustrating Life Cycle Cost Breakdown

| Acquisition | | Operation and Support | |
|---|---|---|---|
| **Basic Engineering** | **Recurring Production** | **Logistics and** | **Operational** |
| • design | **Costs** | **Maintenance** | • supply |
| • reliability, | • parts and materials | **Support** | management |
| maintainability | • fabrication | • pipeline spares | • technical data |
| • human factors | • assembly | • replacement spares, | • personnel |
| • producibility | • manufacturing support | organization, | • operational |
| • component | • quality control | intermediate and | facilities |
| software | • inspection and test | depot | • power |
| | • receiving | • on-equipment | • communications |
| **Test and Evaluation** | • in process | maintenance | • transportation |
| • development | • screening | • off-equipment | • materials, |
| • R growth | • burn-in | maintenance | excluding |
| • R&M demonstration | • acceptance | • inventory entry and | maintenance |
| • R screening | • material review | supply management | • general |
| • R acceptance | • scrap rate | • support equipment, | management |
| | • rework | including | • modifications |
| **Experimental** | | maintenance | • disposal |
| **Tooling** | **Nonrecurring** | • personnel training | |
| • system | **Production Costs** | and training | |
| • R program (Mil-Std- | • first article tests | equipment | |
| 785) | • test equipment | • technical data and | |
| • M program (Mil- | • tolling | documentation | |
| Std-470) | • facilities | • logistics management | |
| • cost | • system integration | • maintenance | |
| | • documentation, | facilities and power | |
| **Manufacturing and** | including maintenance | • transportation of | |
| **Quality** | instructions and | failed equipment to | |
| **Engineering** | operating manuals | and from depot | |
| • process planning | • initial spares, | | |
| • engineering change | organizational, | | Legend: |
| control | intermediate, depot, | | R is reliability  R&M is |
| • QA planning, audits, | and pipeline | | reliability and |
| liaison, and so on. | | | maintainability. QA is quality assurance. |

Measurement of life cycle costs, and many other measures advocated for the proper management of projects is not really about *creating* systems. Such measurements are a proper and necessary part of the management and control of development, manufacture, operation, support, disposal, and replacement. They may be important to program management and project management. It is perhaps inappropriate to overburden the creative systems and systems engineering processes with them, and so to risk systems engineering taking its "eye off the ball."

The breaking down of categories into lists of constituents is a time-honored activity, often referred to as the "laundry-list" approach to unraveling problems. It can be used as a precursor to systems thinking [4], and it works because it is the way in which our minds have been trained when recalling factors relating to something. Those with the systems thinking approach might then observe that many of the factors in the laundry list were mutually interactive, and that they therefore represented aspects of an overall system.

Someone with a systems view of the lists above might observe that spending more money on basic engineering—design, reliability, and maintainability—would reduce costs under test and evaluation, experimental tooling, engineering change control (due to fewer unreliable component replacements), recurring production costs, spares requirements, and so on. So spending more money up front may reduce overall life cycle costs, and a dynamic model would show both the interactions between the various cost centers, and whether or not there was a minimum cost condition. That would be systems thinking. The list appears to represent systematic accounting rather than systems thinking, systems design, and systems optimization.

The U.S. Department of Defense (DoD), having issued military handbooks [5] and standards on all sorts of issues, including the U.S. Air Force Mil-Std-499A for systems engineering, over many years, curtailed the practice in 1994. Toward the end, systems engineering is reputed to have been reduced to the filling in of countless forms, and had the acquired the epithet "paper engineering." Apocryphal, or true, this is a far cry from expectations of creativity and innovation with which the advent of systems ideas and methods was, and still is, heralded. The trend at the time of this writing is to develop and field commercial systems engineering standards, the latest being a new International Standards Organization (ISO) publication, ISO 15288.

## Entropy

Entropy, the degree of disorder in a system, is an extensive measure, addressing the whole system at once. The term "systemic" is sometimes used as a synonym for extensive.

A particular case is configuration entropy—the degree of disorder in a pattern. As we have seen, a system-in-the-abstract is a pattern. An idealized crystalline solid at $0°K$ has zero entropy, and this offers a mental anchor to ground our view of configuration entropy. In addition, there are several well-established and provable features of entropy:

- In a zero entropy system, all internal energy can escape: a zero entropy system converts all its internal energy to external work.

- Conversely, in a maximum entropy system, no internal energy can escape. A maximum entropy system may contain much internal energy, but cannot convert it to external work. So, a complex, hierarchical bureaucracy can achieve little work as output, in spite of the fact that individuals in the hierarchy expend great amounts of energy on (internal) work and mutual interaction.[6]

- Increasing the energy in a system increases its entropy. Boiling a liquid increases the agitation, turbulence, and disorder. Putting more people on a committee or team creates factions within that committee or team, making decisions harder to reach.

- Entropy tends toward a maximum in an isolated system. Isolation implies an inability to transfer mass or energy across the boundary of the system. Entropy is additive. If a system has two parts, with entropies $S_1$ and $S_2$, then the entropy of the whole is give by $S_1 + S_2$.

Entropy is about the number of ways things can be arranged:

$$S = k \cdot \ln (W) \tag{2.3}$$

where:    $S$ = entropy;
         $k$ = Boltzmann's constant $(1.38 \times 10^{-23} \text{ J K}^{-1})$;
         $W$ = number of ways things can be arranged.

It is possible to measure the entropy of a system both as a set of interacting subsystems and as a process or set of interacting activities. Appendix A explains how such measurement may be achieved, with examples.

**System Efficiency**

Systems with their several subsystems do not offer straightforward opportunities for the mathematical treatment of efficiency, where classically:

Efficiency = Work out / work in x 100%                    (2.4)

Most real processes often comprise several activities; each of which may operate with individual efficiency. Altering the efficiency of the upstream system in Figure 2.2 may upset the downstream systems by either reducing essential input or overloading them.

Consider three sequential processes, say, in a factory. Each process is set initially at 50% energy efficiency, to achieve the maximum rate of doing work.[7]

---

[6] "Bureaucracies, like aircraft carriers, exist to defend themselves" may be another statement of the same observation. Government committees may agonize for months at many different levels about whether to go ahead with some project or other. In the end, the result is a "yes" or a "no"; a decision that could have been achieved instantly by flipping a coin, although that would not have satisfied the many people who wanted to have a say.

[7] In accordance with the maximum power transfer theorem.

**Figure 2.2** *Downstream effects of upstream changes.*

The overall energy efficiency for the three machines is $(0.5)^3 = 12.5\%$. Suppose that efficiency of the overall process is measured, not as energy efficiency, but as Material efficiency[8] = Goods out/(goods out + inventory + work in progress).

Figure 2.3 shows the effect on material efficiency of holding the first and last processes constant at 50% energy efficiency, while changing the energy efficiency of the center process. Either increasing or decreasing the energy efficiency of the center process above or below 50% reduces the overall material efficiency. Why? Because altering the efficiency from the 50% optimum slows the center process, causing a build up in items waiting to be processed, while starving the third and final process of work input.

It is tempting to increase the seemingly poor efficiency of systems and process, and there are those who advocate such actions across a wide range of industrial and human endeavors. Where extravagant waste is evident, it may be an appropriate action, but—as the example and Figure 2.3 illustrate—it may be more sensible to concentrate on enhancing effectiveness.

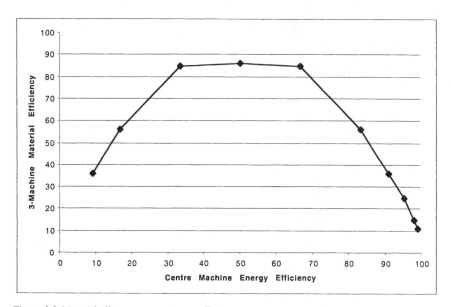

**Figure 2.3** *Material efficiency versus energy efficiency.*

---

[8] Such a measure indicates that it is inefficient to stock an assembly line or manufacturing line with inventory that costs money, but does not bring in revenue.

**System Effectiveness**

For manmade systems, system effectiveness is the effect a system has on other systems, that is, generally a measure of the degree with which it fulfils its purpose. It is practicable to measure system effectiveness only when that system is operating in its intended environment, and interacting with those other systems with which it is intended to operate. Predicting effectiveness for some system as yet to be built is therefore not possible directly. Future effectiveness can only be assessed either by reference to some similar system presently operating in similar circumstances, or by modeling and simulation.

One approach to modeling and simulation is for a simulated representation of a future system to interact with the real systems with which it will interact in the future. This is done with space models, to see if the future system will fit physically, and with emulation models, which can be made to behave as the future system.

New railway carriages may be "mocked up" to see how potential fare-paying passengers like them. Carriages may also be prototyped, and used for short periods as part of a live railway, to gauge customer reaction and to highlight potential operational shortcomings.

Effectiveness in medical terms may refer to the outcome of some procedure or medication. Effectiveness in education[9] may similarly refer to future outcome, perhaps several years after the events. While systems efficiency is generally about cost of operation, systems effectiveness is about value.

**Efficiency, Effectiveness, and Entropy**

In general, system effectiveness is reduced when system efficiency rises above an essential minimum. Measuring entropy gives a guide to the optimum configuration, where efficiency is at a maximum consistent with greatest effectiveness.

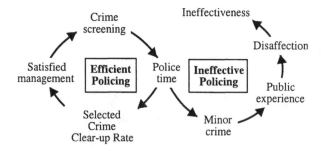

**Figure 2.4** *Police crime screening.*

---

[9] It is noteworthy that few educational pundits seem interested in measuring effectiveness as outcome. It is so much easier to guess or assume.

Figure 2.4 illustrates the manner in which efficiency and effectiveness can run counter to each other. At left, a police service has introduced crime screening. The idea is seductive. Examine the clues associated with each reported crime; allocate scarce police resources preferentially toward those crimes with the most clues, since these are most likely to be solved; hence, make the best use of police time, that is, reduce wasted effort; be more efficient.

At right, the down side emerges. By emphasizing the most easily solved crimes, the police de-emphasize minor crime, such as street violence and muggings, that most people come across in their everyday lives. This de-emphasis leads to public disaffection and a public belief in the ineffectiveness of the police. Worse, public disaffection may lead to public noncooperation, and without such cooperation, the crime clear-up rate is likely to fall, too.

In this example, terms such as efficiency and effectiveness have been used qualitatively; that is, not in any strict mathematical sense. This is more often the way with human activity systems, where complexity makes simple calculation of "work in/work out x 100%" simply impractical.

With efficiency and effectiveness being, at best, only partial measures of a system, and, at worst, infeasible to measure, entropy may look more attractive. Reasons for using entropy as a systems measure include:

- System processes waste the least internally, within the system, when they are direct, linear, and without recursion, deviation, rework, delay, error, and so on; that is, entropy is low.

- System processes do the most work external to the system when system entropy is lowest.

- Minimizing system entropy therefore arranges a system so that it has the efficiency it needs to deliver maximum effectiveness.

Nonetheless, entropy is an abstract, even an opaque concept, and is seldom used as a direct measure. However, see Appendix A, which shows practical ways in which entropy, or some function of entropy, can be used to great effect.

## GENETIC ALGORITHMIC METHODS OF ADJUSTMENT

If, as shown above, it is inappropriate and incorrect to optimize or maximize the individual parts of a system because that will disturb other parts and de-optimize the whole, then an apparent problem emerges. How can the whole system be optimally configured at all?

Actually, it is worse than that. The corollary to the first principle of systems, on page 27, states:

*Altering the properties, capabilities, or behavior of any of the parts, or any of their interactions, affects other parts, the whole system, and interacting systems.*

So, trying to adjust the overall system to achieve any particular degree of entropy, efficiency, or effectiveness would seem to present a level of impracticability similar to that of nailing jelly to a wall.

However, there is a way, using so-called genetic methods. Adjusting any one part (subsystem) or interaction (interface) will affect the emergent, properties, capabilities, and behaviors of the whole system by measurable amounts. If all of the parts and interactions are varied in turn, they will each vary the overall system measurably. If the parts and interactions are varied, not singly, but all together using a random pattern of variation, then the output will also vary by measurable amounts.

Continuing to vary the set of parts and interactions randomly will produce a pattern of overall system emergent properties, capabilities, and behaviors—an $n$-dimensional landscape with peaks and troughs. The ultimate peak determines the optimal performance and, at the same time, determines the corresponding emergent properties, capabilities, behaviors, and configuration of the internal parts (subsystems and interfaces) to afford optimum performance of the whole.[10]

To undertake this practice in the real world would be impractical in many cases: it necessitates, instead, a suitable, behavioral model of the whole system. Given such a model, the process of optimizing the overall system design and, at the same time, establishing the requisite emergent properties, capabilities, and behaviors of each of the component subsystems, is relatively straightforward. It is not, however, widely practiced.

## OPTIMIZING MEASURES

Associated with ideas of optimizing system configurations is the employment of optimizing measures. Optimizing is associated with an organismic view of systems, that is, there is some overall constraint on the end system such that internal subsystems may be balanced, adjusted, and traded in the interests of meeting the overall constraint.

### Value for Money, and Cost Effectiveness

A popular example of an optimizing measure is value for money. The expectation is that the quotient, value/cost, has a maximum value such that cost is within

---

[10] An optimum system is comprised of suboptimal parts. Were each subsystem optimized in its own right, then subsequent integration would not result in an optimum overall system. Instead, each subsystem and interaction has to be adjusted to achieve an optimum configuration for the whole system.

sensible limits and value is acceptable; at this point, it is believed, the customer will receive maximum value per unit cost.

We have seen that systems effectiveness is an expression of perceived value for the whole system-in-action. Hence, cost effectiveness is virtually synonymous with value for money; see Figure 2.5.

However, effectiveness is not a simple parameter to assess or estimate; particularly in advance. As we have seen above, it depends on environment and on interactions with other systems, so not only is effectiveness difficult to predict, but it also may prove to be a variable, rather than the constant that customers may anticipate.

In practice, although cost effectiveness, or value for money, is a sensible measure to aid system optimization, the assessment of effectiveness/value may be qualitative. Indeed, it is not unknown for the assessment to be less than impartial for major projects, where politics becomes a factor. Those who call for value for money may also overlook the shortfall in effectiveness that such optimization invokes, as illustrated at the top right of Figure 2.5; cost effectiveness and value for money are organismic, compromise criteria.

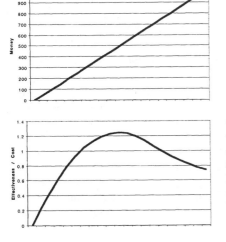

Top left. Project cost rises typically with time.

Top right. Effectiveness follows a logistic or sigmoid curve–there is a limit to how effective something can be.

Bottom left. Dividing effectiveness by cost shows a maximum condition–best value for money.

The maximum condition occurs at less than maximum effectiveness.

**Figure 2.5** *Cost, effectiveness, and cost effectiveness.*

## Cost Exchange Ratios

For two systems locked in competition or combat, an alternative approach is to form a ratio expressing the cost of operations to each party. For many systems, the ultimate challenge is to maintain operations in the face of mounting costs. If one party can keep its costs down, yet cause the other party to raise its costs then–given enough time–the second party will run out of funds and resources.

This measure implies a war of attrition, and is associated less with optimization, more with endurance.

## Casualty Exchange Ratios

A particular approach used during periods of conflict or war is the calculation of casualty exchange ratio. It is a particularly emotive subset of cost exchange ratio, where cost is not measured in money, but in lives. Again, this parameter is not associated with optimization, but more with minimization. However, it may be used to configure an optimal system that results in casualty minimization.

In warfare, while the military may accept the inevitability of casualties, politicians and the media may be less fatalistic. In democracies,[11] even one casualty may be sufficient to weaken political resolve to continue a war, since the politician will realize that his or her prospects for re-election may be prejudiced.

## SUMMARY

Measuring systems is not straightforward; many of the measures that are used consider only parts of a system. Sometimes it is better to measure what a system does rather than what it is. Some measures, such as entropy, are extensive, or systemic, and address the whole system. Reductionist methods such as ranking, and weighting and scoring, are widely used but are suitable for mind clearing, rather than for any valid numerical analysis.

Entropy is a useful measure, since it relates directly to the ability of a system to convert its internal energy into useful external work. Minimizing entropy, particularly in sociotechnical systems, provides a route to improving system effectiveness, while at the same time avoiding the risk of increasing efficiency too much. At high efficiencies, physics' maximum power transfer theorem dominates system behavior, slowing the rate of doing work to a halt at 100% efficiency.

Measures are used as a means to optimize systems; value for money and cost effectiveness are popular examples, especially with bureaucrats and politicians charged with looking after public money. Value, however, may be in the eye of the beholder, while effectiveness can seldom be measured at the time when public money is being spent to achieve it.

## ASSIGNMENT

You are a member of a group whose task is to determine how public revenues should be spent on services available within society such as medical, police,

---

[11] Perhaps this has something to do with Michigan University's ongoing correlates of war (COW) project, which has determined that, to date, no two democracies have ever waged war on each other.

library, and educational services. The group agrees that money should be spent/invested in proportion to the value provided by different service groups to the tax-paying society as a whole.

How would you value the services provided to society by hospices; those places where the terminally ill are cared for as, and until, they die? How would you compare this value to those of other social service provisions?

---

## REFERENCES

[1] de Neufville, R, and J. H. Stafford, *Systems Analysis for Engineers and Managers,* New York, NY: McGraw-Hill, 1971.

[2] Hitchins, D. K., *Putting Systems to Work,* Chichester, England: Wiley, 1992, pp. 86 – 108.

[3] Lacy, James A., *Systems Engineering Management,* New York, NY: McGraw-Hill, 1992.

[4] Richmond, Barry, *An Introduction to Systems Thinking,* Hanover NH: HPS-Inc., 2001.

[5] *Systems Engineering Fundamentals,* Fort Belvoir, VA: Acker Library On-line Catalog, Defense Acquisition University, www.dau.mil, 2001.

# Chapter 3

## The Human Element

*I must Create a System, or be enslav'd by another Man's*
*I will not reason and compare: my business is to Create*

William Blake, Jerusalem, *1757 - 1827*

Perhaps we humans can be forgiven for believing that we are rather fine examples of systems; products of a long evolutionary process. Humans are systems, and associate with each other to form systems: civilizations, social systems, organizational systems, political systems, and economic systems. Humans make physical systems: mechanical systems, electrical and electronic systems, vehicle systems, and so on. Humans form part of sociotechnical systems. Some analysts, however, seem reluctant to view humans as systems, or to view humans as part of a system; to some, systems are manmade artifacts, and humans stand outside. People, they might say, employ such systems, but are not part of them.

One reason for this attitude is the widespread belief that human behavior is uncertain and unpredictable. In consequence, putting a human in the system makes the combination also uncertain and unpredictable. Paradoxically, we take the opposite attitude about airliners, for instance, where the pilot is evidently part of the system. The aircraft will not move without the pilot. The pilot also monitors the automatic flight facilities, providing the final back up when all else fails. In such cases, it seems, we are prepared to believe in, and rely on, the absolute reliability and predictability of human behavior over technological automation.

In some circumstances, some humans are unpredictable. In most circumstances, humans like to believe that they are unpredictable. In reality, most of the time we humans are highly predictable. Given knowledge of culture, background, recent experiences, and current situation, psychologists are astonishingly good at predicting human behavior.

We tend to overlook the fact that we all share our humanity. We may start to see our humanity when we see chimpanzee mothers and infants behaving just as humans do. The observation of similar behavior makes us aware that the way we behave may not be unique, and is observable and predictable—at least in some circumstances.

An example of our innate humanity can be gained from game playing. Imagine that, instead of having evolved from mammalian forebears, we had somehow evolved from reptiles and, as a result we laid eggs, and left them to hatch unattended, as some reptiles do. What effect might this have on behavior? We would not know our offspring, so there would be no suckling; no filial or parental bonds. We would not feel protective, we would not nurture, and we would not provide care. So, a threat to some of our species' eggs, perhaps from flood, would not necessarily bother us. Our decisions as adults would, to a mammalian human, appear quite unfeeling and irrational.

Such bizarre role-playing exercises can be revealing. In some cases the behavior of, and decisions made by, the supposed alien, mimics that of a brutalized gangster; in others that of an uncaring, remote monster. Often it seems coldly, calculatingly logical.

Psychologists have been surprisingly successful in unraveling the basis of behavior in humans. There is much still to learn, but some things seem to be clear.

## CATEGORIZATION

In common with all living things, we have the ability to categorize. At birth we can tell the difference between food and nonfood, mother and not mother, hungry and not hungry, comfortable and uncomfortable.

Categorization is vital. Without the ability to categorize new sensations as "like that previous one, and therefore essentially in the same category" we would be condemned to examining every sensation, every situation, every artifact, every person from scratch, every time. We would be overwhelmed.

We categorize everything. As we walk down the street we categorize the houses (new, old, detached, wooden, quaint, dull...), the people we see (smart, ugly, threatening, pretty...), and the things people do (postman, shopkeeper, policeman, deliveryman, shopper, man waiting outside a bank...).

Everything we encounter is immediately put into mental boxes. We are excellent at categorization, probably because it has in the past been essential to survival. For example, in the bracketed lists above, notice "threatening" and "man waiting outside bank." In prehistory, not to recognize the man with the club as an enemy was likely to end your genetic line. Those who survived to breed went on to categorize and stereotype.[1]

All animals categorize. The garden worm categorizes the ground as either damp (OK) or bone dry (dangerous). The humble worm will skirt round dry patches of ground to reach some goal safely.

Psychologists suggest that we humans create mental "world maps" that contain within them features of the world about us that we accept as normal. For

---

[1] Stereotyping is often decried as socially inappropriate. However, stereotyping is a form of categorization, and people can no more avoid stereotyping than they can stop breathing. How the individual responds to such categorization "alerts" is another matter.

instance, if you observed a helicopter hovering upside down you would immediately sense that something was out of the ordinary; the observation would jar your world model in which helicopters hover the right way up.

World maps help us to categorize on a large scale, so that we can take in broad swaths of information and highlight only those features that jar; another manifestation of perceived entropy. Along with world maps, we also develop from birth a mental store of "tacit knowledge." This is ordinary everyday knowledge that everyone possesses. Grass is green, sky is blue, heavy things drop to the ground, water is wet, and so on. It turns out that the amount of tacit knowledge we each possess is vast. This has become evident, curiously, from efforts to create autonomous robots. Programmers discovered that enabling a domestic robot to undertake cleaning activities around a home necessitated providing the robot with large amounts of this low-level information that humans take for granted.

Tacit knowledge relates to world maps, in that objects and entities in our world map identify with our tacit knowledge. We might recognize a country garden as consistent with our world model until, that is, we notice that the green lawn shrinks when touched. Upon closer inspection, we may observe that the lawn is not made of grass but of sensitive mimosa. Having observed this unexpected phenomenon, it will enter our mental repertoire, updating both our tacit knowledge and our world model. The next time we observe this form of lawn, we will not be surprised, as it will not jar our (updated) world model. Our categorization capability has been enhanced.

## MOTIVATION, INSTINCT, AND INHERITANCE

Psychologists tell us [1] that there are two kinds of motivation: achievement motivation and conformance motivation. Achievement motivation is the incentive to reach some goal or target. Conformance motivation is the incentive to be like others. Teenagers can find these two in opposition. Thus, in some school cultures it is not acceptable within the peer group to be too clever. Yet the bright individual wants to do well in class and in examinations. For some, the tension between the two motivations can be difficult to resolve.

Humans like to believe that they are creatures of independent and original spirit, and that instinct is for dumb animals. So-called dumb animals exhibit remarkable instincts. Some sea-going birds nest and lay eggs in the northern hemisphere, leaving the hatchlings behind while the adults fly south. The hatchlings spend their first year in the north, before they too fly south, to the same spot where their parents migrated. How do the young birds know where to go? With no parents to guide them, and no practical experience, it seems on the face of it to be instinctive behavior employing knowledge somehow inherited from the parents. Perhaps the parents have the ability to convey a route map to the offspring. Perhaps the offspring follow some kind of trail; an odor perhaps. Whatever the explanation, and it is currently not understood, the humble birds exhibit a remarkable instinctive behavior. Humans like to believe that humans do not.


## INTELLIGENCE AND BEHAVIOR

Research into the behavior of genetically identical twins separated at birth reveals that humans are much more instinctive than we like to believe. Some identical twins have grown up in quite different socioeconomic environments, and yet 30 or 40 years later they exhibit startling similarities in behavior. The way we dress, the music we like, the foods we prefer, the partners we choose, the age at which we marry, even the number of children we have, and our political inclinations, as well as our basic intelligence, appear to be genetically influenced or determined.

That human behavior is largely genetically predisposed does not sit well with the widespread belief that we are "free and independent spirits."

Further research is confirming earlier ideas that the two halves of the human brain have complementary, but quite different, roles [2]. For most right-handed people, the right brain stores visual imagery, which is retrieved into the conscious mind by "parallel access." We recall faces directly, without having to go through some sequential index. The left brain stores textual information, names, and so on. Here we might find ourselves going through some alphabetic mental index when trying to recall a name: "I know it starts with 'B'." Somehow, the left and right brains associate the recalled name with the recalled face, respectively, via the bundle of nerves (the corpus callosum), joining the two halves.

The left brain, it seems, is suited to establishing sequence. In recalling a situation, we may recall a series of snapshots, and then piece them together in a logical sequence that satisfies us. The drive to complete the logical sequence is strong; we even reward ourselves when we complete the sequence. This is why we feel a quiet surge of satisfaction when we put the last word into a crossword puzzle, or the last piece into a jigsaw puzzle [3]. The sense of closure, of satisfaction with completeness, is strong within us, and the self reward[2] encourages us to repeat the exercise.

So strong is the sense of satisfaction, in fact, that we will go to some lengths to achieve it. There is some evidence that police witnesses "construct" parts of their evidence without being aware of it. They may recall a number of glimpses of some situation, but, in piecing the record together, their left brain fills in missing pieces, or perhaps arranges the pieces in the most logical sequence, even though that is not what they truly observed. The worrying aspect is that the witnesses may be totally convinced that their evidence is true and complete; in no sense are they consciously trying to deceive. They will swear that their evidence is true, and

---

[2] The level of dopamine in the brain increases, as do endorphins in the rest of the body, giving us a happy feeling of pleasure and enjoyment.

honestly believe it to be so, even in the presence of compelling evidence to the contrary.

Behavior is response to stimulus. Behavior may be the same response to the same stimulus, or perhaps a changing response to the same, repeated stimulus. Some behavior is reflex. If a doctor taps a person's knee in the right place, the knee jerks–every time; uncontrollably. Reflexes are reliably predictable. If, on the other hand, a person gets burned when they try to pick up a hot saucepan, they may use a cloth or glove to pick up the saucepan next time; this is learned behavior.

The distinction may become a little blurred. If a woman slaps a man in the face, he may step back, look (and feel!) surprised, put his hand to his face, but not otherwise react. If a man had slapped him, the response may well have been an immediate returned blow. For some men, to return a blow from another man is virtually a reflex action.

An octopus, placed in a glass jar on a table on the deck of a yacht, has been observed to climb out of the jar, slither to the corner of the table, slide down the table leg, cross the rough, dry deck, and drop over the side into the sea. This is strikingly intelligent behavior, particularly since the octopus had to go into the open air, thereby risking its life, and had to cross the dry deck without the aid of any skeleton.

So, intelligent behavior gives evidence of thoughtful planning, and perhaps a willingness to take risks, or to avoid threats. Another characteristic of intelligent behavior is that the same stimulus, repeated, may not evoke the same response.

Humans in groups may exhibit group behavior. Group behavior is an emergent phenomenon, in the sense that the group behaves in a way that none of the individuals would subscribe to at a personal level. Carl Gustav Jung, the noted Swiss psychiatrist, observed as follows:

> *It is a notorious fact that the morality of society as a whole is in inverse ratio to its size...Any large company composed of wholly admirable persons has the morality and intelligence of an unwieldy, stupid and violent animal. The bigger the organization, the more unavoidable is its immorality and blind stupidity...The greatest infamy on the part of...a man's...group will not disturb him so long as the majority of his fellows steadfastly believe in the exalted morality of their social organization* [4].

If it seems that Jung was exaggerating, think of the rise of fascism during the 20<sup>th</sup> century, or, currently, consider the regime that operates within some industrial organizations. Companies may lie and cheat to conceal problems or protect profits. Some companies sell products, knowing them to be harmful, even lethal, yet do not feel a need to warn customers, let alone withdraw the product. Companies pollute the environment rather than clean up their waste. And so on. It is not that the individuals in the companies are unworthy. This is emergent group behavior, where individuality is, to quote Jung again, "driven to the wall."

Other psychologists have taken up Jung's work; worthy of note is the Myers-Briggs Type Indicator [5], a procedure for testing individuals to determine their "personality type." Types are classified in mutually exclusive pairs, as:

- *E*  Extroversion—people who focus on the outer world of people and the external environment; or
- *I*  Introversion—people who focus more on their internal world.
- *S*  Sensing—people who use their senses to appreciate the realities of a situation; or
- *N*  Intuition—people who look at the big picture and seek to grasp the essential patterns.
- *T*  Thinking—people who work things out logically, objectively, on a cause-and-effect basis; or
- *F*  Feeling—people who make a decision in personal terms, using personal-centered values.
- *J*  Judgment—those who take a judging attitude, structured, organized; or
- *P*  Perception—those who prefer to live in a flexible, spontaneous way.

With four pairs, there are sixteen personality types: a person could be deemed *E-N-F-P*, or *I-S-T-J*, and so on. Some organizations test potential recruits to determine their personality type, using Myers-Briggs.

## BELIEF SYSTEMS

Human behavior is strongly influenced by belief. With experience and education, we come to believe that some aspects of our lives, culture, and situation are open to interpretation. We establish in our minds a set of beliefs that allows us to interpret the way that our world works. These different beliefs may be called our belief system.

Our beliefs are important to us. In primitive times beliefs might have been the difference between life and death: beliefs enabled early man to categorize situations as advantageous or dangerous. Today, our beliefs allow us to assess situations easily and quickly; thereby reducing any psychological uncertainty in the mind of the believer.

Figure 3.1 below shows how beliefs might impact people; both at an individual and at a cultural or societal level. In the upper loop, beliefs substitute for the believer's need to investigate events and situations, allowing instead a straightforward explanation based in the belief. This reduces the believer's psychological uncertainty, and so reinforces the belief system–if only because it works.

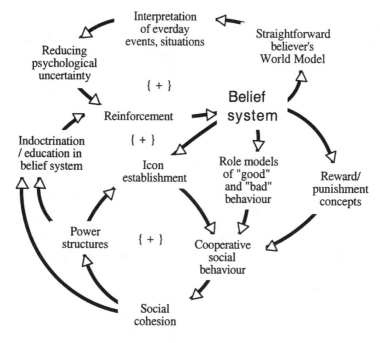

**Figure 3.1** *The human belief system.*

In the lower loops, a belief system may present three aspects: role models of good or bad behavior; reward and punishment concepts for being good or bad; and icons, which represent the belief. An icon may be a figurehead of an organization, a national flag, an award, or almost anything that may represent the belief. During World War II, Field Marshall Montgomery's cap badge became an icon, representing the leader, who was himself an icon; representing the indomitable spirit of the Eighth Allied Army that he led.

Beliefs persist so long as they work; they need have no basis in ground truth whatsoever. If the rains come to the parched land only after human sacrifice, then it may be believed that human sacrifice will bring the rains in the following year; indeed, more sacrifice may bring the rains even earlier. Should the rains, by chance, come earlier, the belief will be reinforced, and more human sacrifice may follow in later years.

Established beliefs, in which numbers of people subscribe to the same beliefs, result in social cohesion, which leads in turn to the development of power structures that are likely to establish and sustain an icon of the mutual belief (perhaps a temple, a book, or a mode of dress). The belief system will also be spread by indoctrination, politics, education, and inheritance.

As Figure 3.1 above shows, a belief system may serve not only to reduce perceived entropy in the believer, but also to proselytize on a wider scale, promoting social order–reduced entropy–in the process.

It is possible to recognize in the figure many of the great beliefs, such as world religions, political dogmas, and so on, but it is also true that beliefs arise and propagate continually at much more modest levels. Fashion, and being in fashion, follow similar rules, as does identifying with a particular culture.

## RESISTANCE TO CHANGE

One notable aspect of human behavior is resistance to change. The causal loop model above suggests one of the reasons. People form into groups with shared beliefs. Such groups develop and exhibit characteristics that may be referred to as "culture." Cultures tend to sustain themselves by continual reinforcement. Cultures provide people with a view of their world and a ready explanation of situations and phenomena that reduce uncertainty (perceived entropy), offering reassuring psychological closure. As we have seen, such explanations need have nothing to do with what might be called reality, or real truth. People of different cultures may encounter the same situation and may explain it to their own satisfaction in mutually contradictory terms.[3]

Cultural beliefs are propagated not only within and between communities, but also through time, as generation succeeds generation. Icons of belief may also persist through time, together with the power structures and indoctrination practices. These combine to create cultural resistance to change, where the term "culture" may refer not only to some great faith or political dogma, but also to local or regional attitudes and behaviors.

Culture develops and persists in companies and organizations in commerce, manufacturing, clubs, families, sports, and entertainment, too. Attempts to

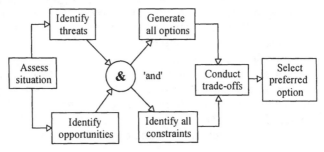

**Figure 3.2** *Naïve decision making.*

---

[3] Aztec sacrifices, including cutting out the still-beating human heart, were apparently designed to ensure that the sun rose each morning. Ancient Egyptians also worshiped the Sun as the source of all life, but seemingly never resorted to human sacrifice.

introduce change management into industry are frequently met with less than total success, leaving would-be "change managers" mystified. But then, they may not have realized that cultures, rather like aircraft carriers, exist to defend themselves.

## DECISION MAKING

We are all experts at making decisions. After all, we are humans, we make frequent decisions, so we must understand how humans make decisions. Research suggests that there may be more to it than we subjectively imagine.

Decisions are choices between options. We make decisions with a view to achieving, or at least influencing, some outcome. Decisions, then, are ultimately based on some predictions about the future: how things will turn out if we choose A as opposed to B or perhaps C. Rationally, we know that predicting the future is fraught with error, but this knowledge does little to diminish our need to make decisions. See Figure 3.2, which shows how analysts thought people made decisions until fairly recently. This so-called naïve approach is widely used for making public decisions—in systems engineering, and in command and control.

One reason why we feel obliged to make decisions is our cerebral drive to reduce perceived entropy. We make decisions to overcome or escape from uncertainty. Uncertainty is psychological entropy. Making decisions gives us the feeling that we have come to a conclusion, that the matter is resolved, that we have achieved closure. We may be wrong, but we nonetheless reward ourselves with a shot of endorphins, believing we have done well, we have completed the jigsaw puzzle, or the crossword puzzle.

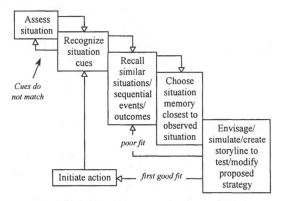

**Figure 3.3** *Satisficing: finding the first acceptable solution.*

The satisficing model of decision making in Figure 3.3 is quite different, and is one that experts may employ under time pressure. Instead of generating all the options and all the constraints, the expert knows from experience [6] what will work. He or she recognizes situations from cues, and recalls appropriate strategies from experience. The expert can then run a fast mental simulation of a putative strategy, perhaps modify it according to the situation, and then initiate action. Unlike the naïve decision maker, the expert does not cling to first decision, but may continually review and change the strategy as events turn out, until a successful conclusion is achieved. Decision making by satisficing is more a continual process rather than a single event.

The generic decision making model of Figure 3.4 shows how decision making appears to operate in general. The essence of some situation or issue is set against a number of strategies, seeking a pattern match that indicates that a particular strategy would have a good likelihood of success. According to the quality of the process, the pattern-matching process might include world models, tacit knowledge, training, experience, beliefs (e.g., cultural viewpoints), and doctrinal factors.

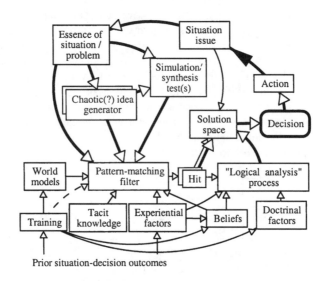

**Figure 3.4** *Generic decision making model.*

For satisficing, a reasonable hit would offer an initial solution and lead to an action, and the process would repeat if the initial solution turned out to be less than satisfactory. In this way, satisficing "homes in" on the best solution.

For naïve decision making, the "logical analysis" process would be employed, bringing many possible strategies together in a trade-off process. The process might, but is less likely to repeat, compared with the satisficing process. It is the mark of the naïve decision maker that he tends to make a decision and stick to it, even in the face of mounting evidence that it was wrong.

## PERCEIVED ENTROPY REDUCTION

Overall, it seems that human behavior is aimed at reducing perceived cerebral entropy. Entropy reduction also occurs in another way: as people form into groups, the group diminishes the individuality of the person, making the behavior of group members more alike, and hence more predictable. Training encourages individuals to conform to behavioral patterns. Military personnel may operate and behave as one to provide the group with more focused power; this conformity not only reduces risk and danger to the group, it also reduces behavioral diversity (perceived entropy) and makes the group more cohesive–and predictable.

Behavioral conformance is encouraged in every walk of life. Road signs tell drivers and pedestrians how to behave. Driving schools teach people the "right way" to drive. People worshiping in churches, mosques, and temples follow strict rules of procedure and behavior. Schools teach a variety of subjects, but within each variety, they tend to teach only established and accepted views. Students are examined so that only those who conform will pass and succeed.

Although some individuals rebel against these "rules of behavior," most do not. Group membership depends upon adhering to the cultural and behavioral rules, and since most people wish–need–to belong, most conform. The rules set out "good" and "bad" culture. An individual may be a member of two different groups. Within one group, he may be an upright member, a pillar of the community. Within the second group, where theft and murder are considered "good," he may be equally esteemed for stealing and killing. The individual may have no difficulty in reconciling these two positions: we all, it seems, have almost schizophrenic abilities to switch between cultures on demand.

## SUMMARY

Human behavior may be complex. Generally, however, behavior can be predicted; given enough information about the individual. People can be trained to behave in like manner, to cooperate, and to limit the range of their activities. People in groups are likely to be more predictable as a group than they would as individuals. Trained people in groups may be highly predictable in suitable circumstances. Different cultures also influence the way people behave, making people of particular cultures more predictable.

## ASSIGNMENT

A soldier, patrolling as part of United Nations peacekeeping duties, is talking at a road junction with local civilians, when they all hear a dull thudding sound coming from a nearby wall. The soldier immediately runs for cover, while the civilians look around to see where the noise came from. Explain and describe the different mental processes occurring in the soldier and the civilians.

---

## REFERENCES

[1] Tajfel H., and C. Fraser (ed.), *Introducing Social Psychology: An Analysis of Individual Reaction and Response*, Harmonsworth, UK: Penguin Books, 1978.

[2] Gazzaniga, Michael S., "The Split Brain Revisited," *Scientific American*, Vol. 279, No. 1, 1998.

[3] Kuhn, Thomas S., *The Structure of Scientific Revolutions*, Chicago, IL: 2nd Edition, University of Chicago Press, 1970

[4] Campbell, Joseph (ed.), *The Portable Jung*, New York, NY: The Viking Penguin Inc., 1971, pp. 100-101.

[5] Briggs Myers, Isabel, and P. B. Myers, *Gifts Differing: Understanding Personality Type*, Upper Saddle River, NJ: Prentice Hall, 1995.

[6] Klein, G.A. "Recognition Primed Decisions." In *Advances in Man-Machine Research*, Vol. 5, pp. 47-92, W.B. Rouse (ed.), Greenwich, CT: JAI Press, 1989.

# Chapter 4

## Systems Engineering Philosophy

*All philosophies, if you ride them home, are nonsense, but some are greater nonsenses than others.*

First Principles, *Samuel Butler*, 1912

### HOLISM AND EMERGENCE

Systems engineering may have existed–at least as a philosophy–for thousands of years. The fourth dynasty of the ancient Egyptian culture created enormous pyramids in relatively short times some 4,500 year ago without any significant tools other than ropes. The Great Pyramid of Khufu, the last remaining wonder of the ancient world, is still the largest manmade stone building on earth, and it was constructed in about 20 years.

The ancient Egyptians drew up careful plans, and built the many parts of the pyramid complex in parallel; a practice that today's systems engineers would call concurrent or simultaneous engineering, intended to reduce project duration. The Egyptians had a goal, and they had a system for building so that upwards of 20,000 men could work simultaneously: leveling, marking out, quarrying, hauling, raising, and constructing causeways, pyramid sides, valley temples, walls, chambers, and so on.

This outstanding feat illustrates some of the features expected of systems engineering: well planned, well managed, effective, efficient, systematic, and organized. (The same might also be said of the ancient Egyptian civilization at that time which came to be looked upon as the Golden Age.) However, these features are not exclusive to systems engineering; project management might justly lay claim to some or all of them. What is it that is special or unique about systems engineering?

The motives for conceiving modern systems engineering are to be found, at least in part, in past disasters. Arthur D. Hall III [1] cites: the chemical plant leakage in Bhopal (1986); the explosion of the NASA Challenger space shuttle (1986) and the Apollo fire (1967); the sinking of the Titanic (1912); the nuclear explosion in Chernobyl (1986); and the disaster at the Three Mile Island power plant (1979). He cites, too, the capture of markets by Japan from the U.S., the decline in U.S. productivity, and the failure of the U.S. secondary school system. He identifies the millions of people dying of starvation every year while other nations stockpile surplus food; medical disasters such as heart disease, while governments subsidize grains used to produce high cholesterol meat, milk, and eggs; and many more. One implication is clear. Systems engineering faces challenges well beyond the sphere of engineering.

The modern philosophy–the "why" and the "how" of today's systems engineering–developed largely at NASA in the 1960s and 1970s; see Table 4.1.

**Table 4.1**
Early Developments in Systems Engineering Philosophy

| Developing Philosophy | Exemplifier |
|---|---|
| Systems engineering requires a clear, singular mission and goal. | "…to put a man on the Moon and return him safely to Earth by the end of the decade." (President John F. Kennedy) |
| There should be a sound concept of operations (CONOPS) from start to finish of the mission. | The mission will be executed in phases, marked by transitions. The first phase will place three men in earth orbit…. |
| There should be an overall system design that addresses the whole mission from start to finish. The full CONOPS should be demonstrably realized in the design. | An overall system design is needed to ensure there are no weak links in the chain… and that the concept of operations can be completely realized. |
| Overall system design can be partitioned into complementary interacting subsystems. Each subsystem should have its own clear mission and concept of operations…. | Each subsystem must be clearly defined in terms of fit, form, function, and interface to ensure overall system integrity upon integration. |
| Each subsystem may be developed independently and in parallel with the others, provided that fit, form, function, and interfaces are maintained. Where any emerging deviations are unavoidable, whole system redesign may be revisited. | As part of development, each subsystem should be rigorously tested in a representative environment, such as would be presented by the extremes of the mission and by the behavior of other subsystems. Deviations at subsystem level will affect other subsystems and the overall mission system. |
| Upon integration of the subsystems, the whole system should be subject to tests and trials, real and simulated, that expose it to extremes of environment and to hazards, such as might be experienced during the mission. These would include full mission trials where recovery from defect was possible. | The whole system, including the operators and crew, should be subjected to rigorous tests and trials in representative environments, including hazards and emergencies. |

Working on high-profile, high-risk projects such as Apollo gave NASA clarity of view. Every ounce of payload counted. Every interface had to match. Every function had to contribute. Nor was it sufficient for the designer and developer of any one subsystem to state that his part was good, and that it was someone else's problem. The whole mission system might be developed as separate subsystems, but they were all to be developed correctly to the overall plan, or they were all subject to change.

On the other hand, it became clear that the detail of what was inside the various subsystems was of secondary importance at system design level, provided the subsystem could be depended upon to operate as it should, and meet the criteria of compatible fit, form, and function.

Although the term may not have been used, theirs was–had to be–an organismic approach, constrained as they were by hard limits in rocket lifting power. The overall payload consisted of men, and several craft that fitted together like Russian dolls, including a moon-lander with sufficient fuel both to soft land and to lift off again into moon orbit, a return craft with fuel, and a re-entry vehicle. Changes to any one part of the payload that affected mass, moment of inertia, interface, form, volume, function, capability, and so on, was likely to impact other parts of the payload, which would necessitate rebudgeting, redesign, and redevelopment. The whole payload was, therefore, analogous to an organism (see page 4), in that the various parts were mutually interactive, interdependent and combined to produce emergent properties, capabilities, and behaviors.

It was not possible for one mission system designer to know the intimate details inside each and every subsystem, but then he did not need to know. There was a system design team for each subsystem, whose task was analogous to that of the whole-mission system designers. These subsystem designers were vitally interested in detail down to sub-subsystem level, and so on.

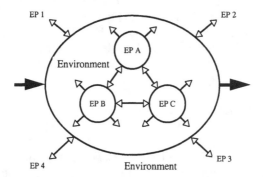

**Figure 4.1** *Systems engineering philosophy and emergence. The emergent properties, capabilities, and behaviors of the whole system, EP1 to EP4, derive exclusively from those of the subsystems, EPA, EPB, and EPC, and from their interactions, interchanges, inflows and outflows, and environments. There is no other source possible.*

So it became evident that the complexity could be encapsulated inside each subsystem, and that each subsystem, no matter how complex it might be on the inside, could be represented by its external fit, form, function, and interface. These properties were emergent, and so there arose the association between systems engineering and emergence. Although it may not have been stated as such, the emergent properties, capabilities, and behaviors of the whole system derived from the emergent properties, capabilities, and behaviors of the contained subsystems, and from their interactions and interchanges. There was no other source, as Figure 4.1 illustrates.

So the philosophy of systems engineering was, and is, fundamentally *holistic*, in the sense that any system should be conceived, designed, and developed as a whole. The basis for this philosophy is to be found in the pathology of systems that have failed in the past, in the poor performance of systems that have been "cobbled together" from available or separately developed parts, and from the failure of many complex projects to complete at all. Only by considering the whole problem, the whole issue, is it feasible to conceive and create a whole, balanced solution.

The philosophy of systems engineering is also *organismic*, in that the whole system-to-be-conceived/designed/created/operated is viewed as an open system, and as analogous to an organism. The various subsystem parts are seen as interactive and mutually interdependent, such that constraints on the whole system necessitate both complementation and compromise within the parts and their interactions. Only in this way, and from this standpoint, is it possible to create optimal systems; that is, systems that satisfy limiting criteria such as value for money, cost effectiveness, overall mass/volume/form/moment, and so on.

Finally, the philosophy of systems engineering is *synthetic*, in that systems are built from parts that are themselves systems, interconnected in such a way that the whole delivers requisite emergent properties, capabilities, and behaviors. Synthesis is the opposite of reduction [2]. Reduction looks into a system; synthesis looks out of a system. Reduction breaks down; synthesis builds up. Analysis, looking into things, yields knowledge; synthesis, looking outwards, gives understanding [3].

## PROBLEM SOLVING

Systems engineering philosophy recognized, too, that there might be many solutions to a problem, or issue. (Thus, there were many different ways in which NASA could have put a man on the moon). Some of these potential solutions will inevitably be better than others, where "better" means that they match the mission objectives and CONOPS more precisely.

So, there emerged a systems engineering problem-solving paradigm (SEPP), the purpose of which was to find the best solution to any problem. Figure 4.2 shows the process. The intention is to avoid seizing upon the first idea that comes to mind, but instead to conceive a range of possible solutions. At the same time, but independently, the criteria for a good solution are generated. There is a trade-off process, in which the optional solutions are matched against the criteria, and the best match is selected. Variations on the theme allow good features from losing options to be included into the preferred solution, making it even better [4].

All parties within a project, business, or enterprise may use the systems engineering problem-solving paradigm at all levels to address all kinds of problems from the everyday, mundane to the grand scale.

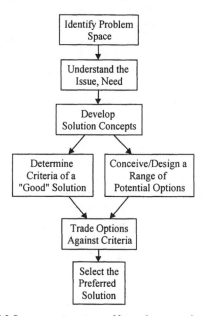

**Figure 4.2** *Systems engineering problem-solving paradigm (SEPP).*

Problems arose at every turn. For NASA, some were big problems, as when rocket control tests failed. Some problems, on the other hand, were not so big, as when some process was too fast. Using the SEPP at each and every decision point throughout a project enabled rapid, effective decision making and problem solving. It also involved a number of people, usually in a team, so it had the side benefits of informing and involving team members. Today's projects, businesses, and industries face the same situation. Problems arise at every turn, necessitating rational, effective problem solving and traceable, justifiable decisions.

Using broad-based methods such as the SEPP has become an integral part of the systems engineering philosophy, to the extent that some organizations have adopted the SEPP almost as a way of life for making all decisions throughout the company.

## PERCEPTIONS OF CONNECTEDNESS

A theme within systems philosophy concerns "connectedness," in that all things in the world are connected, either directly or indirectly, through other things. The parts of a system in particular are connected by definition, since the parts influence each other and the whole. This notion has permeated through systems engineering philosophy. In solving a problem, a system solution will interact with other systems in the solution space, as a stone thrown into a pond creates ripples that reflect off other stones in the pond.

The realization of "universal connectedness" has generated the need to create boundaries, to prevent systems engineering from attempting to create systems that are too big or that address too much. It has also generated awareness that introducing a new system, or altering the emergent properties, capabilities, and behaviors of an existing system, will affect other systems in the solution space; that, after all, may be the intention. These other systems will react, influencing and changing the new system, so that its emergent properties, capabilities, and behaviors will change with time after introduction. (Although obvious, this is an organismic viewpoint; one not readily accepted by reductionists.)

Recognition of this dynamic interactivity has led to a belief in simulation and modeling as the key to system conception, design, development, and proving. System engineering seeks to create a system that will exhibit requisite emergent properties, capabilities, and behaviors when it is in operation; interacting with other systems, but not necessarily when it is in isolation, without feedback from the environment and other systems. By simulating the future system interacting dynamically with other systems in the future environment, actions and reactions can be observed and accommodated.

Dynamic simulation reveals counterintuitive behavior [5] in particular. This is unexpected system behavior that may arise within interacting systems, even relatively simple ones. Counterintuitive behavior may be beneficial, adverse, or even chaotic, but, once predicted, it can be anticipated and, if need be, avoided or exploited. Dynamic simulation and modeling have also become part of the philosophy, practice, and even the touchstone of systems engineering.

## SYSTEMS OF SYSTEMS

In recent years a new philosophy has appeared on the scene calling itself "system of systems." To those steeped in system theory, science, and engineering, the title is tautological. In principle all systems are comprised of subsystems that are themselves systems, so every system is a "system of systems."

The term is being applied to the creation of new systems by bringing together existing operational systems under a single umbrella and, presumably, creating or adapting links and interactions between the operational systems, which become subsystems of the higher level umbrella system.

An illustrative example of a commercial system of systems might be the creation of an integrated public transport system, where there previously existed separate, uncoordinated railways, ferries, air services, and bus services. To make such an amalgamation worthy of the title "integrated system," there would need to be harmonization of timetables and of capacities, and the introduction, or enhancement, of "interchange stations," where passengers could transition smoothly from one mode of transport to another; all of which would take time and money to create, but which would, it is believed, solve a problem.

Forming a system of systems may be pragmatic and it may be necessary, but is it systems engineering? Certainly, it appears inconsistent with the philosophy created by NASA in the early days of Apollo, although that philosophy evolved in the first instance to address new, unprecedented systems. An integrated transport system is new in one sense, but it would be more reasonable to view it as an adaptation of existing, separately developed subsystems, or piecemeal development as Gwilym M. Jenkins aptly described it [6]. Missing are: initial, overarching conception; a complete CONOPS; and an end-to-end balanced system design optimized to provide, say, value for money.

It is not unreasonable to predict that, in consequence, this kind of system of systems will be less than optimal (that is, will not provide good value for money), and may exhibit counterintuitive behavior.

## Bottom-up Integration

While the concept of "system of systems" may be poorly defined, creating a new system from existing subsystems is not. In some industries the practice is to buy complete subsystems from subcontractors, or to build a new system using some bought-in and some legacy subsystems; that is, subsystems that are left over from a previous system. The practice has its merits. Subsystems are available, they may have performance and reliability track records, and their cost is known. However, the bought-in or legacy subsystem rarely does precisely what is needed; in particular, it may provide unwanted functions (paid for, but not used), there may be some essential functions it does not provide, and its interfaces may not be as required by other subsystems. Moreover, there is an implicit risk that unwanted functions within bought-in items may create unwanted effects, or security risks, in the final system. Verifying and validating such composite systems may prove difficult; even impracticable.

The answer is frequently a custom interface device; one made especially to match the bought-in/legacy item, and to provide the missing functions. The custom interface device may overcome the immediate interconnection problem, but it may take time to conceive, design, and create; it costs money; and it has to be tested, trialed, and maintained. There is evidently a breakpoint beyond which the penalties of introducing the custom interface device outweigh its value. In practice, this approach to creating systems may be pragmatic, but it is unlikely to create an optimal solution, and it risks being viewed more as piecemeal cobbling together, than as effective systems engineering.

Systems integration is advocated as a simpler and cheaper alternative to systems engineering. It is a mechanistic, building-block approach, as opposed to a holistic, organismic method, and, as such, is unable to accommodate the internal subsystem trades necessary to satisfy overall system constraints, such as value for money, life cycle cost effectiveness, and so on. Moreover, it will not anticipate counterintuitive behavior upon introduction into operation. However, it is a form of synthesis and as such can be used to create systems.

The basis for advocating systems integration, as opposed to systems engineering, is one of pragmatism. Bringing together systems that already exist, or that can be readily obtained, is a faster, cheaper way to create an integrated system. Advocates propose that it amounts to the same thing as systems engineering in the long run. It does not. Bringing together existing systems under the aegis of some umbrella system in the hope and expectation of creating an effective solution to a problem, or bottom-up integration as it is often called, risks creating inappropriate, ineffective systems, deficiencies, and even disasters, such as those cited by Arthur D. Hall III at the start of this chapter.

Bottom-up Integration: the Integrated Transport Dilemma

As an illustration of the limitations that are inherent in bottom-up integration, consider the integrated transport issue. The drive for integrated transport usually arises because of road traffic congestion around and in big cities. Analysis of traffic patterns shows that congestion peaks during times of commuting into and out of the city, in the mornings and evenings, and also in the suburbs, particularly when mothers take children to and from school. An integrated public transport system is expected to prove such an effective and attractive alternative to people using their cars, that the traffic peaks will be reduced, traffic flows will be eased, and expensive major new road schemes can be shelved.

So, the root problem is traffic congestion caused by synchronized commuting and taking children to school. Studies and experiments all over the world have shown that people using their cars to commute into and out of work in the cities are unlikely to be persuaded out of the car and into the bus or train, no matter how integrated these might be; many people just want to be in their cars,[1] on their own, or with a colleague. Mothers take children to school by car out of fear for their child from traffic, pedophiles, molesters, drug pushers, violent children, and so on. More available public transport does not answer their concerns.

Is integrated transport the answer? Systems engineering philosophy dictates that we look at the wider picture, before moving down to detail; that is, analyze the problem top down, and not integrate a supposed solution bottom up. The wider picture looks at the reasons for the congestion. People commute into work because there are more jobs of greater variety in the city, because city jobs pay more, and because they like being at the center of things; the seat of power.

An alternative approach to an integrated transport system would be to take work to the workers, rather than bring the workers to work. This can be done in at

---

[1] Even the Los Angeles driver on the Ventura Boulevard who, over an extended period, shot other drivers for lane changing, only managed to improve lane discipline. Drivers still stuck to using their cars, even at the risk of being shot.

least three ways. In Paris, France, the population density outside of working-hours is over five times that of London, England. Why? Because a large proportion of the people that work in Paris also live there. Most Parisians live in large apartment blocks with inexpensive, government-regulated rents. So there is less need to commute into the center of Paris from the outlying suburbs. Schools also exist in the city, reducing–although not eliminating–the daily school delivery run.

In the City of London and in Manhattan, by contrast, most people commute to and fro, so the large office blocks, filled by day, are mostly empty outside of working hours.

Another way to take the work to the workers is by teleworking; that is, to have staff working at home using personal computers connected to the organizational hub by telephone, cable, satellite, or radio. This approach is not for everyone, but many people are happy to work at home two, three, or more days a week, which allows the organization to rent less expensive city space. Where home working is not the ideal, groups of staff may elect to set up a small local office in the suburbs or the countryside, and telework as a group, thereby maintaining the gregarious work pattern that some people require.

Government departments and large companies may move lock, stock, and barrel to some town or "green-field site" remote from the city, taking as many of their staff as are willing to go, and recruiting new staff in the new location. If just one major company moves out of a city, the impact on road congestion and on public transport can be significant–for a while. New organizations, new commuters, and new mothers will soon move in to use up any spare transport capacity–unless there is either a disincentive or a more attractive option; a stick or a carrot.

Organizations may be persuaded to telework, move out of the city, and so on, by governments making it economically worth their while. Such potential solutions to the problem of traffic congestion are not mutually exclusive, either. So, there are several cogent options to an expensive, possibly ineffective integrated transport system.

Unless system conception starts at the top, however, objectives, options, and alternatives are likely to remain unrecognized or undervalued. In this particular example, chances are that a combination of all the approaches may be necessary. In addition, chances are that an integrated transport system on its own would *increase* the numbers commuting daily, not only by bus and train, but by car too, as new companies took up city center residence in this now more attractive city. Overall, the issue is so complex that a simulation model would seem to be necessary to represent the many subsystems, flows, interchanges, and influences....

### Systems of Systems: Volume Supply

For decades, electronics manufacturers have built stocks of standard modules: power supplies, preamplifiers, amplifiers, tuners, modulators, demodulators, and so on. These have been made in such a way that they can be "bolted together" almost willy-nilly, making the design and construction of a new radio or high fidelity system, straightforward and fast. The characteristic of the various modules is that they have standard interfaces arranged such that each module is largely unaffected[2] by the absence or presence of other modules. This appears, prima facie, to be bottom-up integration.

The use of standardized parts is an indicator of a different kind of systems engineering: mass production or lean volume production systems engineering. The overall system is made up from a number of smaller process subsystems arranged in pipelines—a system of systems—with upstream processes manufacturing parts, subsequent processes incorporating parts into subassemblies, and the final process assembling the subassemblies into a variety of finished products that then go into the marketplace to be sold.

Systems engineering concerns itself with overall conception, design, and balancing of the end-to-end process, and with emergent properties, capabilities, and behaviors of the subsystems; for example, their capacity, material efficiency, throughput rate, and on. Only by balancing the many interactions between all the subsystems within the overall mass production or lean volume supply system is it possible to optimize the performance of the overall system. Changes in any subsystem impact other subsystems, causing blockages, queues, shortages, incompatibilities, and so on. Systems engineering in this context is, again, holistic, synthetic, and organismic.

## SYSTEMS OF SYSTEMS IN DEFENSE

Systems engineering has been employed for over half a century in the defense industry, notably in the U.S., but in many other countries, too, largely to reduce development risk, and to promote thoroughness.

Perhaps not so obviously, military[3] formations are also systems of systems. Figure 4.3 shows a simple example of a military group or task force. It could represent an army battle group, for instance. The fighting elements might be infantry, armor, and artillery. The battle group would have its own command and control, with intelligence to provide information about an enemy's current activities and future intentions. Support units provide logistic support, ammunition, food, transportation, medical treatment, and casualty evacuation. Communications enable the various elements to coordinate their actions. Depending on the theater of conflict, there may be surface-to-air missile defenses, electronic countermeasure facilities, psychological operations, and many more.

The whole makes up a classic system in which the internal parts interact in a coordinated manner to create overall system emergent properties, capabilities, and

---

[2] For example, some modules are designed to exhibit infinite input impedance.
[3] In this context, military includes army, navy, air force, marine, and related categories.

behaviors; these can only be defined sensibly in the context of an operational environment that, in the case of a military group, would detail the theater of war and the opponent.

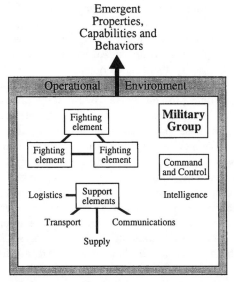

**Figure 4.3** *A military group as a system of systems.*

As Figure 4.3 suggests, the performance of an individual platform (tank, ship, or aircraft) within any of the fighting elements is not necessarily paramount in the context of the whole group. Platform-to-platform automated data communications, such as the Joint Tactical Information Distribution System (JTIDS), may enhance the capability of a group of platforms, for instance, forming a mosaic area situation "picture" from platform sensors, thereby enabling the weapon from one platform to be guided by another's sensors and controls, triangulating on jamming sources, and so on. This real-time data sharing constitutes a so-called "force multiplier," where the capability of the group exceeds the sum of the individual fighting elements. The use of such communications systems, which provide not only communications, but also navigation and identification (hence called CNI systems), too, has led to the introduction of the term "network-centric" warfare.

It has become the practice to describe the *capability* of a complete military group. "Seize and hold an opposed bridgehead anywhere in the third world" might describe a capability; the description carries within it implications of environment, climate, and opposition capabilities. For a particular nation's forces, the ability of different fighting elements to form and work together quickly and effectively is the subject of much training and practice. It is also important, in network-centric warfare, for the fighting elements to be able to network; that is, to be fitted with compatible, interoperable data links; to exchange the requisite information across those links; and to employ appropriate and compatible procedures.

Bringing together different force elements with their different support facilities creates different capabilities. Less obviously, the same force elements would have different capabilities if their doctrine of war, their concept of operations (CONOPS), and their rules of engagement were changed: Figure 4.3 represents a complex sociotechnical system; not just an array of weapons. Similarly, battle-hardened infantry might not perform too well in a peacekeeping role. The "construction" of a group capability from the various fighting and support elements is holistic, synthetic, and organismic[4].

A potential difficulty arises in the case of international coalition forces deployed to address some problem on the world stage under, for instance, the auspices of the United Nations. Coalition partners contribute their own parts of the overall military group: these may be fighting elements ("teeth arms"), logistics, communications, intelligence, and so on. Partners may also wish to retain some level of control over their own forces, although there may well be a need to coordinate the activities of the whole coalition force as a single entity.

As a result, building a coherent overall capability is less than straightforward. Looking at the issue from an abstracted systems viewpoint, there are two potential routes: build the overall capability by choosing lower level elements from each and every coalition partner and combining them as in Figure 4.3; or coordinate the capabilities of each coalition member's contribution to the force, as implied in Figure 4.4.

In the figure, Capability X might be a multiarm (e.g., air and land) force capability, or it might be an airpower capability, say, and Capability Y might be a naval power capability. Capability Y, for instance, might include carrier-borne early warning aircraft, ship launched missiles, antisubmarine, and mine clearing capabilities; all from different coalition partners.

Potential issues are evident in Figure 4.4. Coordinating the activities of the coalition necessitates a singular point of command and control; it also requires an excellent communications infrastructure, so that appropriate information is available where and when needed, and not otherwise. Single-point, high-level planning ensures that various force elements coordinate their actions; do not interfere with each other's actions (deconfliction); do not attack a target already destroyed by another force element; use the appropriate weapons to avoid collateral damage, for example, where that is an issue; do not cause so-called blue-on-blue casualties (that is, kill other coalition personnel), and so on.

The concept of network-centric warfare also raises issues. To qualify as a coalition partner, it may prove necessary for the fighting, command and control elements, and others to form part of the network. Only those nations with equipped platforms, the appropriate automated information handling facilities, and the correct procedures would be acceptable coalition partners.

---

[4] Many military personnel would neither recognize these terms, nor–if they did–wish to be associated with them. That does not detract from the veracity of the statement.

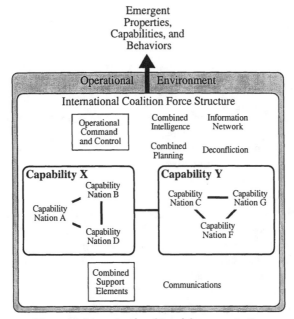

. **Figure 4.4** *International coalition defense organization.*

An overall coalition force, with its emergent properties, capabilities, and behaviors, might be described as a system of systems. The creation of such a force would appear, however, to be a classic systems engineering process, starting with an overarching concept of the force capability to be created in the appropriate context and environment. A concept of operations would follow, coupled with an overall, end-to-end design. This design would then be partitioned into subsystems; each of which would have its corresponding capability, together with interfaces, infrastructure, communications, and so on, to enable the whole to work as one system.

There are many ways to partition an overall design. For an international coalition force, the simplest way might be to choose partitions that corresponded to, or were consistent with, the likely contributions from the various nations. However, it is possible to partition in many different ways; each of which presents different interface, infrastructure, information exchange, and communication needs (partitioning schemes).

By pursuing a number of different partitioning schemes it is possible to: a) compare and contrast different schemes in terms of effectiveness, political acceptability, and so on, and b) establish a scheme that would accommodate many of the more acceptable portioning schemes. The complexity of this process would be managed using simulation models showing operational effectiveness with different partitioning schemes and with different networking capabilities, for instance.

Systems engineering in this context, too, is evidently holistic, synthetic and organismic, as a comparison between Figure 4.4 and Figure 1.3, on page 12, confirms.

## SYSTEMS ENGINEERING PHILOSOPHY IN METHODS

Systems engineering methods and practices have emerged that are consistent with systems engineering philosophy, and that may be employed successfully without the philosophy being necessarily evident to the user. The following is a short list of tenets applied to concept formulation:

- *Highest level of abstraction.* When approaching a new issue or problem, maintain the highest level of abstraction for as long as practicable. For example, consider a sports car, not a Jaguar 120; a means of air transport rather than a Boeing 777; a system for selling new products rather than a sales force; an operating system rather than OS 10.2 Jaguar, and so on. Being too specific too soon leads to premature assumptions[5] and missed opportunities.

- *Disciplined anarchy.* Maintain a high level of abstraction by generating many options and many criteria, and by questioning tacit assumptions. Brainstorming is an example of disciplined anarchy.

- *Breadth before depth.* Analyze the whole problem space before "drilling down" into parts of the potential solution. It is all too easy to generate overwhelming detail so that trees hide the wood. Identify and describe the entire first level of elaboration, interactions, external interactions, and environments before partitioning or elaborating any subsystem.

- *One level at a time.* This is complementary to the previous tenet. Complete each level of elaboration before moving down to the next level. This prevents imbalance, with some parts/subsystems receiving undue prominence, or being neglected.

- *Functional before physical.* Elaboration of a purposeful system necessarily generates functions and activities, which may be grouped into various subsystems for physical realization.

These and other tenets [7] embed systems engineering philosophy into the conception of a system solution.

Many systems engineering tools exhibit consistency with systems engineering philosophy and the tenets bulleted above. The simple, but powerful $N^2$ chart (see Appendix A) encourages a wide system overview, enabling the user to see all of

---

[5] There is an old adage: "If the only tool you have is a hammer, pretty soon everything starts to look like a nail." Maintaining a high level of abstraction seeks to anticipate and prevent that kind of thinking.

the entities and their many interactions at once. By its design, it prevents the user from going into too much detail–there is insufficient room–and it also reveals the patterns that underlie the structure of systems.

Other computer-based tools invite users to enter, not only the requirements for a new system, but also the environment and wider system context in which that putative system will exist and operate. Some of these tools allow the user to run dynamic simulations of the requirements to see if they are complete, mutually consistent, dynamically sound and will result in the requisite, end system properties, capabilities, and behaviors. This process, too, is entirely consistent with systems engineering philosophy.

## SUMMARY

While systems engineering may have been around for hundreds, if not thousands, of years, it has been codified only in modern times. NASA's development of the Gemini and Apollo programs exemplified the new philosophy of systems engineering.

Systems engineering philosophy emerged in response to many major disasters (often attributed to piecemeal developments) as holistic, synthetic, and organismic. In this, its roots are to be seen in the models and ideas from General Systems Theory, but without reliance on its mathematics.

System of systems, supposedly a new philosophy, presents in different forms. In the commercial environment, it implies the bringing together of extant, operational systems under a single, higher level systems umbrella. This practice seems on the surface to be unsound if the objective is to create an overall optimal balanced system, with no counterintuitive behaviors.

In the defense arena, the same term, system of systems, is associated with network-centric warfare. This involves the careful bringing together of complementary force elements using communication networks as binding agents and as a mean of sharing and enhancing resources. There is an overarching defense system concept, design, and CONOPS. In this defense context, it is difficult to distinguish system of systems from classic systems engineering philosophy.

## ASSIGNMENT

What is meant by "systems engineering philosophy?" Explain, justify and illustrate the following concepts:

i)   Holism;
ii)  The "organismic analogy;"
iii) Top down; and

iv)   The building-block approach.

The practice has grown of customers undertaking their own concept formulation, CONOPS development, and overarching design, before presenting a commercial systems engineering organization with a set of requirements; created either by the customer or by a separate group employed by the customer. The organization is then expected to create the solution and to signify that it is "fit for purpose."

Comment on this arrangement. In particular give your views, with substantiation, on the following:

1.  The consistency of the practice with systems engineering philosophy, and the implications of any supposed inconsistencies.
2.  If the customer is always right, it is conceivable that his requirement will be: a) inadequate, b) incorrect, c) a poor solution to the problem, d) overly expensive, e) beyond the laws of physics, or f ) so expensive as to bankrupt your organization. For each of the possibilities a–f, suggest how likely the possibility might be and what might be done about it–supposing that you wish to retain the customer!
3.  The reasonableness and implications of accepting a "fitness for purpose" condition of contract in such circumstances.

## REFERENCES

[1] Hall, Arthur D., III, *Metasystems Methodology*, Oxford, England: Pergamon Press, 1989.

[2] Ackoff, R. L., *Creating the Corporate Future*, New York, NY: Wiley, 1981.

[3] Ibid.

[4] Lacy, James A., *Systems Engineering Management*, New York, NY: McGraw-Hill, 1992.

[5] Forrester, Jay W., *World Dynamics*, Williston, VT: Pegasus Communications Inc., 2nd Ed., 1971.

[6] Jenkins, G. M., "The Systems Approach," *Systems Behaviour*, Milton Keynes, England: Open University Press, 1972.

[7] Hitchins, D. K., *Putting Systems to Work*, Chichester, England: Wiley, 1992, pp. 246-248.

# Chapter 5

## A Theory of Complexity

*It is the nature of an hypothesis, when once a man has conceived it, that it assimilates everything to itself as proper nourishment; and from that first moment of your begetting it, it generally grows the stronger by everything to see, hear, read or understand. This is of great use.*

*Laurence Sterne, 1713-1768*

### CHARACTERIZING COMPLEXITY

Complexity is subjective. When a trainee technician first comes across an integrated avionics system, an integrated circuit diagram, or a process plant layout, he or she may well perceive complexity. After a week or two of learning about the overall system, its subsystems, and their various roles, and how the subsystems work together, the trainee will no longer perceive the same degree of complexity.

If familiarity reduces the perception of complexity, then it is not a fixed and arbitrary thing. Nonetheless, it is possible to characterize the elements of complexity. Complexity, or the perception of complexity, appears to be generated by three factors working in combination: variety, connectedness, and disorder.

In the context of complexity, variety is seen as a commodity such that one can add or take away variety without having to specify what constitutes the variety. Adding variety is adding a new category.

Connectedness, or connectivity, is the degree of linkage between elements. A little mathematics can be of assistance here. Between N elements, the maximum number of one-way links is N x (N-1). So, for 10 items, the maximum number of unidirectional links is 10 x 9 = 90. If the links are bidirectional, the number of physical connections might be half, that is, 45.

For some systems, the number of links may exceed the logical maximum: this could occur when links were replicated, perhaps for reasons of integrity, reliability, or aesthetics. For most systems, the number of links is likely to be less than the maximum possible; however, this raises the possibility of measuring connectedness as a proportion of the logical maximum.

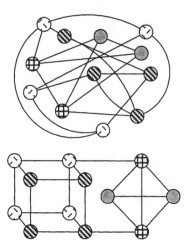

**Figure 5.1** *Reducing the perception of complexity.*

Patterns of connection describe and delineate architecture. Some configurations seek to reduce physical connectedness. A star-connected architecture has one system at the hub, connected to others on the rim. For 10 items, the number of unidirectional physical links is only 18 (or nine bidirectional links). The number of logical bidirectional links between items may still be 90, provided the hub system is transitive; that is, it is capable of transferring whatever an incoming link carries onwards to the appropriate rim system destination. The number of physical links, and hence the perceived complexity, may be reduced by such architectures, but only at the cost of greater complexity and capacity in the hub system.

The third element of complexity in the trio is disorder, perhaps more simply thought of as tangling. Where the various subsystems or parts of a system are arranged so as to cause the various linkages to cross, interweave, and overlap, the resultant tangle of interconnections increases the perception of complexity.

While it was possible to see the different varieties in the upper diagram of Figure 5.1, the lower diagram reveals two connected sets of entities: a cube at left, and a rhombus at right. There is even, in the cube, a perception of three-dimensionality. (The 3-D impression appears only on the cortex of the viewer since the diagram is, of course, only two dimensional.) Sets containing the same varieties could have been observed in the upper diagram; sets with different but related varieties can be seen in the lower diagram. Systems are unlikely to be comprised of identical parts. They are much more likely to be comprised of different, complementary parts, again as in the lower diagram.

Since the degree of variety is unchanged between top and bottom diagrams, the reduced complexity must be due to the untangling of the many interconnections, which was engendered by rearranging, or reconfiguring, the entities. It follows that the perception of systems within a complex set of many interconnected parts will be encouraged and simplified by reconfiguring[1] the parts so as to reduce overall configuration entropy. Such methods require the organizer to: a) evaluate the configuration entropy of a pattern of interacting systems, and b) reconfigure the pattern so as to minimize the disorder.

## ELABORATION

Disaggregating or decomposing is the process of breaking up complex systems into smaller, simpler parts. It is the tool of Cartesian reduction, with all its inherent limitations, and as such is inappropriate to systems ideas and methods. Nonetheless, it is often necessary to examine and analyze systems in some detail. The process of "looking inside" a system is elaboration. Unlike decomposition, elaboration does not disconnect parts, but acts rather like a magnifying glass, enabling the user to see and express more detail while that detail remains in situ; connected, dynamic, and interactive.

Thus elaborating a radio receiver design might reveal a preamplifier feeding a mixer, which mixes the radio-frequency signal with that from a local oscillator, leading to a multistage intermediate frequency (IF) amplifier followed by a detector, and finally by an audio amplifier and loudspeaker. Any, or all, of these items may be elaborated down to a further level of design detail in which individual components and interconnections are shown. Unlike reductionist methods, none of the elaborations would present parts or components on their own.

The radio receiver example is straightforward; the various parts are matched to each other in such a way that changes in one part do not materially affect the performance or behavior of other parts. On the other hand, decomposing, say, the architecture of a cathedral, or the behavior of a psychotic killer would not only be difficult, it would produce incorrect results. Why? Because the various parts/subsystems mutually influence each other; it is possible to understand any one part only in the dynamic context of all the others.

The spire of a cathedral creates a down thrust on the central tower that buckles stone pillars and walls; see left-hand image in Figure 5.2. Flying buttresses are necessary to prevent walls from bowing outwards, and the arches of these buttresses are themselves liable to spring outwards and collapse. To counteract this, a weight, in the form of a stone pinnacle, is often seen mounted on top of the buttress arch, at the point where it would spring out; see the right-hand image in Figure 5.2.

So, all of these parts mutually interact. Remove the pinnacle, and the buttresses buckle, the pillars bend further, the cathedral walls bow out, the tower

---

[1] This approach forms the basis for a simple system analysis and design tool, CADRAT©, which will be introduced and used later.

tilts, and the steeple leans or falls. Decomposing the components ignores the interactions. Elaboration looks at the components in situ and in context. Because the different parts work together, or complement each other, they form a system; in this instance, a system for maintaining the integrity of the tower and spire. Different parts complementing each other are a key indicator of a system.

**Figure 5.2** *Medieval systems engineering in stone. Stone pillars bowing (left) under the weight of Salisbury Cathedral spire are supported (right) by orthogonal flying buttresses, capped by a pinnacle.*

Similarly, the behavior of a psychotic killer is not a simple cause-and-effect, or stimulus-response affair. To understand and perhaps even predict behavior, as the so-called criminal profiler seeks to do, would require a deep knowledge of the killer's background, history, treatment, childhood, and more recent experiences.

A killer's behavior is determined not just by the present situation and stimulus, but by recent events and longer-term influences, too. Behavioral signposts emerge as aberrant behavior tends toward some outburst, although these signposts of disturbance may be discovered only after the event. They nonetheless suggest that there is a dynamic, a process that the psychotic mind goes through, that exhibits similarities between quite different individuals.

Again, decomposing the elements contributing to behavior ignores their mutual interactions. Elaboration looks at the various elements as part of the dynamic whole. The different experiences that the killer has had, combined with his condition, conspire to generate a broadly recognizable pattern of behavior. The idea of a system is at work here, too, with different parts mutually interacting to produce a whole behavioral paradigm.

Figure 5.3 elaborates on part of the concept. At center, an individual seeks to join a social group. He may join a group that finds him acceptable, shown at left.

If so, he will be influenced by the group. The individual's social behavior is, then, likely to be "grounded" by group norms. Most of us belong to many groups: family, work, sports, clubs, and so on, and we develop appropriate patterns of behavior for each of them.

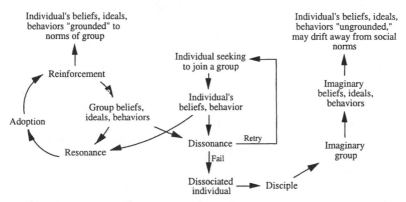

**Figure 5.3** *The development of group behavior?*

If an individual fails to find himself acceptable to one group, then he will search for another group that is more amenable. Should he fail altogether, then – as the figure indicates at right–he may create his own mythical or imaginary group. If the individual should become immersed in this fantasy world, he may adopt its "norms" of behavior, which he may have imagined, read about, or seen at the movies.[2] Since these norms are not grounded in reality, his behavior may drift toward the unusual, bizarre, or sociopathic.

The path toward sociopathic behavior may be a long one, with many stages along the way; simply looking at individual parts of the journey would not be sufficient. On the other hand, since so few people follow that path, there must be a suggestion that certain individuals are susceptible, perhaps genetically, perhaps due to early environment or traumatic experiences. Yet again, not everyone with a genetic predisposition and/or disadvantageous early environment becomes sociopathic. Figure 5.3 does not address these fundamental issues, but it does point to the potential dangers of social exclusion, and to the advantage of social inclusion.

## ENCAPSULATION

One of the issues facing the task of elaboration is that of too much detail. How far need the process be taken? "About two levels down the hierarchy" seems to be the

---

[2] Some "loners" create secret shrines to a mythical or remote figure, seeing themselves (they are usually men) as disciples. Some reveal their inner broodings in dark, stylized paintings and drawings.

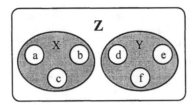

**Figure 5.4** *Encapsulation.*

surprisingly simple answer. The reason has to do with encapsulation, the opposite process to elaboration.

If elaboration is the magnifying glass, which looks in more detail at things without taking them apart, then encapsulation is the placing of containers around sets of entities such that a single label–one that describes their emergent properties, capabilities, and behaviors–can represent them. Containers do not interfere with interactions, which continue unabated "underneath the covers."

Consider three sub-subsystems; a, b, and c; see Figure 5.4. Suppose they complement each other in such a way that they can be reliably described as a single set, or system, X, with singular properties. Then we may not need to know about a, b, and c; we can simply use the label, X, and the list of X's (stable, predictable) properties. Similarly, we may not need to know about d, e, and f, but we may find that the label Y describes the properties of the set quite sufficiently.

It may be that a, b, and c are parts of a radar receiver, while d, e, and f are parts of a radar transmitter. Together they form a radar, Z. If we were comparing radars, to see which was the more suited to a new airfield, we would be interested in the performance of the overall radar, in terms of range, performance through rain, susceptibility to jamming and noise, and so on. These are properties of the whole system; not of the parts. On its own, the transmitter has no range; neither does the receiver. So, for comparison purposes, we may not be interested in a, b, c, d, e, or f. We may even have little interest in X and Y, but only in Z–until, that is, it comes to specifics about installation, electromagnetic compatibility, cooling, and ventilation, and so on, which we might call the engineering aspects.

The process of encapsulation hides unnecessary detail and reduces perceived complexity in the process; it does not, however, alter the underlying structure or connectivity. The "ground level" architecture, on the left in Figure 5.5, is the only substantive architecture. Note from the diagrams how elaboration effectively constitutes the sequential removal of the capsule covers. Note also that groupings are not unique. The left-hand eight entities could have been encompassed in a single capsule, for instance.

Complexity is often viewed as something to be avoided. Complexity is neither good nor bad, however. The human brain is undoubtedly more complex than that of a mayfly, but then the human brain is capable of more. In manmade systems, complexity may go hand in hand with capability, reliability, integrity, and survivability. Complexity is something to manage rather than avoid, reduce or increase. Encapsulation allows the concealment of complexity; as such it is one tool of complexity management.

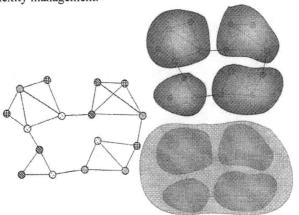

**Figure 5.5** *Progressive encapsulation.*

## VARIETY

Addressing complexity as a subject allows that variety, connectedness, and disorder may be treated as commodities, and this is so far outside of our usual experience that it can appear strange at first. For example, the implications of "not enough" or "too much" variety may not be instantly obvious. We will explore the following variations on variety:

- Minimum Variety;
- Useful Variety;
- Limited Variety; and
- Excess Variety.

### Minimum Variety

The Western world, with its many mouths to feed, has developed agriculture based on large-scale monocultures. Western farmers grow vast fields of a single species of wheat, or they farm salmon spawned in great quantities, or they grow a single species of apple. Singular varieties are known to be susceptible to disease, or predation, and a complementary industry exists to provide artificial fertilizers, pesticides, and a continual supply of new, disease-resistant varieties. Genetic engineering is coming to the aid of farmers who are finding that nature is rather good at evolving diseases and predators to damage or feed off their mono-crops. Genetic engineers hope to keep ahead of the game by conferring disease resistance, say, from one species to another. The approach is necessary to support vast populations, but it is also highly energy intensive.

South American farmers have farmed using polycultures for many hundreds of years. Polycultures bring together complementary plant varieties, with mutual nitrogen fixation, natural pesticides/herbicides, wind brakes, climbing frames, and so on. The different plants, working as a set, combat a wide range of threats/potential deficiencies. In the Western world, these techniques are well known to domestic gardeners, too, with marigolds and cabbages (for instance) being planted together, since the marigold protects against some cabbage pests. Polycultures require much less energy in the form of artificial aids. Their disadvantage, of course, is the unsuitability for mechanization that militates against large-scale cultivation.

Polyculture farmers probably identified suitable combinations of plant variety by trial and error in the first instance, which raises an interesting question—how much variety is enough? Clearly there could be inappropriate variety, but can there be too little, or too much, as well?

The various systems within our bodies are complementary, or mutually supportive. With the exception of the reproductive system, little could be removed without serious consequences  such as immediate death. The viability of any system is determined, not only by internal complementary parts, but also by the threats that it has to meet. Figure 5.6 shows an N-squared ($N^2$) chart showing a typical air defense organization.

In an N-squared chart, component systems are shown on the leading diagonal, with the other rectangles representing interfaces, interactions, relationships, and interflows between the systems. So, interceptors give "some protection when airborne" to both the Air Defense Ground Environment (ADGE, or ground radar and command network) and to the air bases and other vital points. The bottom right rectangle, representing surface-to-air missiles (SAMs), is highlighted to show that it is needed only if the airborne threat to the other members is significant.

Were a tactical ballistic missile to be considered a threat, then an antimissile missile would be needed in addition to a SAM. The minimum system variety to ensure viability is partly determined by the external threats to the system. The greater the environmental diversity or threat, the greater must be the minimum variety essential to viability. This idea suggests that there may be a basis for determining minimum variety, based on cause and effect.

**Useful Variety**

The concept of useful variety is distinct from that of minimum variety. Useful variety makes an essential, complementary contribution. For example, the variety of clothes and sizes in clothes shops makes a useful contribution to the ability of the shop to satisfy its varied customers. Beyond the smallest and largest likely clothing sizes, variety would not be useful. Similarly, a word processor or drawing

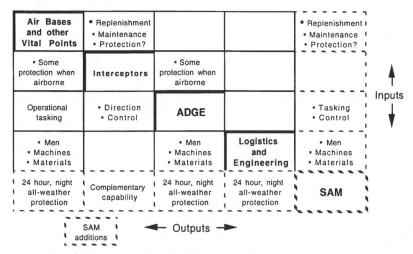

**Figure 5.6** *Minimum variety in air defense.*

program has varied capabilities for different people, tasks, and environments. Beyond some limit, more variety in the capabilities of the program is superfluous, and could make the program difficult to understand and to use. This would be excess variety.

Determining the sensible limit of usefulness may not be straightforward. In the clothing shop example, it might be possible to analyze the likely population and to stock only sizes that would sell well, although the existence of outsize, short, and tall specialist shops testifies to the difficulties of meeting this goal. The ideal variety of capabilities in a word processing package depends upon the tasks it will need to face after purchase, and these may not be so easily predicted. How about the skills that a company should retain during a recession to enable rapid start-up once the recession lifts? Holding redundant skills against a future need is tantamount to predicting the future, which may not be knowable. Not holding the skills may be an act of delayed corporate suicide.

Useful variety, then, is less amenable to rational determination than minimum variety, since the former implies prediction. Nevertheless, we often choose in everyday practice to hold on to something, or to buy a product with variety that we may not need directly, as a hedge against the uncertain. Useful variety is therefore associated with risk management.

### Limited Variety

Limits to variety can be both perceived and real. First, consider perceptions. If we try to list the major religions of the world, for instance, we find a fairly short list. If we then take any one of those religions and try to list the sects within it, we again come up with a very short list. We can try the same technique with major languages, types of jokes, or different film genres.

Each category can be broken down into a relatively few divisions; often about seven. As we increase the resolution, we seem to see the same broad constraint on variety emerging, no matter what the general topic. Why should there be a limit? Why is it so small?

Psychologists and anthropologists propose that there is some built-in, simian inheritance in our brains that tends to make us confused when we try to contain more than about seven [1] mental categories at the same time. Experiments with dots randomly scattered on a sheet of paper, and shown for only half a second or so, suggest that the perception of numbers of dots up to about seven is simple for most people. Beyond seven, there is a sharp cut-off, with most people responding with "lots of dots."

Could it be that our reaction to things or situations as being complex is an inherited mental processing limit? Perhaps during our evolution, it was unnecessary to address more than seven or eight categories of anything at the same time in a life-threatening situation; perhaps we evolved an ability to perceive up to seven things in parallel under "fight or flight" conditions. If "Miller's seven" (as the phenomenon is called) is simply that, a subjective and inherited limit of human parallel processing, then perhaps complexity itself is simply our mental processor balking when we try to process too much variety at once.

There are real limits to variety also; these will be explored below.


## GENERATION OF VARIETY

For open systems to be stable, they may need a minimum variety; to adapt, they may need useful variety. Excess variety may change the nature of system behavior.

Looking to the natural world, we see that the profusion of species, plants, and animals, or "speciation," is greater in more energetic systems. Biodiversity is much greater in a rain forest than in cold tundra areas. Is it too great a leap to note that car variety, the profusion of car makes and models, is greater in New York, London, Paris, or Tokyo than in Moscow? In each case, the common theme is the amount of energy available; in nature, energy may be measured in joules, while in economies it may be measured in wealth. We may note similarly that the variety of weapons in the respective arsenals of the U.S. and the U.K. favors the U.S.; not only does the U.S.—the wealthier nation—possess more weapons, but a greater *variety* of weapons, too.

It is tempting to deduce that variety is somehow generated or "pumped" by energy, although we might have to be a little free in our interpretation of energy according to the situation.

## ENERGY AS A SOURCE OF VARIETY

It has long been established that species evolve to be efficient users of energy [2]. If two species compete for the same resource then, during times of shortage, the more efficient will survive. On the other hand, during times of plenty, when speciation occurs—as it does naturally owing to genetic inheritance from two parents—the environment may nurture a weaker variety to propagate and evolve further. Subsequent reduction in available energy may see new species sufficiently changed to be able to shelter and flourish in new niches.

It is tempting to equate these ideas to recession and the rise of a variety of businesses during times of plenty, only to see many of them—but not all—go to the wall when times are harder.

Both the natural and the business worlds behave, in this respect, in a way that is consistent with the observation of living on the "edge of chaos." To a detached observer, species/businesses generate and multiply, then fall back again, filling the constantly changing ecological/economic space in a seemingly haphazard manner. From this perspective, variety seems to come and go, in a complex, continually shifting pattern.

Behind these phenomena and perceptions we see, not control and regulation, but complementation and connected variety as the bases for open, interacting systems stability. It is also evident that open, interacting systems stability breaks down when connected variety reduces in a changing environment, owing to an inability to adapt to a changing environment.

## OPEN SYSTEMS INTERACTIONS

We can make some general observations of open system interactions:

- *Energy creates disorder*–If you heat a substance, its molecules vibrate, and there may be a change of state. If energy is injected into an argument, disorder may result. If energy is injected into an ecology, speciation occurs, activity levels rises, and entropy increases.
- *Variety interacts to form systems*–Sometimes interaction is uneventful, sometimes inimical, but sometimes the parts find mutual benefit. On those occasions, separation would be unnecessary or disadvantageous, and the varieties may form a system.
- *Systems reduce entropy*–Because of the complementary nature of the parts from which a system must be formed, both the parts and hence their intraconnections form a pattern. That pattern represents reduced (configuration) entropy; a localized ordering against a disordered backdrop.
- *Systems break apart* sooner or later–This increases entropy back toward the state before the system formed.

The bullet-points above present a concept of continual change over time. The mean level of disorder is increased by energy, and the rate of change from order to disorder to order is increased by energy.

Energy, it seems, does not only create disorder. For open, interacting systems *it also creates order from that disorder.* A powerful example of this open, interacting system life cycling is illustrated in Figure 5.7. It shows the surface of a pool[3] that contains a submerged circulating water pump, causing the water to well up to the surface. The resulting surface is turbulent, with swirls and vortexes forming and reforming, some lasting, some ephemeral, some large, some small; the whole presenting self-similar patterns, driven by the energy from the pump. Were the pump to stop, the patterns on the surface of the water would die away. Were the pump to increase its flow, the surface would become more turbulent; more systems would form and reform, and at a higher rate. At a still higher level of energy, different types of pattern would occur, with bubbles, spouts, and fountains.

A similar example could be offered in the form of a meteorological weather map, with cyclones, depressions, ridges, fronts, isobars, and so on. As the weather changes, these many features form, develop, last for a while, and then decay. Global warming, we are incessantly advised through the media, will result in more turbulent weather, with deeper depressions, higher speed winds, rising tides, a faster change of climate, and so on.

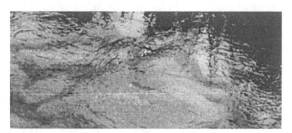

**Figure 5.7** *Open, interacting system analogy.*

This seems to be the case with all open interacting systems. Ecologies employ energy from the sun, which causes growth, activity, and formation of floral and faunal systems, decay, and reformation. While solar energy lasts, life cycling continues. Where solar energy is greater, the number of species and their rates of growth and decay increases. Typically, rates of growth and decay are five times greater in a tropical rain forest than in a tundra.

Civilizations behave similarly. The ancient Egyptian civilization grew rapidly using energy from the sun and fertility from the Nile River. With such exceptional and dependable energy, the civilization produced a profusion of open interacting agricultural, building, cultural, and transcendental systems. Dynasties came and

---

[3] In Disneyland, Paris (!)

went. Eventually, however, the civilization decayed as other lands grew stronger, and as its internal political and economic systems became moribund.

It is possible to formulate a hypothesis, or empirical law of open systems dynamics as follows:

*Open, interacting systems' entropy cycles continually at rates and levels determined by available energy.*

This law of open systems dynamics applies to all open, interacting systems. It does not seek to operate within any boundary, unlike the Second Law which, by referring to isolated systems, implies a boundary. The law of open systems dynamics applies to a never-ending network of systems, providing a basis for understanding parts of this infinite network without bounds or preconceptions. Instead of isolation, the law of open systems dynamics embraces openness and interchange, with energy entering and leaving any part of the infinite network that may be of interest.

At first reading, the law of open systems dynamics may appear a disappointment. After all, how can it be used in connection with, say, a social system, an economic system, or an ecological system? The first reading of the Second Law of Thermodynamics may engender similar reactions, but that Law is one of the cornerstones of physics and engineering. The law of open systems dynamics has to be seen in context. It applies to open, interacting systems which are in a state of continual, dynamic change. The law of open systems dynamics is scale independent and type independent: it applies to large and small systems of any kind. The law proposes that systems, far from decaying with time, recycle themselves forever, provided they continue to receive sufficient energy to sustain the processes.

How might we apply the law? We can use it to understand and predict open system behavior, remembering the following:

- Complexity can be measured as (configuration) entropy; the greater the variety, connection, and tangling, the greater the disorder in pattern.
- Energy may take many forms, including wealth, ideology, belief, and many more, according to the types of system of interest.
- A reduction in energy inflow to zero will result in zero entropy, eventually; this is consistent with the Second Law.
- An influx of energy will increase the general level of entropy, that is, create more disorder; this is also consistent with the Second Law.
- A steady flux of energy will result in *continual* entropy cycling, but not necessarily *continuous* entropy cycling, that is, the rate of change of entropy need not be smooth.
- A rise in the energy flux to a new level will increase the rate of entropy cycling.

- There is a balance at all times between the energy influx and the total energy contained, utilized, and dissipated by systems as they traverse the life cycle. The conservation of energy applies for social and transcendental systems, too.

Using these precepts, new viewpoints about social systems behavior are possible, concerning, for example, criminality and the law, education, committees, power structures, organizational behavior, and so on. Similarly, new viewpoints emerge about economic systems, engineering systems, transcendental systems, and even physical systems; particularly at quantum level.

## SUMMARY

Complexity is relative, not absolute. It seems to emanate from three related factors: variety, connectedness (of that variety), and disorder, or tangling, of the connections. Variety is seen as a commodity, with more being created by energy.

Elaboration, which magnifies rather than decomposes, looks into things, and may increase perceived complexity. Encapsulation, on the other hand, reduces perceived complexity.

A scale-independent and system-type independent law of open system dynamics is proposed; whereby *open, interacting systems' entropy cycles continually at rates and levels determined by available energy.*

This law, equivalent to the Second Law of Thermodynamics, but applicable to unbounded open systems, suggests that interacting open systems create order as well as the disorder predicted by the Second Law, and that they will continue so to do while energy is available.

## ASSIGNMENTS

1. The economy is entering an uncertain period of downturn. Your company, a creator of cutting-edge technology products, is concerned at the drop in sales. Corporate accountants propose that all technical and engineering staff not actively concerned with production should be dismissed immediately as no longer necessary in the circumstances. You are tasked with mounting a robust challenge to this viewpoint by making a more considered proposal that will retain essential skills but will not ruin the business. Make your case.

2. Companies and corporations are often faced with trimming the workforce. The process frequently goes too far, with those who retain their jobs after the cutbacks losing morale and actively seeking new jobs in a more secure environment. Propose a rational method of slimming

down the numbers in an organization that leaves adequate variety and capacity, and does not prejudice company viability.

3. Explain why there is a greater *variety* of jobs in a large city than in a small town.

4. You are planning the so-called "terra-forming" of the planet Mars. The plan is to create an elementary biosphere, and to encourage its growth and spread across the planet. How might global warming assist in, and accelerate, this process? How might you go about warming the surface of Mars, what would be the most likely effects, and how—once the biosphere was established—do you think that global warming might be reined in?

5. A commercial manufacturing company is experiencing poor sales because its products lack appeal. Using the law of open systems dynamics as a guide, suggest how the company might improve itself and its performance.

---

## REFERENCES

[1] Miller, George A., "The Magical Number Seven, Problems of Perception," *The Psychological Series*, Harvard, Vol. 63, No. 2, 1956.

[2] Lotka, A. J., "Contribution to the Energetics of Evolution," *Proc. Natl. Acad. Sci.,* Vol. 8, pp. 147-155, 1922.

# Chapter 6

# Systems Life Cycle Theory

*"Begin at the beginning," the King said, gravely, "and go on till you come to the end, then stop."*
Alice's Adventures in Wonderland, *Lewis Carroll, 1865*

## INTRODUCTION

The empirical law of open systems dynamics, presented in the last chapter, concerned itself with an infinite network of open, interacting systems continually creating order and disorder, dissipating inflowing energy. Systems life cycle theory proposes how systems come into being within such an infinite network, grow, exhibiting stability and maturity, before eventually decaying and collapsing.

An understanding of the stages in the life cycle of a system can be valuable. It may enable us to prolong the useful life of a system; alternatively, it may explain the implosive collapse that many complex social and sociotechnical systems exhibit, from the domino collapse of the former Soviet Union, to the slow decline of traditional industries.

The search for such a theory, concerned as it is with undefined, conceptual systems, invokes ideas and principles that must be system-type independent and system-scale independent. The following principles will be introduced individually before being synthesized into a unified proposition.

## THE SEVEN PRINCIPLES OF OPEN SYSTEMS [1]

### The Principle of System Reactions

*If a set of interacting systems is at equilibrium and, either a new system is introduced to the set, or one of the systems or interconnections undergoes change then, in so far as they are able, the other systems will rearrange themselves so as to move toward a new equilibrium.*

The principle is unexceptional for physical systems, to the point that it may seem axiomatic; Le Chatelier (1850-1936) expounded the principle in 1888 in the context of chemical reactions. Newton's Third Law is also relevant for physical systems, if not for the more general case. The principle applies equally to interactions between economic, political, ecological, biological, stellar, particle, or any other aggregations that satisfy the definition of a system.

The principle says nothing about the manner of the movement toward equilibrium: it may be linear, nonlinear, periodic, catastrophic, or chaotic.

The contention is not necessarily as bold as it might seem at first. If an open system is introduced and connected to other open systems, interchanges will occur that will affect both the new and the existing systems and, due to the openness, these effects will be felt in systems that may be connected only indirectly to the new member.

In the general case, open interacting systems may be changed in some way by the fluctuations in interchanges; encouraging a variety of potential responses that, according to the principle, will settle to some state after an indeterminate time. We are aware of such effects in terms of share price movements after a devaluation, changes in ecological balance following the introduction of a new species, drug users becoming inured to particular drugs, insect populations becoming resistant to insecticides, and so on. Other examples include the relations between unions and management, which may be disturbed either by bad management or undue union demands; adjustment of attitudes to new ideas; meeting new people; and the reaction of a beehive to invading wasps.

### The Principle of System Cohesion

*Within a stable, interacting, system set, the net cohesive and dispersive influences are in balance.*

For physical systems such as a star with its planets or the quarks in a neutron, such a statement is axiomatic, another way of looking at Newton's Third Law. The proposition, however, is that the principle applies to all interacting systems: social, economic, political, and so on.

Honeybees present a social system example. Research has shown that there is a natural tendency among bees to swarm that is suppressed by a pheromone expressed by the one active queen and shared between hive members to calm them. As the hive flourishes, workers have to forage progressively further from base, until it starts to become uneconomic, due to time delays and to the energy expended by individual bees. At the same time, the number of bees is growing, and the share that each bee receives of the calming pheromone is reduced. Eventually the pheromone level per bee drops below a critical threshold and swarming occurs. Swarming is a classic example of the balance between cohesive and dispersive influences breaking down to produce a sudden event.

## The Principle of System Adaptation

*For continued system cohesion, the mean rate of system adaptation must equal or exceed the mean rate of change of environment.*

The principle is evident within biological systems, and is underpinned by Darwin's theories. The extinction of the dinosaurs may be a spectacular example. The proposition is that it applies to all interacting systems, including manmade systems. The idea is not novel; manufacturers regularly upgrade hardware and software, to keep it up to date, for instance, and eventually ditch such things when updating becomes ineffective, too difficult or too expensive. The need to update is engendered by a changing environment, and as that change accelerates, so must updating to keep pace.

## The Principle of Connected Variety

*Interacting systems stability increases with variety, and with the degree of connectivity of that variety within the environment.*

To understand this principle, qualitatively at least, consider Figure 6.1. The figure represents part of a set of interacting systems. The organic shapes may represent any systems: biological, economic, social, transport, stellar, ecological, and so on. As the outflow from any system alters, it causes a change to the inflow of others. These changes migrate throughout the set; some returning to affect the initial system.

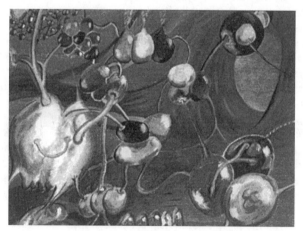

**Figure 6.1** *Abstract open systems.*

The greater the variety of systems, the more likely it is that, across the set, outflows from some of the set will satisfy the inflow needs of others, through a variety of mutually redundant sinks, sources, and transforms. In the limit, a point would be reached at which there was so much variety, all of it cross fed between all systems, that it would become increasingly difficult to invoke any change.

Figure 6.2 illustrates connected variety diagrammatically. The $N^2$ chart shows five open systems: A-E, on the leading diagonal. The inflows to each open system are shown in the corresponding column. The outflows from each open system are shown in the corresponding row.

|   | 1 | 2 | 3 | 4 | 5 |
|---|---|---|---|---|---|
| 1 | A | abc | bcd | bcde | abcde |
| 2 | ab | B | bc | | |
| 3 | bc | | C | de | |
| 4 | cde | | be | D | e |
| 5 | de | a | b | c | E |

← Outputs → (column header)

Inputs (↑↓, right side)

**Figure 6.2** *Stability and connected variety.*

The interface (5,1) shows composite variety *abcde* passing to open system E, which decomposes *abcde* into *de, a, b,* and *c*. Each of these varieties passes to the other open systems as shown in row 5. Checking the column and row for each of systems A-E shows that each system both receives and emits, or ingests and exhausts, the same varieties, albeit sometimes as composite varieties, and sometimes as discrete varieties. So each of the open systems A-E both provides for, and is dependent on, the other open systems in the set. Each open system may also retain some part of the inflowing varieties for construction or maintenance.

Open system A ingests and exhausts all varieties *a-e*. Open system B ingests and exhausts only varieties *a-c*. Open system C ingests and exhausts only varieties *b-d*. Open system D is a decomposer. Open system E is a decomposer. Open system A is a composer, or integrator.

The members of the set are self sufficient, except that work is done[1] in transferring the varieties between systems, in constructing and maintaining the systems, in joining up varieties, and in decomposing composites [2]. A continued supply of energy is therefore necessary to enable the work to continue. The whole set is also an open system; able to receive discrete and composite varieties from outside, and also able to attach suitable new systems to the existing set, provided that they fit in with the set rules of mutual contributory interchange.

---

[1] Figure 6.2 is a parody of life, with the model emulating an organism, or ecology. It has been suggested that life itself is "a Byzantine energy-shredding machine, to reduce the Universe to a state of bland uniformity" [2].

Eliminating any one system, system E for example, need not debilitate the overall set, which would continue to maintain its identity through alternative source, sink, and transform interactions. This is stability, or resistance to change. These notions lie behind the principle of connected variety, which finds its most obvious example in the ecological stability found in nature between flora and fauna that depends upon variety and interconnectivity.

Examples of stability through connected variety are also common in business and international trade. A company would be foolish to make only one product, or to rely on only one source for a vital raw material. Instead, manufacturers seek a variety of sources, and manufacture a range of products that sell into different market areas. The manufacturer seeks and promotes variety in his process of synthesizing products: the greater the (connected) variety, the greater the resilience, or stability, of the company.

Economies, too, are founded on variety of sources, manufacturers, transportation systems, and consumers. Money may be viewed as the principal variety exchange medium. Both nature and economies illustrate the potential for stability emerging from complex interactions between essentially chaotic systems. In both instances, planned systems and interactions can prove disastrously unstable.

**The Principle of Limited Variety**

*Variety in interacting systems is limited by the available space and the degree of differentiation.*

The principle is axiomatic once "space" and "degree of differentiation" have been established. Consider a guitar string. It can vibrate in a variety of modes limited by the need for nodes at bridge and stop. This maximum set of modes is the available space; the degree of differentiation is limited by the need for each mode to comprise waves in integer half-wavelengths only. There is (believed to be) a finite number of fundamental particles, because there is limited space within the framework of a logical set of particles.

Specialization [3] increases as the environment becomes more benign. Large cities have greater specializations among tradesmen and craftsmen, rich countries have a greater variety of cars, and tropical rain forests have a greater variety of species in a greater variety of niches. In every case, however, there is a limit.

**The Principle of Preferred Patterns**

*The probability that interacting systems will adopt locally stable configurations increases both with the variety of systems and with their connectivity.*

As the weave of interactions between systems becomes more complex, it is increasingly likely that feedback loops will arise, some perhaps acting through many successive systems and exchanges. Figure 6.1 illustrates the phenomenon qualitatively. If the various systems and their interconnecting tubes in the figure were filled with some working fluid and, say, the large system at lower left were squeezed repeatedly, then some of the reactions from other systems would occur almost immediately, while some would reverberate through other systems before returning to the source. The prospect increases for nonlinear interacting system behavior. Oscillation due to positive feedback is to be expected, if only because of resulting delays, and leads to the principle of preferred patterns.

Locally stable, interacting systems abound. Cities, computer giants, international conglomerates, thunderclouds and tornadoes, molecular micro-clusters, ecological niches, bureaucracies—all are instances of positive feedback, or mutual causality leading to stable configurations.

### The Principle of Cyclic Progression

*Interconnected systems driven by an external energy source will tend to a cyclic progression in which system variety is generated, dominance emerges, and suppresses the variety, the dominant mode decays or collapses, and survivors emerge to regenerate variety.*

This is a restatement of the law of open system dynamics, with additions. Empirical observations indicate that, although variety is generated by energy, not all varieties are created equal. Some systems may overshadow others, or be dominant.

Dominance denotes substantial imbalance in favor of one system at a given hierarchy level. The weight of evidence suggests that there may indeed be a repeating pattern in systems where variety, the mediator of stability, is suppressed by dominance, which in turn leads to vulnerability through an inability to change.

Such patterns are evident in business; human organizations such as committees and parliaments; ecologies, where one species dominates, eliminates rival species, and in so doing damages the environment on which they all depend; car racing, where one manufacturer/driver combination may dominate, eliminating competition, and damaging the business viability of the sport; overbearing leaders, who eliminate rivals, surround themselves with sycophants, and progressively lose touch, both with reality and with those they lead; and so on.

### SYNTHESIZING THE SYSTEM LIFE CYCLE

The seven principles can be combined as shown in Figure 6.3. The principles are arranged and interconnected with signed digraphs (positive or negative arrows) that indicate the influence that one principle has upon another. The figure illustrates how systems are formed, how they come together to reduce entropy, how the systems grow, and how they become compromised and eventually decay or collapse to regenerate higher entropy.

To understand the process, start at Energy, at the upper right-hand part of Figure 6.3. Energy promotes differentiation, which generates variety (speciation); Adaptation occurs, reinforcing the generation of variety. This top loop indicates that environmental change is a continuous process, provided only that energy flows into the environment.

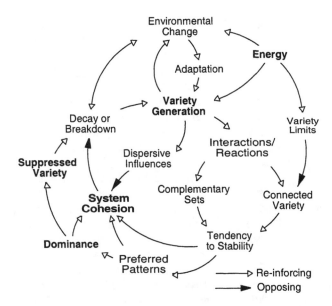

**Figure 6.3** *The system life cycle map.*

The generation of variety results in interactions between the varied elements. Interactions and reactions result in complementary sets, and in connected variety; both of which promote stability. The tendency to stability is confined by limits to the variety, so that stability cannot increase indefinitely. Figure 6.3 says nothing about the *manner* of the tendency to stability; it may be linear, nonlinear, catastrophic, or chaotic. Nonlinear interactions may produce multiple points of stability. Catastrophic systems may switch rapidly from one state to another. Chaotic system interactions may be short-term unstable and long-term stable.

The tendency to stability promotes system cohesion, which may arise with or without the generation of preferred patterns. Preferred patterns lead to, and are a nucleus for, dominance, by which some members of the set of interacting systems become larger, stronger, and/or consume more resources than others. Preferred patterns are an indicator of system cohesion; interacting systems with dominant members may remain cohesive for periods of time.

Two factors disturb this increasing order:

1. All variety that is generated is not taken up into interacting system sets; some variety remains free to act as dispersive influences, or pathogens, potentially degrading system cohesion.
2. Dominance may tend over time to suppress variety, which results in interacting systems' rigidity and inflexibility, creating thereby an inability to adapt to subsequent environmental change.

A positive feedback (reinforcing) loop connects decay or breakdown, variety generation, dispersive influences, and system cohesion. Once the process of decay or breakdown is initiated, it may be difficult to restrain, and is more likely to gather momentum or be catastrophic. The end result is total dispersion of the previously cohesive interacting system set into separate parts that add to the changing environment and to the pool of variety–and so the cycle progresses.

On a larger scale, a negative feedback (opposing, or counteracting) loop can be traced from variety generation, through a tendency to stability, through system cohesion, through decay or breakdown, and back to variety generation. The negative effect of this loop indicates long-term stability; in other words, the process will continue, although the time taken to traverse the loop is uncertain.

A similar loop goes from variety generation through the tendency to stability, through dominance and suppressed variety, to decay or breakdown, and back to variety generation. This loop may appear to reinforce, but it contains the negative term, suppressed variety, which reverses the sign logic. Suppression of variety does not *cause* decay or breakdown, but it sets the stage for it. Interacting systems with suppressed variety may persist until their environment changes sufficiently to trigger decay or breakdown. In effect, then, both of these larger scale loops are long-term stable, indicating the inevitability of the cyclic processes, but without prescribing cycle times.

Thus we have moved full cycle from high configuration entropy, through the formation of lower entropy, ordered, structured, interacting system sets; through the flowering of dominance, and dispersive influences, until the ordered structures break down with a flurry of increased entropy to return to a state which, although not the same as that at the start of the cycle, contains much the same ingredients. Cycling expends work and dissipates energy. Cycling is maintained, therefore, by energy pumped in from an external source. A balance may evolve in which the energy pumped into the interacting systems environment matches that expended in the cycle; this is long-term stability incorporating paroxysms of configuration entropy. The paroxysms are not a separate condition; they are fundamental to the long-term stability.

There is an implication, evident from the model, that chaotic systems may have essentially stable features. The same variety that provides the basis of stability can, under suitable conditions, provide the basis for chaotic behavior. Chaotic behavior is not associated with dominance; almost by definition. Hence, it is reasonable to deduce that chaotic interacting systems may be *more* enduring, *more* robust, than interacting systems which behave in a more ordered manner. Free market economies are an example.

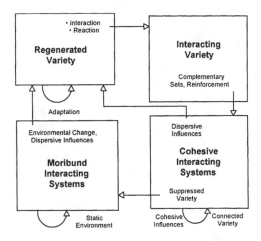

**Figure 6.4** *System life cycle: state-transition diagram.*

Figure 6.4 shows the system life cycle in state-transition form, revealing implications not always evident from Figure 6.3. Interacting systems may remain in particular states for indefinite periods if the appropriate conditions prevail. Variety in open systems requires renewal. We renew the cells in our organs, nerves, and blood by absorbing a variety of essential substances from the food we eat. If our diet lacks variety, we may weaken and die; the essential variety has not been sustained. Organizations respond similarly; employees grow old and must be replaced upon retirement, or they simply need their skills updated, to "learn new tricks."

Open system life cycle theory offers an understanding and explanation of the behavior of open interacting systems. It encompasses all kinds of systems, tangible and non-tangible, linear and chaotic, natural, social, and manmade.

## APPLYING LIFE CYCLE THEORY

### Economic Systems

Economic systems exhibit chaotic behavior in several ways. When a New York stock market guru makes comments about the value of stocks, the effects of the comments reverberate around the markets of the world. This phenomenon is redolent of Edward Lorenz's so-called butterfly effect [4], suggesting great sensitivity to starting conditions. The eponymous Russian discoverer of the Kondtratieff cycle observed a longer term cycle in international economics, on the scale of 50 to 60 years. He deduced that capitalism was long-term stable, that is, it might have its ups and downs, but it should always bounce back. Such periodic undertones are characteristic of chaotic systems

There is sufficient evidence, then, to suggest, not only that free-market economics may be chaotic, but that they are inherently long-term stable *because* of that chaos. This places it at the lower end of the life cycle map in Figure 6.3, where it stays because chaotic systems, as we have already observed, tend to avoid dominance and maintain essential variety.

The former Soviet Union economy, on the other hand, was organized to a preferred pattern by a dominant ideology in such a way that the variety was deliberately suppressed. Quasilinear behavior was carefully planned by setting production quotas, and stability was sought by cross connecting between member systems, requiring particular products to be supplied to particular markets, often in other Soviet bloc countries. The cross connection did not compensate for the lack of variety, leaving the overall economic system vulnerable to the domino collapse that eventually occurred.

**Political Systems**

Figure 6.3 can be used as a template or map on top of which the progressive states of a complex system may be superimposed. For example, it is possible to map out happenings in Eastern Europe since 1900 A.D., and identify a number of cycles since that time, with czarist, Bolshevik, Stalinist, and post-war dominances; each with transient stability characteristics of their own, each collapsing fairly spectacularly and none more so than the domino fall of the Warsaw Pact and the U.S.S.R.

Individual principles can then be used as probes to seek understanding of the reasons for the changes that cause moves from state to state clockwise around the cycle, as perceived on the map. Pursuing the Soviet example, Stalin's approach to promoting system stability may be explained using the principle of connected variety. His methods may be characterized as suppressing variety (by imprisoning or killing dissidents), while at the same time greatly increasing connectivity to compensate, using various kinds of police to continually reinforce ideology and to regiment behavior. Such systems are both dominant and highly stressed. They may persist while the environment remains steady, but are compromised by changing environments, to which they can respond only by collapse as the stress is released.

Stability is not always the perceived ideal. Proportional representation (PR), in which politicians are elected in proportion to the number of votes cast state- or nationwide, encourages the proliferation of smaller parties, which tend to form mutual associations. This makes for a political environment that is expected to be more stable than the more traditional simple majority approach. Some pundits declare PR to be turgid rather than stable; essential change is very difficult to introduce.

**Business Systems**

The life cycle map of Figure 6.3 suggests explanations of counterintuitive behavior in systems. The map shows that preferred patterns may emerge such that anticipated change does *not* occur. Attempts to "unwind" bureaucracies may be confounded; left alone, they rapidly repair any damage, rebuilding to their former state. Bureaucracies exemplify the principle of preferred patterns, since they sustain themselves by a positive feedback mechanism.

The principle of system reactions indicates that perturbations within an interacting system set may make other set members respond, even reverberate, so that the system may take some considerable time to settle to a new stable state—itself unpredictable owing to uncertainties in the participating systems.

Introducing a new management information system (MIS) to a company may cause ripples, even shock waves. Reactions within the company to the new MIS change the environment, rendering the original specification conditions invalid. Individuals and groups may adopt the new system readily, while others cling to previous ways, causing at best uneven take-up, and at worst downright intergroup hostility. The reverberations and hostility to change may settle down, either to an uneasy truce or to genuine détente; however, they may not do so. The short lifetimes of MISs in corporate businesses, and their uncertain contribution to return on investment indicate that, more often, they do not settle down.

**Organizational Systems**

A phenomenon in some organizations is the periodic switch between project and functional organization. Consider an organization with several project teams, each semi-autonomous. Cooperation may develop between like functions in several of the projects, and they may share ideas, methods, and tools; even resources. After a while this cooperation becomes institutionalized, and individuals undertake the lateral cooperation as a formal task. Gradually, lateral ties increase until it becomes apparent that a de facto functional organization is emerging. Then the organization may switch formally to a functional structure. Functional structures encourage corporate learning, but are less obvious in their direct impact on a particular end product of an organization. The effect is more difficult to trace back to cause, and, before long, the impression emerges that overheads are too high. The switch back to project organization is imminent.

From a life cycle theoretic viewpoint, this continual switching is long-term stable and is best represented by cycles around the life cycle map. Unfortunately, it results in an organization that is, at best, slow to learn, since the periodic upheaval to project organization is likely to be accompanied by a reduction in functional skills, tools, methods, and knowledge. The organization passes right around the life cycle loop, experiencing only transient decay (rather than collapse) before reconfiguring into the next stable state.

## Changing System Behavior

System life cycle theory indicates several ways in which the behavior of a system might be modified, according to the state it is in at the time. Adding variety encourages stability, not only to a fledgling system, but also to a moribund system where variety has been previously suppressed.

Conversely, it is possible to create a situation in which a system will decay and collapse; simply by starving it of variety. The timeframe for the collapse is uncertain, since it necessitates a change in environment.

Accountants can unwittingly cause companies to collapse by cutting back on variety in an attempt to hibernate through an economic winter. Come the spring, the company may be too weak to revive itself.

## Predicting System Behavior

Predicting system behavior is fraught with problems. Homo sapiens seem deeply committed to prediction, however, and no matter how often predictions fail, attempts are repeated. Pundits predict weather, economic situations, military outcomes, global freezing/warming, and so on. No one seems to notice that they are consistently wrong. There is a belief that, if only we had more information and better models, we would *then* be able to predict.

We may be able to predict chaotic system behavior under two circumstances only: in the immediate near term, but with decreasing accuracy as we seek to extend the time horizon; and in the long term, but only general trends, and on the (gross) presumption that the environment remains broadly constant throughout.

For example, it is possible to predict the weather tomorrow with some accuracy, next week with less, and next month with very little, depending on the general climate. A turbulent climate means a shorter term horizon of "knowability." Three months from now, the weather conditions at a particular site may not be knowable. It is not that we presently lack facilities to calculate; they simply are unknowable, in any detail. On the other hand, we can be reasonably sure that next summer will be warmer than next winter, and we might be able to predict trends in how warm and how cold—but not that a particular winter or summer will have a precise day-to-day temperature profile.

Chaos is evident in football matches. Although all matches may start under similar conditions and we may know the rules, it would be impracticable to predict the location and vectors of all players and the ball at, say, the twentieth second of the fourteenth minute. Yet there are patterns and consistencies. Football, like the weather, seems to be short-term unstable, but long-term stable.

System life cycle theory offers insight for short-term and long-term prediction. It also affords more help in the understanding of why predictability is so limited.

## Systems Engineering

Classic systems engineering seeks to understand systems so that new systems can be introduced coherently to improve a problematic situation. Over the last 50 to 70 years, ad hoc rules of thumb have been adopted that have been found to be successful in some cases. Certainly, when some of these time-honored practices are not observed, as in not testing the Hubble space telescope in one piece on the ground prior to launch, spectacular failures can occur.

System life cycle theory introduces a new set of ideas for systems engineering. If a newly created system is launched into an environment, perturbations will occur as the pre-existing systems react to the newcomer. Some of these perturbations may be undesirable, and may even militate against the operation or stability of the newcomer. Complementary systems, referred to in the approach below, are additional systems to those being considered, which should be provided at the same time to anticipate or reduce perturbations caused by their introduction. A new paradigm for systems engineering presents itself, outlined as follows:

- Introduce complementary systems to neutralize unwanted perturbations;
- Introduce complementary systems to link the output of a new system back to the input of a new system;[2]
- Increase internal variety;
- Intraconnect that variety to enhance internal stability;
- Enhance cohesive influences, and diminish dispersive influences;
- Interconnect the variety to promote external stability.

These new rules are very much outward looking compared with traditional, more introspective, systems engineering: System life cycle theory is primarily concerned with systems as part of an interacting set of systems. System life cycle theory approach is thus complementary to, rather than an alternative for, classic systems engineering rules and methods. Using system life cycle theoretic methods enables a system to be designed first in terms of its ability to integrate with other systems in its environment, after which its internal characteristics may be conceived and created conventionally. The approach is also consistent with GST, which established that putting systems in dynamic opposition is the primary

---

[2] N.B. The second bullet anticipates counter-intuitive responses from the interacting system set, by ensuring that it will exist in a positive feedback environment to ensure self-sustainment and growth.

mode of systems regulation, rather than cybernetic, feedback approaches. (See page 5.)

## SUMMARY

Seven principles of open systems contribute to the synthesis of an open system life cycle theory and model. The theory, which is system-scale independent and system-type independent, is exemplified in application in a wide variety of situations: political, economic, and business. It also explains limits to the prediction of system behavior, and suggests ways to modify systems behavior. Using life cycle theory, an alternative paradigm for systems engineering is created, consistent with general systems theory.

## ASSIGNMENTS

1.  A political leader, renowned for having a dominating character, ridicules opposition and includes only sycophants in the inner sanctum of advisers. Using system life cycle theory as a guide, predict the fate of that leader.
2.  There is a view that species extinction is associated with meteoric impact. The extinction 64 million years ago coincides with a layer of iridium in the geological record, said to have come from a meteor impact and explosion. There have been many species extinctions, however, and too few meteoric impacts to explain them all. We are presently in the grip of a major species extinction with about five species being lost per day; the woolly mammoth's loss is part of the current extinction. Species extinctions take hundreds/thousands of year to occur.

    Establish an alternative extinction theory; one not dependent on external events such as meteors. Using only the system life cycle theory and principles, plus your imagination:

    - Explore the possible causes of species collapse;
    - Postulate one or more theories to explain the catastrophic collapse, comparing each as necessary;
    - Consider the build-up to extinction—its nature and cause; and
    - Also consider what happened after the collapse. While wholesale species extinctions occurred, at no time was all life extinguished (as far as is known):

        o   Thus crocodiles, turtles, sharks, and many land and sea species, including early mammals, survived the extinction 64 million years ago;

- o If the theory that today's birds are an extant branch of dinosaurs is true, then they, too, survived;
- o The meteoric impact theory finds some difficulty in explaining the selectivity of extinction.

- Compare your theory with conventional (meteorite/climate shift) theories.

3. State hospitals and schools may find themselves administered by councils and by civil servants sitting in some remote regional capital. For hospitals, resources, training, recruitment, and many other features are therefore administered–chosen, ordered and supplied–largely remotely. Using system life cycle theory as a guide, propose ways in which such hospitals and schools might be made more responsive and responsible. Identify both the potential benefits and the risk in your proposals.
4. In forming a new team to tackle a novel problem, would you select people with similar or dissimilar backgrounds? What other personality traits would you consider essential, and why?
5. Why do fish farms run such high risks from disease?
6. Why do many companies switch periodically between project organization and functional organization? Map the switch on to the life cycle map? Is it chaotic, catastrophic, inevitable, or advisable?
7. The Vance-Owen plan for former Yugoslavia proposed a patchwork quilt of interspersed Moslem and Serbian domains. Why could this never be acceptable to either combatant?

# REFERENCES

[1] Hitchins, D. K., "A Unified Systems Hypothesis," *Systems Practice*, Vol. 6, No. 6, 1993.

[2] Minkel, J. R., "The Meaning of Life," *New Scientist,* Vol.176, No. 2363, 2002.

[3] Odum, Howard T., *Environment, Power and Society*, New York, NY: Wiley, 1971.

[4] Gleick, James, *Chaos – Making a New Science*, London, England: Macdonald & Co., 1989.

# Chapter 7

## The Social Genotype

*Hardened around us, encasing wholly every notion we form,*
*is a wrappage of traditions, hearsays, mere words*
Thomas Carlyle, 1841

### INTRODUCTION

Genotype describes the genetic make up of an organism or group. Phenotype refers to the observable features of an organism as a result of interactions between the genes and the environment. Both are terms from biology.

The so-called organismic analogy (see page 4), suggests other systems as having genotype analogs, or exhibiting phenotypes. In particular, it is useful to consider the analog of a social genotype for social and sociotechnical systems.

Complex organisms develop from the pattern stored in their respective DNA: a helical chain of nucleotides and hydrogen bonds. The pattern expressed in the deoxyribonucleic acid (DNA) is unique to a given organism. DNA is passed down through the generations, changing only slowly over time as organisms evolve.

Social groups develop cultural patterns of behavior that survive changes of individuals in the groups. Analogously, the pattern of group behavior, locked in a convoluted web of roles and relationships, is inherited, transcends individuals, evolves slowly over time, and exhibits immune response; that is, it rejects would-be intruders.

This, then, is the idea of a social genotype. A social group (company, organization, family, political party, religious group, hive, clone, and so on) forms a pattern of group beliefs and behaviors that is effectively stored in the web of social roles and relationships that binds the group together.

The pattern of beliefs and behaviors transcends time. As a member of the group retires or dies, he or she is replaced by someone who either fits the role perfectly–and can therefore maintain the relationships with other role holders–or who does not, and is summarily ejected, to be replaced by someone else who does. Role beliefs and behaviors, then, are defined and affirmed through the relationships with other role holders. Newcomers adapt and fit, or leave; the role-relationship structure, the social genotype, persists.

By further analogy, the social genotype interacts with the environment to exhibit a social phenotype. Group behavior and, to a lesser extent, group beliefs, may change according to the environment in which they finds themselves.

Examples of the social genotype abound:

- *Patterns of belief and behavior in transcendental systems*– The Council of Nicæa, summoned by Constantine the Great in 325A.D., laid down the pattern of services in the Christian church. The same council set the date for Easter. Both the pattern of service, which detailed such matters as when to stand, kneel, and sit, and the date for Easter, which relates to the phases of the moon, are still in use today. The language, Latin, survived until the 20th century.

- *Traditions, rituals, and rhymes*– Parents pass on behaviors to their children; sometimes long after the original cause has been forgotten. Children recite "ring-a-ring of roses," unaware that it recorded the great plague of London, where a ring of roses was supposed to ward off evil. Subsequent lines include: "atishoo, atishoo, we all fall down," at which point the children fall down on the ground, unaware that they are feigning sneezing followed by death from plague.

- *Guy Fawkes celebrations*– Each year the English celebrate Guy Fawkes who attempted to blow up the Houses of Parliament in the 17th century. Quite why this is cause for celebration is unclear.

- *Historical events*– The Irish remember the battle of the Boyne, in 1690, when William III–William of Orange–defeated James II at the Boyne River. This event, over 300 years ago, is a focus for celebration and hostility today, as though the event imprinted itself on the social genotype of the people.

A social genotype can form and take effect in much shorter timeframes. Start-up companies, initially without much real internal structure, soon develop roles and relationships, and soon exhibit a "corporate style," their way of doing business. Older companies can become so set in their ways that they are unable to respond to changes in their market. For example, IBM, a world leader in large computer manufacture, found itself painfully slow to realize that the world had moved on, and that it desperately needed to reinvent itself if it were to stay in business, let alone recover lost ground.

## CHANGE AND THE SOCIAL GENOTYPE

A characteristic shared by biological and social genotypes is resistance to change. Perhaps an examination of the way a social genotype manifests itself within an organization might throw light on the manner in which change management works in organizations.

For new organizations, roles are initially fluid, and not determined by relationship. Start-up companies, for instance, are characterized by a freedom to think and act, a lack of bureaucracy, and a lack of structure. It is not so much that there are no titular heads, it is more that everyone can, and does, turn their hand to every task. This exciting, creative, and innovative phase gradually gives way over time to a more structured organization, with hierarchy and divisions, territories and boundaries. The role-relationship structure forms, determining the way the company operates internally, who does what to whom, how things are done, and so on. At this point,[1] company culture is established.

The time it takes for this "ossification" process depends on the size of the group. For committees, it can happen in a few hours. For start-up software or electronics companies in Silicon Valley, it seemed to take several years. Once established, there will be many roles and relationships; mutually reinforcing each other to resist change.

What about the newcomer? When joining a new organization, it takes time to learn the ropes and fit in. This is learning the job (role); part of learning the job is to learn about the relationships. Newcomers may be trained to help them fit in. In an existing organization, any given role is determined (in part) by pre-existing relationships to other roles.

During the newcomer's period of adjustment, his or her opinions[2] will be sought and valued, particularly with respect to things outside the group's immediate experience. Once fitted in, the newcomer, having adjusted his or her way of behaving and performing to those of the group, will no longer be considered a font of knowledge. Instead, he or she will be a relative newcomer who still has a bit to learn. This change of attitude toward the newcomer arises because it is perceived that he or she has become synchronized and resonant with the culture, beliefs, and behaviors of the group, and so has nothing new to offer. Instead, such newcomers may have to be literally "kept in their place." (Such familiar expressions take on a new significance in the light of the social genotype.)

## Quantifying the Social Genotype

Because the social genotype is about roles and relationships, it is possible in principle to count up the numbers of each and to perform numerical analysis.

---

[1] Company culture is often marked by the issue of a company handbook defining the way things are done in the company and, in effect, restraining further uninhibited innovation.

[2] The willingness of organizations to listen to and accept the advice of a newcomer, is of particular concern in the case of consultants, who are generally new by definition. Many organizations would rather accept external advice, even from a highly questionable source, than listen to considered advice from a group of their own staff with many years of direct, relevant experience and expertise. Good consultants, aware of this danger, resist advising and, instead, act as catalysts to help organizations reach their own solution, using their own experience and insights.

Since the role-relationship structure underpins group beliefs, behaviors, and practices, and since these are not themselves quantifiable in any useful manner, numerical analysis shows only the magnitude of the network, rather than any intensity of cultural feeling. Nonetheless, it may prove a useful guide, especially when comparing situations, or when comparing alternative strategies for change management.

Consider a corporation of, say, 1,000 employees. Each employee is contained in several (containing) systems at once: the work hierarchy; family; union, perhaps; and informal groupings (e.g., the secretaries' network). Each containing system influences behavior in real time. So, an employee at work may be thinking about union matters, an argument with a partner, a forthcoming social event, or a member of the opposite sex who has just passed by. Each employee has several roles within each containing system: boss, subordinate, peer/colleague, confidante, father figure, worker, leader, and so on. Each role is determined by interactions with other roles, reinforced by the appropriate belief system (see Figure 3.1 on page 69), and by perceptions of status (or loss of status).

- A corporation of 1,000 employees might be organized into, say, four divisions; perhaps of 250 each. Each division might be further organized into 25 groups of 10 each.

- A group of 10 people has up to $10 \times 9 = 90$ interpersonal relationships, making 9,000 relationships across all $4 \times 25 = 100$ groups.

- Within each division there may be $25 \times 24 = 600$ intergroup relationships, making 2,400 across all four divisions.

- Within the corporation there will be $4 \times 3 = 12$ interdivisional relationships.

- Total = 11,412 relationships; each complex, and multifaceted.

- This web of relationships connects 1,000 employees, each with an average of, say, 3 roles, making 3,000 active roles.

- The total role-relationship product then is $3,000 \times 11,412 = 3.42 \times 10^7$.

The size of the role-relationship product may seem surprising, yet the sum is incomplete. Individuals also form relationships with others who happen to work in the adjacent offices, or perhaps in the staff cafeteria, even though there may be no direct organizational relationship between such people. Such ad hoc relationships are indicative of a friendly, open culture in many organizations.

When an organization decides to introduce some major changes, it may underestimate the size of the role/relationship complex – or corporate social genotype. The large numbers imply at least a degree of organizational inertia, and that time will be needed for evolution. One supposed method of change management–posting a new organization, with new office layouts, and people allocated to new jobs with immediate effect–has a limited prospect of working;

even with a willing and enthusiastic workforce. Change management has to be driven, it seems, from the top of the organization, or it runs out of steam.

## Change by "Budding Off "

There are many potentially successful strategies for change management; there are also many failed attempts, and apocryphal tales to embroider them. One approach is to start up a new division or new enterprise in a new location, away from the influence of the older outfit. If the new enterprise is "seeded" with a few people steeped in the desired culture, they will quickly adopt a suitable role-relationship structure, or social genotype, within which the desired beliefs, culture, and behaviors–a social phenotype–will be both expressed and maintained. Newcomers will then fit into the new role-relationship structure, and will adopt the new culture without question. In this way, a new company or corporation can be gradually "budded off" from the old, with an entirely new social genotype.

## Progressive Conversion

An alternative approach, shown in Figure 7.1, does not necessitate budding off. Instead, without altering the social genotype, it transfuses the organization with new ideas, methods, and thinking.

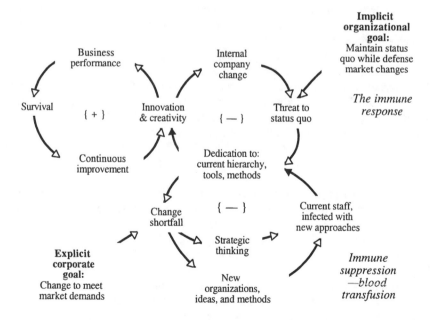

**Figure 7.1** *Change management strategy.*

In the figure, the top left loop shows corporate policy for a defense corporation, business, or enterprise. (Open arrowheads indicate a positive, synergistic influence; filled-in arrowheads show an opposing, contrary influence.) The policy is one of continuous improvement, or *kaizen*. This causes internal resistance (see the top right loop), with the workforce preferring to stick with older, trusted processes, ideas and methods, and resisting innovation. In the organismic analogy, this is Immune Response: invading new ideas are neutralized and rejected.

Bottom right is the means of Immune Suppression; a continuous transfusion of personnel retrained into strategic thinking, new methods, ideas, and organizations. This is a relatively slow, but effective, process, which leaves the underlying social genotype unscathed. In effect, organizational stability is being restored by the infusion of fresh, connected variety. (See The Principle of Connected Variety on page 109.)

**Continuous Revolution**

A radically different approach is to never let the social genotype have sufficient time to set in place. Chairman Mao's cultural revolution in China typified this radical approach.

Recognizing that a social genotype locks in patterns of belief and behavior, a continuous revolution aims to prevent the formation and setting of the social genotype by continual upheaval. In principle, this is expected to maintain a group–if it can be called a group in such circumstances–in a high state of entropy, excitement, innovation, and creativity.

Some organizations bring together ad hoc groups to address problems; perhaps setting up more than one such group to address the same problem, so that different potential solutions can be compared. Some organizations turn over staff at such a rate that there is always a ready supply of "new blood" with new ideas, not yet inculcated into organizational beliefs and behaviors. Other organizations employ continual streams of external consultants to achieve the same ends.

A more extreme approach is to continually reorganize the company in such a way that individuals are dissociated from previous colleagues. This might seem tantamount to throwing a pack of cards up in the air and seeing how they land, but it is feasible where individuals have specific skills and roles. The technique is more common on a smaller scale where people work in teams or syndicates. This mixing up of teams, or syndicates, transfers new ideas across syndicate boundaries, resulting in richer analysis, synthesis, and solutions.

One aspect of continuous revolution to consider is the increased entropy that it invokes. Higher entropy systems are, by definition, less able to convert their internal energy into useful external work. Or, in simpler language, the group has to get organized to get serious work done, which takes time and, in the process, starts to form a social genotype. Perhaps *continual*, rather than continuous revolution, would be a more workable compromise.

## THE MILITARY AND THE SOCIAL GENOTYPE

There are situations in which the durability of the social genotype may be advantageous. Human beings are, by nature, creatures of habit, ritual, and routine. This reduces the perceived entropy in our daily lives; making us individually more comfortable.

Military groups are formed from regiments or squadrons that have existed sometimes for hundreds of years. The regiment has traditions. Recruits are trained, not only in their military skills, but also in the traditions, beliefs, behaviors, and practices of the regiment. Training is intense and continuous; after a while the recruit's trained behavior supersedes his or her natural behavior. The trained and disciplined soldier's habits, rituals, and behaviors are now those of the regiment.[3]

This is important for several reasons. The soldier's instinctive reaction to threat and opportunity becomes controlled, predictable, and consistent with group capability and survival. Allegiance to the regiment breeds esprit-de-corps, which gives soldiers the energy and drive needed to work together as a team in the face of threat, danger, and casualty. Self discipline becomes second nature. Knee-jerk reactions give way to considered judgments and decisions. And, when the soldier finds himself or herself isolated, the disciplined rituals and behaviors maintain the soldier's mental and physical equilibrium. Each trained soldier effectively carries a copy of the regimental social genotype in his memory.

This is the beneficial side of the social genotype that formed in regiments many years ago and that persists today in tradition, ritual, ceremony, dress, training methods, loyalty, and so on. The general public sees aspects of this in parades, with soldiers marching in precise synchronization and wearing identical, distinctive uniforms to indicate membership in a special group. However, the public does not see the culture and beliefs shared by group members, and maintained in the invisible, but palpable, role-relationship structure, or social genotype.

## SYSTEMS ENGINEERING AND THE SOCIAL GENOTYPE

Systems engineering is, by design, intended to be creative and innovative, yet with a well controlled management of risk. Systems engineering teams develop a social genotype, however, which might be expected to inhibit innovation.

The apparent paradox is addressed by introducing deliberate disorder (creative entropy) into the systems engineering process. This is generally done in two ways: first, by the extensive use of brainstorming and similar sessions early in a project; second, by continual use of the systems engineering problem-solving paradigm (SEPP) throughout the project life cycle. See Figure 4.2 on page 79.

---

[3] The example here is of soldiers and regiments. The concepts apply equally to sailors, airmen, marines, special forces, and so on; that is, to all disciplined military, and quasimilitary groups, and disciplined emergency services.

Both processes benefit from the juxtaposition of experience and youth; the latter injects new ideas and creative approaches into the discourse, while the former is able to sift the viable from the nonfeasible proposition.


## SUMMARY

As a model of social role-relationship bonding, the social genotype is one of several ways that can be used to view social behavior and social resistance to change. It is useful in that it offers a chemical bonding analogy that can be quantified, although the strength of the bonds is not accounted for by simply counting their numbers.

The substance of social genotype bonds is shared beliefs, culture and behaviors. In effect, a culture becomes embedded in the role-relationship structure, transcending individual role holders, and becoming potentially eternal, unless the genotype is shattered or starved into nonexistence.

The social genotype explains difficulties experienced in managing change within organizations, and suggests alternative approaches that are likely to be more successful. Continuous revolution is seen as a way to overcome and minimize the change resistance; military tradition, on the other hand, is seen as a valuable instance of the social genotype in operation. Finally, creative entropy is presented as a valuable compromise within systems engineering, preventing local stagnation without prejudice to the overall systems engineering process.


## ASSIGNMENT

We have all experienced attending a reunion at work, school, university, and so on. The years fall away and we each find ourselves, almost irresistibly, adopting the same roles and relationships that we had all those years ago. We avoid the one-time bully, we embrace and associate with old friends, we exchange secret signs, we like and dislike as before, we envy as before, we recall the nicknames we used for one-time authority figures, we still fear those figures, and so on. We recall long-forgotten events with ease. We even find ourselves behaving as we did before, although we may now disapprove of such behavior.

Explain these reactions and behaviors in terms of the social genotype. Explain what is implied by this common experience in terms of our memories, the recall of prior roles and relationships, and the durability of stereotypes. Suggest ways in which the phenomenon might prove valuable.

# PART II

# SYSTEMS THINKING

# Chapter 8

## Tools and Methods for Systems Thinking

*We must dare to think about "unthinkable things," because when things become "unthinkable," thinking stops and action becomes mindless.*

Speech, U.S. Senate, *J. William Fulbright, March 25, 1964*

### ABOUT SYSTEMS THINKING

#### Facing Up to Our Real, Nonlinear, Dynamic World

Systems thinking is not new; it has been around for thousands of years in many different guises. Ancient creation myths were instances of systems thinking. Operations analysis, systems analysis, failure analysis, risk analysis, corporate benefit analysis, financial modeling, quantity surveying, investment appraisal, finite element analysis, civil engineering models, economic modeling, simulations, and many more are all modern ways of thinking about systems. Imaginative visualization should be on the list, too.

What is new, perhaps, is the ready availability of powerful desktop tools that permit and enable us to think about the most complex and complicated issues and systems. Processors allow us to tackle problems of such complexity and magnitude that, without them, we would be obliged to guess. The same tools reveal unexpected complex behavior from simple systems.

To avoid guessing, some confine their activities to a relatively simple, straightforward world of linear, singular cause and effect. Physics and engineering have, in the past, been examples of this classification, associated as it is with a mechanistic view of the world.

Before the advent of the processor, and its ability to repeat experiments hundreds and thousands of times, emphasis rested on the development of definitive equations that uniquely explained how the world worked. Newton's three equations of motion were a masterful stroke of genius: they provided explanation and proof in a compact, readily understood form. Since the processor, it has been possible to consider proof by exhaustive testing; performing so many calculations, each with different conditions, that all–or nearly all–possibilities are explored. This may not be quite so intellectually satisfying, but it is a practical approach to issues where the prospects for simple equations are poor to non-existent.

In particular, it is a practical way to approach nonlinear problems. Nonlinear equations generally cannot be solved and cannot be added together. Yet, at every turn, we find our world to be nonlinear.

Also relatively new is the degree of understanding we are gaining about ourselves and other species, both as individuals and in social groups. The burgeoning of the social sciences is enabling us to think rationally about ecologies, economies, and systems that include human beings; hitherto frowned upon as unreasonable or nonfeasible.

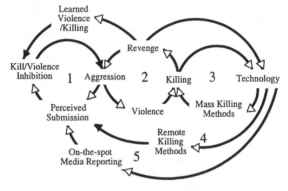

**Figure 8.1** *Kill inhibition in humans.*

Figure 8.1, for instance, shows one of the unexpected side effects that media coverage is having on warfare. The diagram is a causal loop model (CLM) with five principal, numbered loops:

- Loop 1 shows that humans will usually not continue with violence to the point of killing when they see an opponent submit; this behavioral reaction is common in fights between male members of the same mammalian species.

- Loop 2 shows that, when killing does occur, it may promote a continuing cycle of revenge, aggression, and violence that can overshadow the one-on-one kill inhibition, so that killing becomes routine and even acceptable.

- Loop 3 shows how technology enables mass killing methods, such that hundreds and thousands may be killed for revenge.

- Technology has also given man remote killing methods, shown in loop 4, such that the attacker may never see the victim submit, so that inhibition upon seeing submission is never triggered.

- Recently, however, the advent of roving, on-the-spot television coverage of warfare has shown the public the effects of shootings and bombings in stark close-up; triggering kill inhibition in millions of noncombatants; see loop 5. These people may be thousands of miles from the scene, but they also have influence over politicians who control the combat, and they may use that influence to stop the bloodshed.

Figure 8.1 explains a nonlinear dynamic social phenomenon that has political and military overtones, but that is outside the sphere of linear cause-and-effect analysis.

### Chaotic Perspectives

Along with many others, Benoit Mandelbrot [1] and Edward Lorenz [2] have opened Pandora's box, and introduced us to chaos; there is no going back, no resealing the box. Lorenz's classic weather simulations (see Figure 8.2) showed that the real world does not, and cannot, behave in a precisely repeatable way; prediction of the future is not practicable in a nonlinear world. Physicists and engineers may be unhappy with chaos, but it is not going to go away.

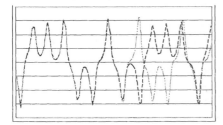 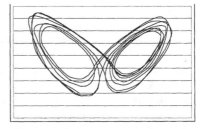

**Figure 8.2** *Icons of chaos. At left, Lorenz showed how even the smallest difference in starting conditions could materially alter the behavior of his (simulated) weather patterns. The lines from two simulations (started with miniscule differences in initial conditions) at first overlap, but then unexpectedly diverge. Their behavior is similar, but different. At right, Lorenz also showed that weather patterns "revolved" around what came to be called a "strange attractor" that was, curiously, shaped like a butterfly. Weather, the archetypal open system, might be long-term stable, but it never precisely repeats itself.*

Systems thinking, then, addresses the nonlinear dynamic systems of the world around us, and systems thinkers are prepared to use powerful tools to test hypotheses that may be quite closed to conventional mathematical analysis.

How does systems engineering come into the picture? Systems engineering is a process, or set of processes, conducted largely by people, who–along with their facilities–form a sociotechnical group. The people in the group are certainly nonlinear dynamic. Because of inevitable feedback and constraints, the overall systems engineering process is nonlinear dynamic, too.

Does this mean that the systems engineering process is unpredictable? Yes, in the same sense that Lorenz found the weather patterns to be unpredictable beyond a certain time horizon (see the left-hand chart in Figure 8.2.) However, the overall process, as well as its many parts, is amenable to simulation and so it is potentially open to solution under some conditions. Such solutions are likely to be stochastic, as the many starting conditions, many optional routes and changes during the processes, and many environmental changes impacting a typical project, will testify.

A well-founded and well-run systems engineering project can never be repeated precisely. Like Lorenz's butterfly (see the right-hand chart in Figure 8.2), successive projects revolve around an attractor, representing the "climate" in which each project exists, and that climate may be long-term stable; environment permitting. All of this will be anathema to project planners, project managers, accountants, and those many others who desperately want the world to be precisely controlled and predictable.

**Minimalist Behavioral Systems Thinking**

Contrary to popular myth, systems thinking generally neither represents nor simulates systems; the representation and simulation of just one sociotechnical system in full detail would be at least very difficult and time-consuming, and at worst impossible.

Systems thinking looks at cause, effect, and the behavior of systems. So, instead of trying to represent complex systems, the models and simulations represent the *behavior* of those systems, that is, the response to stimulus.[1] If one system impacts a second, then the behavior of the second is *caused* by the first. Changes in the emergent properties, capabilities, and behaviors of the second system are the *effect*. So behavior is a cause-and-effect, or stimulus-response, phenomenon. For nonlinear dynamic systems, as we have seen above, behavior will be conditional, and may not be repeatable. Moreover, singular causes may have multiple effects, and a singular effect may not be exclusively attributable to a singular cause.

Working with system behavior, rather than attempting to represent complete, complex systems; enables system thinking to address complex issues using minimalist models. The following topics look at a range of tools and methods for systems thinking. The list is far from exhaustive, and each item is addressed only briefly. The tools presented are mutually compatible, however, so that information from one may be used in another. The set offers a fabric of simple yet powerful

---

[1] Cause-and-effect, or behavioral analysis, is not new. Isn't that just what Sir Isaac Newton did when he established his three laws of motion? He certainly did not concern himself with the detail of what was to be moved, beyond its mass – an emergent property.

tools that will enable the user to tackle just about any problem. There is, however, no guarantee that nonlinear dynamic problems necessarily have a solution.

## CAUSAL LOOP MODELING

There are many different ways of looking at the world, as Figure 8.3 shows. For many people, cause and effect come in discrete pairs. One cause results in one effect, and that is that. It is a convenient viewpoint; one not requiring the holder to question too much. Some expressions of this are: "Poverty is the cause of crime;" "urban decay is caused by the lack of government investment;" and "the project is late because of underfunding."

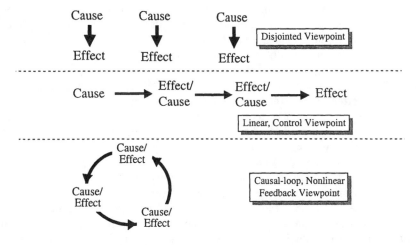

**Figure 8.3** *Cause and effect.*

Politicians of all varieties tend to stick with this disjointed view of the world; it enables them to seek a singular solution—the so-called "silver bullet." However, even politicians may find this stance untenable at times; then, they talk about "joined-up government," by which they mean that coordination between government departments; that is, multiple causal activities; may work where singular ones have failed. The disjointed view of discrete, cause-and-effect pairs is the antithesis of systems thinking.

Planners, project managers, and many engineers take a linear control view of cause and effect. They string activities into sequences such that the output (effect) of one becomes the input (cause) of the next in line. This is typical of program plans, GANTT charts, PERT charts, and simple bar charts. The linear control viewpoint is dominant. It is, however, deeply flawed; it takes no account of reaction. Stimulating any system causes a reaction, which opposes the stimulus.

This is true in physics and engineering in that:

- Any block of conducting material moving in a magnetic field experiences induced eddy currents that flow in such a direction as to oppose the movement.

- Action and reaction are equal and opposite; a paraphrase of Newton's third law of motion.

- Placing blotting paper on the edge of an ink blob encourages the ink to soak into the blotting paper under capillary action. As the capillaries fill, the blotting paper gorges, and the rate of uptake slows and then stops. This is a nonlinear, reactive phenomenon.

Reaction to stimulus is at the heart of science and engineering, yet it is often overlooked when developing plans for scientific and engineering activities: people systems, those processes conducted by people and machines, are somehow exempted from the universal rule. Although planners are constantly disappointed when plans do not work out, they continue to use limited planning methods; this is a continuing triumph of hope[2] over experience.

The lowest diagram in Figure 8.3 comes closer to the real world. Cause results in effect, which feeds back–sometimes through other systems–to interact with cause. This is a causal loop, and it is generally either negative, or positive, in the sense that the feedback either counters, or reinforces the initiating cause. In the bulleted list above, reactive feedback always opposed the cause – were it not so, we would have energy for nothing.

When a stage amplifier system howls, however, it is due to positive audio feedback from the loudspeakers into the microphone. It is more interesting to observe positive visual feedback. In a darkened room, connect a mounted video camera to a television, and point the camera at the screen. Light a match in front of the camera, and the light will travel into the camera, to the television set, back into the camera, and so on. Extinguish the light and the phenomenon persists, creating interesting and bizarre dynamic, nonrepeating patterns, indicative of nonlinearity and constraint within the positive feedback loop. The set-up is so sensitive that simply blowing on the camera may change the nature of the video pattern; this confirms Lorenz again, with a great sensitivity to starting conditions. In both examples of positive feedback, the nonlinear dynamic phenomena occur in systems that are designed by engineers to behave linearly.

Causal loop models (CLMs) have already been introduced (see Figure 1.17, Figure 1.18, Figure 1.19, Figure 2.4, and Figure 8.1). Figure 8.4 shows a further example. In the figure, signed digraphs join rates and levels of relevant factors,

---

[2] This is like Samuel Johnson's assessment of second marriage as "the triumph of hope over experience."

representing the flow of causality.[3] In this book, the signs are implicit by convention: Hollow arrowheads denote reinforcement; solid arrowheads denote a contrary or opposing effect.

The figure shows how body temperature stabilization might work, but only where there is a tendency for the temperature to rise above nominal; for lower than nominal, another CLM or an extension to this one would be needed, since the means used by the body to preserve thermal energy are quite different from perspiration.

The CLM of Figure 8.4 should be self explanatory, or it fails. At bottom left are reasons for body temperature to rise above nominal. The rise causes perspiration (at least in humans). If it can, the perspiration evaporates, extracting latent heat energy from the surface of the skin in the process. Heat energy can also be lost in liquid perspiration droplets, although this is only some one thousandth as effective as evaporative cooling. In both cases, bodily fluid is lost, reducing available body fluids. Should the latter fall too far, consequent dehydration will cause body temperature to rise.

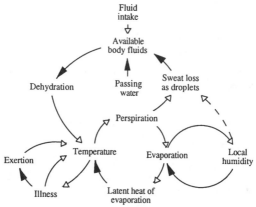

**Figure 8.4** *Causal loop model of body temperature regulation.*

The whole is a temperature stabilization mechanism apparently using negative feedback. The greater the (tendency toward) a temperature rise, the more profuse the sweating, the greater the evaporative cooling, and the less the temperature rise. The CLM also suggests how illness may result in profuse sweating, and how local humidity can affect the issue.

In addressing only heating and not cooling, the figure is incomplete. For instance, nominal body temperature is not represented; the thermostat setting is not evident. Is there a thermostat setting in the body, or is nominal body

---

[3] Causal loop models are not the same as influence diagrams. CLMs restrict themselves to direct, causal relationships, whereas Influence diagrams (as the name implies) include influences that need not be causal, and do not necessarily work in feedback loops.

temperature the result of opposing dynamic influences rather than cybernetic feedback?

For instance, stable core body temperature in humans might be the result of autonomic blood diversion. Blood diversion to the skin surface (and consequent sweating) prevents overheating. Blood diversion away from the skin surface uses surface flesh to insulate against loss of blood heat to the environment, thereby preventing overcooling. Both processes may be going on at the same time in mutual antagonism. The result is temperature stability, but without cybernetic control.

Is the mechanism cybernetic feedback, or antagonistic processes? One value of a CLM is that it expresses openly and very clearly the thoughts and ideas of its originator, enabling dialog and, where appropriate, correction or further development. The CLM is particularly useful for representing the behavior, and only the behavior, of nonlinear dynamic systems. Nowhere in Figure 8.4 is there any mention of bodily structures, pores, sweat glands, bladder, water-retaining tissues, and so on; they may be essential biological structures, but are irrelevant to the exposition.

### Promoting Completeness with CLMs

One powerful use for the CLM is to promote and ensure completeness. The method is illustrated in Figure 8.5.

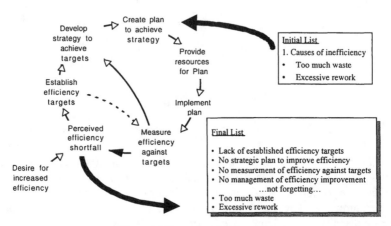

**Figure 8.5** *Promoting completeness.*

The figure considers the issue of inefficiency. A typical reaction, when asked the likely causes of inefficiency in a workforce, might be to draw up a candidate list. This is shown as the initial list in the figure. As is usual in such cases, the items appearing on the so-called laundry list are related to each other. Perhaps there are more. A CLM is formed, using the ideas generated from the initial list. In this case, the CLM represents some ideal world where efficiency is desired, at bottom left, and is achieved by developing targets, strategies to reach the targets, and plans to execute the strategies. Progress toward the targets is measured, and used to back off demand, so that the perceived shortfall is diminished, and the process terminates.

The final list is taken from the CLM by noting that, since the CLM represents an ideal world that is clearly not in place, each element in the loop is effectively missing. For instance, if "Establish Efficiency Targets" should be there, but is not, then a factor contributing to inefficiency is "a lack of established efficiency targets." This is negative inversion. Note how much richer the final list is than the initial list. CLMs are invaluable in this kind of activity. The CLM, in showing causal sequence, adds the dynamic aspects missing from the static laundry list.

The so-called negative inversion approach is useful, too. We humans have a natural propensity to criticize; so much so, that if you ask a group of people to list the good factors about something in their experience, and to separately list the bad features, the list of bad features will invariably be much longer.

CLMs can be built by taking advantage of this phenomenon. Create a list of negative factors, deliberately consisting of a pejorative adjective and a noun. Suppose a string of adverse comments about a hotel included the following: "Low cost efficiency," "poor maintenance," "dirty laundry," "badly cooked food," and "sloppy service." These constitute an initial laundry list.

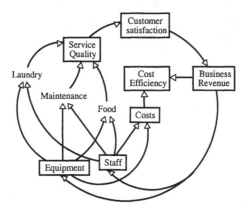

**Figure 8.6** *Cost efficiency.*

Putting the various factors together and dropping the pejorative terms, resulted in the ideal world CLM of Figure 8.6. In this ideal world, improving service quality increases customer satisfaction, which results in more business and more profit. Some of the profit is put back into the business to pay for more and better staff and equipment, thereby enhancing service quality even more, in a never-ending cycle. As the business attracts more customers, cost efficiency goes up by virtue of scale; with better staff and equipment, the same numbers can accommodate more staff with higher service quality, and gain higher business revenues. Hence, there is improved cost efficiency.

Such CLMs are interesting for another reason: they can spin down as easily as they can spin up. All the arrows indicate reinforcement, but that can work in a negative sense, too. If you reduce service quality and customer satisfaction is reduced, so too is business revenue; there is less money to pay for staff and equipment, and so on. Since there is no external reference–no cybernetic control–the only way to guarantee success is to maintain the upward spiral of continual improvement.

The CLM illustrates one other important factor. It is important for all the contributory factors to be in place. Having good food and maintenance but poor laundry would not result in customer satisfaction, and the loop would not "spin up." Also, if it is possible to enhance customer satisfaction without improving the factors shown in the figure, then revenue will increase, and it may subsequently be possible to pay more for equipment and staff. Sometimes the pump has to be primed to get the CLM cycle going.

## Patterns in CLMs

Quite different problems and issues may result in CLMs with similar topography. The following CLM, in Figure 8.7, considers truancy and its possible causes, presented as an intentionally pejorative list. The resulting CLM, having dropped the pejorative adjectives, offers an ideal world solution that combines several ideas. Note how similar in concept it is to Figure 8.5, with a driving influence for change, measurement to assess the change, and an indication of termination when the measured deviation from target goes to zero.

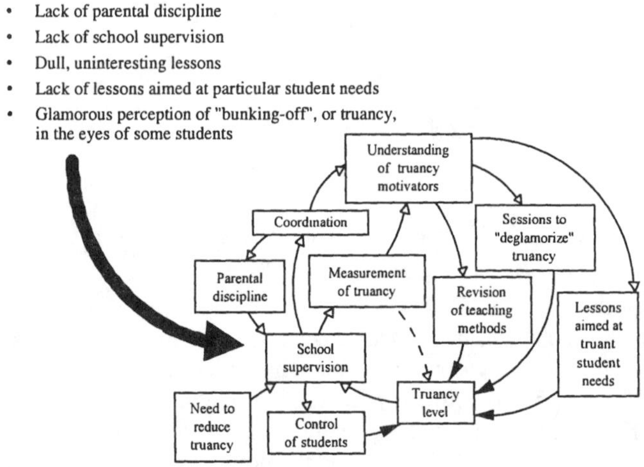
• Lack of parental discipline
• Lack of school supervision
• Dull, uninteresting lessons
• Lack of lessons aimed at particular student needs
• Glamorous perception of "bunking-off", or truancy, in the eyes of some students
Understanding of truancy motivators
Coordination
Sessions to "deglamorize" truancy
Parental discipline
Measurement of truancy
Revision of teaching methods
Lessons aimed at truant student needs
School supervision
Truancy level
Need to reduce truancy
Control of students

**Figure 8.7** *Truancy.*

A contemporary, underlying management paradigm, one that emphasizes measurement, generates these similarities. In many instances, such as school truancy, measurement is feasible and sensible. It may not be so in all cases.

The truancy CLM of Figure 8.7 also illustrates a limitation with CLMs. While the CLM might explain quite nicely one approach to truancy, it gives no indication of the time it might take, or the resources that might be needed over an extended period. However, the CLM forms a very convenient first step toward nonlinear dynamic system modeling, which will be addressed later, and which could assess the likely timeframes and costs.

# $N^2$ CHARTS

## The Hand-Drawn Tool

The $N^2$ chart [3] is a popular tool among systems engineers. It is a table or chart with N rows and N columns; hence N x N = $N^2$ cells. It records entities and the relationships or interchanges between them. Entities are entered on the leading diagonal, while relationships appear in the other cells as appropriate.

← Outflows →

| Commander CEO | Love House | Friendship Information Advice | Orders Instructions Loyalty |
|---|---|---|---|
| Love Support Home | **Wife** | Hospitality | ? |
| Friendship Information Advice | Respect | **Commander / CEO's Colleague** | ? |
| Reports Respect Loyalty | Respect | ? | **Subordinate** |

Inflows ↕

**Figure 8.8** $N^2$ *chart example.*

A simple N-squared ($N^2$) chart, as shown in Figure 8.8, illustrates the idea. A commander or chief executive officer (CEO) has a wife, a colleague, and a subordinate. The commander gives love and a house to his wife, while she gives love, support and a home to the commander. All the outflows from the commander are on the top row, while all the inflows to the commander are in the left-most column. Similarly, outflows from other entities are in their row, and inflows are in their column.

This $N^2$ chart reveals a particular culture. The commander, presumed male, gives his wife a *house*, while she gives him a *home*. The commander's wife gives hospitality to his colleague; the colleague gives respect to the wife, and so on. Some interchanges are unknown. Does the wife have colleagues or friends? None is identified. Does the colleague have any relationship with the subordinate? The subordinate gives respect to the wife, but what she gives him is not known.

$N^2$ charts are very compact, allowing an overview of even the most complex of systems. The interfaces tend to occur in pairs, as we have just seen, potentially forming simple, reactive causal loops.

However, the $N^2$ chart may represent multi-element loops; for example, the commander gives his wife a house; the wife gives the colleague hospitality in the house; the colleague gives the commander friendship, information, and advice. This three-element loop may also be causal. The colleague may give friendship, information and advice because he has received hospitality from the wife in the commander's house. Just as it is possible to simulate a causal loop model, so it is possible to simulate the dynamic behavior of an $N^2$ chart, with all of the various causal loops in simultaneous and synchronous operation.

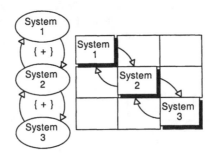

**Figure 8.9** *Representing system interactions in an $N^2$ chart.*

Interacting systems can be represented very simply in an $N^2$ chart; see Figure 8.9. The $N^2$ chart in the figure has six interface cells, and so can represent three reactive causal loops. (Systems 1 and 3 could interact directly.) For a large number of interacting systems, the lines representing interaction complicate conventional layout diagrams, while the $N^2$ chart eliminates crossing lines, thereby significantly reducing perceived entropy.

Where the $N^2$ chart represents a system's internal subsystems, and interconnections, it is important to remember that the whole system exists in an environment with other systems. These other systems may provide inputs to and receive outputs from the system represented by the $N^2$ chart. Not to include these inputs and outputs is to disconnect the system, encouraging a static, linear reductionist viewpoint.

One way to overcome this potential limitation is to surround the $N^2$ chart with another layer, as shown in Figure 8.10, where the shadowy systems X and Y represent the systems that interact with the system-of-interest, represented by A, B, and C. In this manner, system ABC will exhibit and be subjected to the behavior of interacting systems. As a simple example, suppose A was a chassis, B was transmission and C was suspension; then ABC might be a vehicle. System X might then be a driver and system Y might be a vehicle test system. The behavior of the vehicle would then be seen in the context of the driver and the test road surface conditions. Without X and Y, ABC would be static and lifeless.

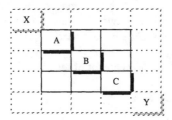

**Figure 8.10** *The $N^2$ chart in context.*

Used as a hand-drawn chart, the $N^2$ chart can be a valuable aid to amassing large amounts of information about a complex system or project. Rows and columns can be interchanged so as to create patterns of interfaces. In Figure 8.11, representing the systems within some overall system, interface patterns are arranged to reveal some of the archetypal interface patterns [4]:

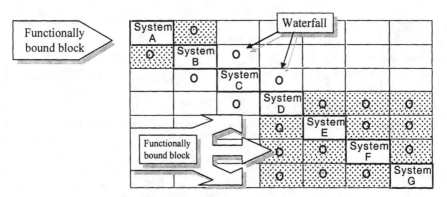

**Figure 8.11** $N^2$ *Chart interface patterns.*

- Functionally bound blocks of interfaces appear as a rectangle of interfaces, indicating that the corresponding systems have many mutual interfaces. This is an indication of tight functional binding, and a suggestion to a system architect that the systems may suit physical collocation and/or special communications to facilitate interchanges between them.

- Waterfall patterns indicate, as the name suggests, that information or a substance is flowing, perhaps down a management chain, or a chain of command. Mirror-image upward waterfalls, representing a reporting or acknowledgement chain, match downward waterfalls.

- Not indicated in the figure is the node, since this arrangement evidences only a partial node. System D is at the center of an orthogonal cross of interfaces, indicating that this system interfaces with systems C, E, F, and G. Were system D vulnerable, damaged, or removed, then the overall system would separate into two disconnected parts ("disjoint sets"). Nodes, even partial ones as shown here, are important in system design and operation, and may need to be replicated, protected or given the additional capability to maintain the behavior of the overall system.

**Encapsulation and Elaboration**

The process of clustering $N^2$ charts to reveal the archetypal patterns, as in Figure 8.11, facilitates encapsulation and elaboration. The functionally bound block, represented by systems D to G, could be capped (encapsulated), and treated as one system, system Q, and the chart would be significantly simplified. This reduces perceived complexity and entropy. Expanding system Q back to its current state would be elaboration, since the environment, interfaces, and interflows would be maintained, but more detail would be revealed in context.

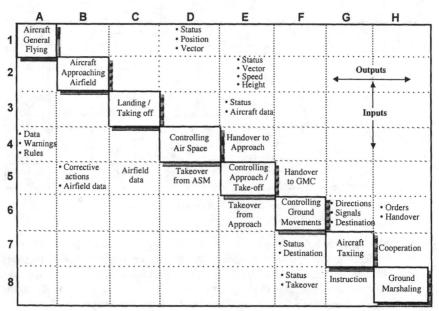

**Figure 8.12** $N^2$ *Chart of air traffic management at an airport (incomplete). Ground movement control is expressed as GMC. Air space management is expressed as ASM.*

Since a system is, by definition, made of connected parts, it should always be possible to progressively encapsulate parts of its $N^2$ chart until only one single entity remains.

## Dynamic $N^2$ Charts

$N^2$ charts can represent many activities occurring in series and parallel, and can reveal activity nodes and potential management overloads. Figure 8.12 is an $N^2$ chart of a notional airport, showing air traffic management (ATM) activities. The entities in the leading diagonal are sets of activities, or system processes.

Using the chart, management of an individual aircraft can be visualized, from its approach (B2) to landing (C3) to taxiing (G7) to marshaling into the appropriate unloading bay (H8). Activities receive and transmit data and information in real time; this is an information decision action (IDA) system. Besides being holistic and synthetic, using the $N^2$ chart in this way reveals the organismic and dynamic nature of the systems in question.

## Automating the $N^2$ Chart

When an $N^2$ chart is formed, it is unlikely that the patterns will be evident, since the entries into the chart may be haphazard. For a large $N^2$ chart (and charts with over 100 entities are not uncommon), the process of rearranging rows and columns to reveal the patterns can be time consuming and labor intensive. Programs may be devised to undertake this procedure under user control, redrawing the $N^2$ chart as necessary. Even then, the process can be time consuming and involved, and results may prove less than optimal.

The solution is to automate the process;[4] this is made practicable by observing that, when clustered as in Figure 8.11, both the perceived and measured entropy are at a minimum. Appendix A shows the evidence and the method. Using a suitable genetic algorithm or hill-climbing routine, the rows and columns of the $N^2$ chart can be rearranged, while still maintaining the system-to-system relationships intact, to reveal the patterns of nodes, waterfalls and functionally bound blocks.

The automated $N^2$ tool is a powerful ally in the processes of concept formulation, system design, and system management. Figure 8.13 shows a typical problem that could be tackled manually, but that becomes easier when an automated tool is used.

The figure consists of five annotated panels. At top left is a simplified view of some tangled system. The view might represent a number of people sitting at desks who go to each other's desks to exchange information. It might represent a set of electronic devices joined by connectors and cables, or hospital departments

---

[4] Programs are available on the market to do this, but it is not difficult to write one's own program, or to automate a suitable spreadsheet for the purpose.

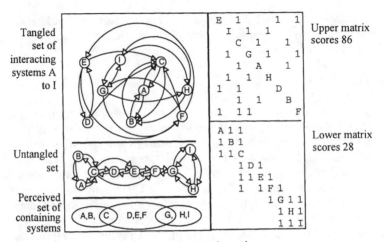

**Figure 8.13** *Automated* $N^2$ *chart[5] in action.*

being visited by outpatients. The important point to note is that the pattern is complex, with overlapping lines and many connections.

At top right, the top-left pattern is entered directly into an $N^2$ chart, using the same order in which the entities present themselves, and indicating a connection between two entities by a "1" in the appropriate space. The upper matrix scores 86; this number is analogous to entropy (see Appendix A for details). Turning loose the genetic algorithm in the automated tool results in the $N^2$ chart at bottom right, with a minimum score of 28. The chart is converted to a diagram of interacting systems at center left, but it is evident from the reformed $N^2$ chart that functionally bound blocks have appeared, together with partial nodes; this enables the bottom left systems diagram to be drawn, revealing three overlapping systems.

It would be difficult to impossible to perceive three overlapping systems in the tangled web at top left of Figure 8.13. Using the automated $N^2$ tool greatly simplifies the practical process; many real-world problems would be impractical without it. Using the $N^2$ chart enables the systems thinker to remain holistic, synthetic, and organismic; the automated tool makes it easier to manage the complexity that comes with variety and tangled interconnectivity.

## INTERPRETIVE STRUCTURAL MODELING

Interpretive structural modeling (ISM) [5], is a powerful graphical method for finding and presenting purposeful relationships between entities. Given a set of entities, the relationships between them can be identified using Saaty's pair-wise comparison technique [6]. Simply, this technique examines a list of entities taken

---

[5] The two small $N^2$ charts shown here were produced using my simple, homemade program, computer aided design relationship analysis tool (CADRAT©).

two at a time and asks questions of the pair. A typical set of questions might be: "does entity X contribute to entity Y; does entity Y contribute to entity X." Each question can be answered yes or no; there are four possible outcomes to the two questions.

With ISM, the results of the questions are entered into a matrix such that yes is recorded as "1" and no is recorded as "0." In the general case, it is likely that, if X contributes to Y (1), then Y does *not* contribute to X (0) or at least not so strongly. ISM matrices tend, therefore, to be sparse. It is possible, as a result, to rearrange the matrix so that the entities are arranged in successive order of contribution: "P contributes to M contributes to D contributes to Z contributes to...." This chain of successive entities and contributions can then be drawn and presented as a structure.

The ISM technique can be used widely. Different questions evoke different results:

- Objectives help to achieve each other; these are intent structures.

- Entities contribute strongly to each other; these are attribute enhancement structures.

- Activities precede each other; these are precedence networks.

- Projects "are more important than" each other; this is a priority structure.[6]

The implications of the rearranged matrix are more readily appreciated in graphic form; see Figure 8.14. The entities appear in boxes, and the relationships as arrows. The arrows are read according to the question that was posed, such as "helps to achieve," "strongly contributes to," "precedes," or "is more important than." The boxes are deemed "transitive;" that is, influences pass through them as behaviors.

Since Figure 8.14 is an intent structure, the arrows in would be read as "helps to achieve." The objectives at the bottom of the graph are those that must be achieved first; hence they are "most pervasive," since their effects will pervade all later objectives. Conversely, the objectives at the top of the "tree", or the ultimate objective, may be equated with the mission of the whole enterprise; all the other objectives help to achieve it, while it helps to achieve none. Objectives 4 and 5 in the figure would appear as they do if both objectives were deemed to help to achieve each other in similar degrees. Objective 8 is nodal, or pivotal, since many objectives combine in helping to achieve it; if it is not achieved, the mission is not achieved either.

ISM is particularly useful for attaining group consensus. Whereas a group of people may have great difficulty in agreeing, for instance, which is the most

---

[6] Project structures are subjective, and must be associated with a set of rules by which "more important" is to be judged.

important entity when faced with 15 or 20 different choices, they can usually agree which is more important on an entity-by-entity pair-wise basis. The final, overall result may surprise the group, but the process of choosing involved them all, and used no information other than they provided–which makes it difficult to argue.

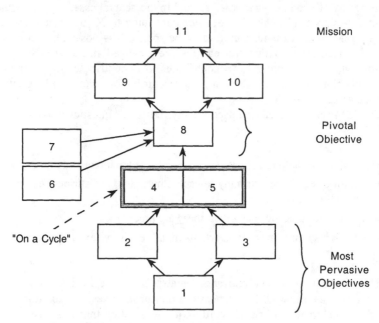

Figure 8.14 *Interpretive structural modeling scheme: intent structure.*

The choice of the questions to be asked is important. In an ideal world, the question might be: "does X *cause* Y to...." Causality is the strongest behavioral relationship in this context. "Strongly contributes to" is the next strongest, with "helps to achieve" being perhaps the weakest. On the other hand, there will be more occasions in which "A *helps to achieve* B" than "A *strongly contributes to* B," which is a more restrictive behavioral category.

Figure 8.15 shows a business intent structure; the various activities were gathered by brainstorming with a group that was considering setting up a new business. Its mission, the group members believed, was survival, rather than simply making a profit, although they would clearly not do one without the other.

Initially unstructured, the group related the activities to each other using ISM, and formed an initial intent structure, from which it became clear that there were potentially not one, but three associated businesses: a business concerned with conceiving, designing, and prototyping new products; a product assembly business; and a product repair business. The distinctions were important, because the three businesses had three distinct rhythms: designing new products is a project-type activity, with a start, duration, and end-point; assembly is a continuous flow-line business; and repair is a spasmodic business generated some time after the products are sold.

The three potential businesses were felt to be synergistic, but separate; a clarity of view not evident before the intent structure was formed. Note in the figure how "Make Profit" (strictly an objective rather than an activity) is included twice, enabling the lines of causality through each of the businesses to become causal loops. This simple addition adds dynamism to the intent structure, and highlights the continuous flow nature of the businesses. Note, too, how the lines of causality have created a minimal entropy structure; the businesses, if organized as the figure suggests, would be simple and straightforward.

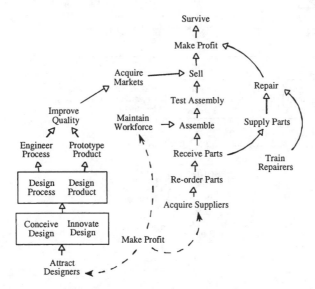

**Figure 8.15** *Business intent structure example.*

Used in the way exemplified, ISM is a powerful tool for conceiving systems and, as in this case, for designing systems. It is invariably synthetic and, if used appropriately, holistic and organismic too.


## R-NETS

Requirement nets (R-nets) [7] are simple activity flow charts, using AND (&) and OR (+) gates to address branching and simultaneity in process. An R-net for a typical company bid is shown in Figure 8.16, from which the similarity to a conventional flow chart is evident.

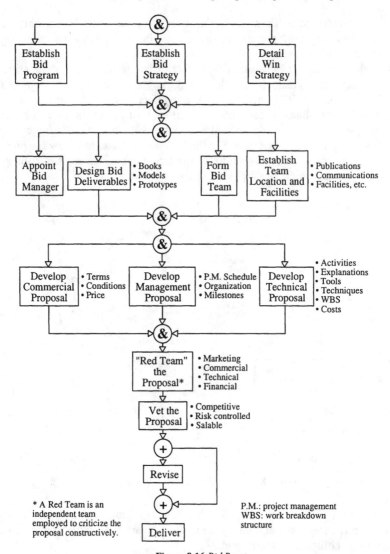

**Figure 8.16** *Bid R-net.*

Where parallel activities are shown between two successive AND gates, the implications are not that the activities must be done in parallel, but that all of them must be completed before moving on past the later AND gate. Whether things are done in parallel or series is left unspecified, just so long they are done. This makes the chart a flexible tool for planning.

R-nets such as the one shown can be useful when estimating. The effort required for each activity can be estimated against the appropriate block, and summed at the bottom. The R-net can be turned on its side and changed into a project bar chart annotated with milestones and deliverable outputs.

R-nets are also so simple to read and so self-evidently logical and rational that they can be used directly in proposals and bids, too. They can also be elaborated and encapsulated to reveal or conceal complexity in context.

## BEHAVIOR DIAGRAMMING

The techniques illustrated so far have looked principally at cause-and-effect, or behavior, synthesis. For its part, behavior diagramming is a particularly powerful, yet simple technique that combines data, function, and dynamic into one persuasive design and presentation technique. The basic elements of the presentation technique are shown in Figure 8.17. There are three columns, organized as input, process and output, in sequence.

An input is required to activate a function that, when performed, results in an output. Since functions logically and sequentially relate to each other, so do inputs and outputs; the output from a preceding function or process becomes the input to the successive function or process. Sometimes a function is changed by an output, as shown.

While the technique is machine supported, it is probably at its most powerful when implemented by hand. The orthogonal nature of the resulting matrix is convincing, as well as self checking. It is difficult, in drawing up such a behavior diagram, to make a mistake and be unaware of it.

Behavior diagrams may be used in a wide variety of situations. For example, Figure 8.18 shows a behavior diagram outlining the basis for a business research program. Note how the symbols associated with R-nets have been used in this

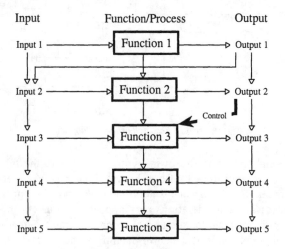

**Figure 8.17** *Notional behavior diagram.*

instance to increase the flexibility of the behavior diagram.

The diagram shows the processes associated with the research program in the central shaded panel. Inputs of data, resources, and management control are shown at left as activators of their respective process. The results from each process are at the right. Note how the outputs form a logical train or sequence: goals, strategies, results, and the presentation of results. The logic of this sequence emanates from the sequential logic of the processes; the inputs at left also form a rational sequence when arranged as shown.

Having developed a high-level behavior diagram such as that in Figure 8.18, it is then possible to elaborate on a horizontal strip-by-strip basis, elaborating the process first by showing the subprocesses/activities within it. Each of these requires its own inputs, which must then be a subset of the inputs to the higher level process. Similarly, outputs from each subprocess, if any, must form part of the output for the whole process.

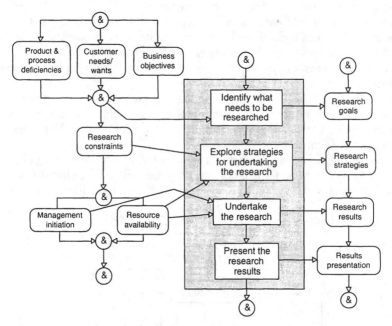

**Figure 8.18** *Research behavior diagram.*

A typical elaboration is shown in Figure 8.19, where the function "identify what needs to be researched" from Figure 8.18 has been elaborated. The elaboration process in this instance is not blind decomposition; in addition to more detail, new ideas and considerations may appear, such as (in this instance) the potential feasibility of manufacture as a research constraint. Either such new ideas must stay within the bounds of the corresponding higher level diagram element, or else that element description must be expanded to accommodate the new idea.

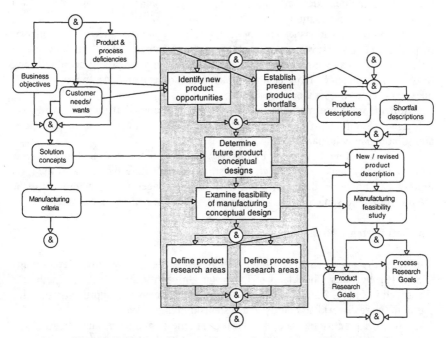

**Figure 8.19** *Elaborated behavior diagram: the first process of Figure 8.18.*

Behavior diagrams will appear later as part of case studies; they are a particularly powerful, yet simple, means of describing system behavior cogently and rigorously.

## SOFT METHODS

### Soft Systems Thinking

Some problems and issues are more complex, abstract, obscure, and intractable than others. Sometimes motivation is unclear; sometimes people lie about their true intent. Often, it is difficult to see how one factor might affect others, or how best to address issues where there are many opposing viewpoints, motivations, and activities.

Such problems and issues are sometimes referred to as soft, to distinguish them from so-called hard problems. Hard problems are not so much hard in the sense of difficult, but more in the sense of firm and solid, with known causation and clear singular objectives. Thus how to design a new bridge is a hard problem,

amenable to mathematical and engineering analysis, and there is probably a best
solution for a given situation.

Soft problems and issues are not so clear. How can we improve the quality of
life of the nation? How can we sustain our environment when we continue to
pollute it? How can we reduce violence in schools? Why are young people turning
to drugs and alcohol? How can we make democracy work? At times, even the
question is doubtful, let alone the answer.

Soft methods are methods based on systems theory that attempt to resolve
soft issues. Solutions, if there are any, are composed of a combination of factors,
and solutions may be partial, temporary or expedient.

Soft systems methods tend, by the very nature of the task, to be broad, vague,
and even wooly, and are subject to criticism because of it. In the hands of the right
practitioner, however, they may work well, raising concerns as to whether it is the
method, or the person employing the method, that is the real problem solver.

**The Soft Systems Methodology**

A noteworthy approach to resolving soft issues is the soft systems methodology
(SSM) [8], by Peter Checkland of Lancaster University, England. The method is
illustrated in Figure 8.20, where it is shown to be a seven step approach, operating
partly in the real world and partly in a systems thinking world.

An SSM practitioner adopts a variety of roles: facilitator, analyst, modeler,
systems thinker, presenter, and–principally–consultant. He, or she,[7] first highlights
the problem or issue by working with a client, finding facts and factors perhaps in
a fairly ad hoc way; "1" in the figure. These facts and factors are brought together
in a so-called "rich picture," often a simple hand drawing showing the various
participants, activities, lines of communication, and–particularly–points of
confrontation ("2").

The consultant then retires to think through the issue. He defines an ideal
system ("3") like the real world one under investigation, except that the ideal
system would not exhibit the issue or problem. He then develops (paper)
conceptual models ("4") of how the ideal system would be structured and would
work, before going back to the client ("5") where together they identify
differences between the way the real world system operates and the way the ideal
world system would operate.

These differences form a so-called "agenda for change," from which is teased
out what would constitute both feasible and desirable change in the real world
("6"). Real world managers take actions ("7") to incorporate agreed changes that
should, it is hoped, go some way to resolve the original problem.

---

[7] Hereafter "he" for brevity, but the consultant could be of either sex, and could actually be a small
team.

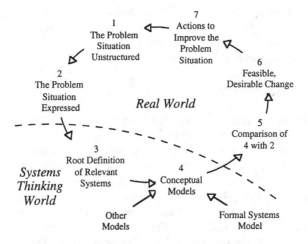

**Figure 8.20** *Checkland's soft systems methodology (SSM).*

The success of this process is dependent on several factors: successful detection of all the facts and factors relating to the issue or problem; correct diagnosis of the many, interwoven issue themes that may exist in a real world situation; the ability to develop meaningful "rich pictures" that successfully highlight the central issue themes; accommodation and management of the large amounts of data and information generated in typical investigations; sufficient domain knowledge and expertise to develop effective conceptual models; and, since these conceptual models are static, the ability to foresee and anticipate dynamic issues such as chaotic behavior and counterintuitive response.

Perhaps these potential difficulties explain why some SSM consultant-practitioners are more successful than others; SSM may provide a framework for systems thinking, but it is the individual consultant who does the thinking, and many consultants do good work without the benefit of SSM.

Checkland included a number of ways of making SSM analysis and synthesis potentially rigorous, in the sense of complete, with all aspects and factors in the issue space accounted for and resolved. Individual practitioners may be less than assiduous in the rigor of their approach; some declaring that the strength of SSM is that its very lack of rigor makes it flexible and adaptable (sic!).

**The Rigorous Soft Method**

There is, however, every reason to ensure that a soft method is rigorous. Without rigor, faulty diagnosis, inappropriate remedial action, and invalid prognosis are inevitable in any complex situation or issue. We would not like a medical practitioner to lack rigor when examining us because he or she might miss something important; such oversights may miss a major problem, may result in a misdiagnosis, may invoke inappropriate treatment, and may, in the worst case, result in death. Surely we should expect no less a standard of rigor from a consultant who is examining our sociotechnical, organizational, political, economic, or ecological body?

RSM - The Medical Analogy

The medical analogy forms the basis for the rigorous soft method (RSM) approach to soft systems thinking. To understand how it works, consider a general medical practitioner working in a surgery or clinic; receiving patients. A new patient declares that she does not know what is wrong with her, but that she "feels out of sorts."

The doctor will then ask a number of questions, perhaps as follows: when did this feeling start? What have you been doing recently that is unusual for you? Have you been in contact with any people from another region/country? What do you do for a living? Are you stressed at work or at home? The purpose of such questions is to find out the patient's environment and what changes in that environment, if any, may have triggered the symptoms.

Next, the doctor will ask questions about the patient's symptoms, and will probably also take some tests: pulse, respiration, eye whiteness, temperature, urine, blood samples, and so on. To use this information effectively, the doctor will have to know what a normal healthy person's test results should look like, for a patient of this age, sex, and circumstance.

A deviation of the patient's test from the norm constitutes a symptom. For each symptom there may be a variety of possible causes; each of which may be associated with a particular organ, or an imbalance between a pair of organs. A symptom of a high resting pulse rate could be a racing heart (tachycardia), infection, low lung capacity, dehydration, stress, stimulants, and many others. Similarly, a pain in the upper left arm could be a symptom of muscle strain, shoulder joint inflammation (pericapsulitis), or angina. As the doctor identifies more and more symptoms, he or she generally finds either that:

- The symptoms form a familiar pattern, one that enables swift diagnosis by experience; or
- Whereas each symptom considered in isolation could implicate several organs, likely causes of *all* the symptoms taken together overlap and coincide, usually on just one organ, or the interaction between two organs.

So, by finding more symptoms, the doctor gradually narrows down the search. At some point in the process, the doctor is able to predict what the diagnosis is most likely to be, and to make a confirmatory diagnostic test. (Interestingly, this approach is able to unravel the potentially confusing situation where a patient is suffering from two quite independent afflictions at once.)

Last, the doctor has to recommend treatment. Effective treatment will be such as to relieve all the symptoms, not just a few; and treatment may have several facets: medication, surgery, rest, change of environment, and so on. The treatment has to be suited to the patient and the patient's circumstances, and it is likely that there will be optional routes for a restoration to full health and quality of life.

RSM–The Approach

RSM approaches soft issues in a manner directly analogous to that used by medical doctors:

Step 1. Appreciate the broad area of concern;
Step 2. Find the symptoms causing concern;
Step 3. Find suspect *implicit* systems;
Step 4. Group suspect *implicit* systems into sets;
Step 5. Highlight set deficiencies compared with the ideal;
Step 6. Propose a remedy; that is, optional system solutions and criteria for a
    "good" solution;
Step 7. Check if the remedy resolves all the symptoms.

(By analogy, implicit systems equate to the body's organs, also implied by the symptom, except that, for the body, the organs are already known and defined. For RSM, implicit systems are detected by their behavior, or misbehavior.)

The process may be represented diagrammatically; see Figure 8.21. Starting at the top, the issue is defined by emerging symptoms, suggesting some unease, concern or dysfunction in what may be a diffuse system. Further symptoms are sought, together with a range of possible causes for each symptom. The repetition of possible causes assists in the logical grouping of problem symptoms into problem themes.

RSM employs a neat idea, also found in SSM: using the differences between real and ideal worlds to drive change toward the ideal. We have already met a means of generating ideal world models (see Promoting Completeness with CLMs on page 140) and of modeling them dynamically. The differences between real and ideal worlds enable the derivation of requirements for a system that would resolve all the symptoms. The bottom four boxes in Figure 8.21 may be familiar, too; they are the heart of the systems engineering problem-solving paradigm; see Figure 4.2 on page 79. Using the SEPP concept, RSM generates both options for a system to resolve the issue and, criteria for an effective solution. As the figure shows, the most important criterion is that the system resolves all the original symptoms. Options in RSM refer to full systems, however, not just to cause, effect, and behavior. A full solution requires the definition of a system with functional, physical and behavioral characteristics; hence the reference model in the figure.

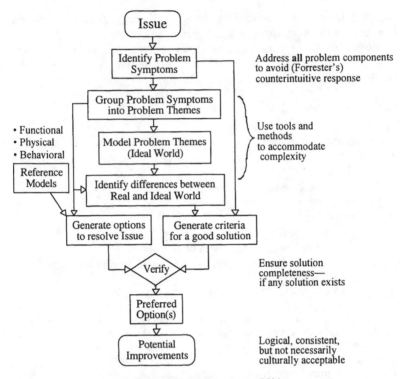

**Figure 8.21** *Resolving issues: RSM.*

RSM is tool supported. In addition to the use of causal loop models to facilitate completeness, $N^2$ charts are used to accumulate the implicit systems; that is, contained systems, or those implied by symptoms. The charts are then clustered to reveal functionally bound blocks, constituting higher level or containing systems. These higher level systems, together with their interactions, represent an ideal world. By virtue of their method of derivation (having been derived from symptoms of what was wrong, missing or dysfunctional), they focus on the ideal solution that would resolve the originating symptoms.

Typical of the types of issue that RSM might address are:

- Concern about an organization's morale, where there is no obvious culprit;
- The urgent requirement for a high-level briefing to refute the detailed criticism of a project;
- The risk of a partnership breaking up owing to the lack of a shared vision;
- Differing views within an organization of the causes for lack of performance, effectiveness, or efficiency;
- Inconsistent fault symptoms emerging from complex equipment;

- Erratic organizational behavior, but no clear strategy for improvement;
- Burgeoning social unrest, but no single cause and no evident remedy;
- Disagreement on the best way to integrate multinational forces;
- Concern over global warming and future sources of energy at the national level.

RSM is a powerful, even heavyweight, method for addressing complex issues. It is at its best when employed by a multidisciplinary team with good knowledge of the issue domain. RSM cannot guarantee results–some problems are insoluble–but it does offer rapid, focused progress, an audit trail of justification, and a provable solution–where one exists–to a soft issue. RSM is holistic, synthetic, and organismic throughout.

An example of RSM, showing how it works in practice, is given later (see National Energy Strategy on page 190). The RSM is presented in Appendix B using sets, where a set theory proof of RSM is also provided.


## THE TRIAD BUILDING SYSTEM

At the opposite end of the scale from soft methods, hard issues and problems tend to have clear, singular objectives; but the means of achieving those objectives in the presence of risks and threats may be uncertain. The TRIAD building system is a sharply focused method of defining a system to achieve an objective in a threat environment.

The TRIAD building system, so called because it employs triangular structures, is shown in Figure 8.22 as a procedural method for defining the elements of a system. At the top, the prime directive (PD) of the system is the ultimate statement of purpose. This is a high-level statement, with one verb only. Semantic analysis elaborates the PD, revealing one or more objectives that must be achieved to satisfy the PD. Threats to achieving each objective are identified. Two kinds of strategy are then formulated: one to achieve an objective, and the other to overcome or neutralize a threat. For each objective there may be several threats; each of which necessitates a strategy. Objectives, threats, and strategies form the first TRIAD. So, there are strategies to overcome threats to achieving an objective necessary to achieve the prime directive.

From this point, the TRIAD building system can be used either to develop a process system, or to define a physical system. If strategies are elaborated into sets of activities, then these activities can be gathered into a process model for the achievement of the PD. Alternatively, if strategies are associated with functions that would perform them, then the functions can be organized, grouped, and structured into a physical system for the achievement of the PD.

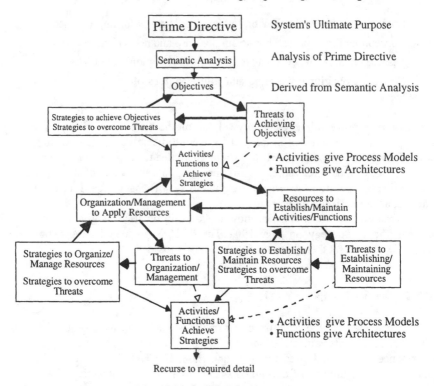

**Figure 8.22** *The TRIAD building system.*

Whether it involves activities or functions, the TRIAD building system continues with a second TRIAD, identifying necessary resources on the one hand, and organization and management of resources on the other.

Two more TRIADs appear because of threats to resources and threats to organization and management; both of which require strategies to overcome them.

The TRIAD building system procedure can generate a large amount of information, either in the form of activities or functions. The activities require integration into a process model in the appropriate sequences, and this can be achieved using the ISM method (on page 148); in particular, the precedence network or structure. Where functions have been accumulated, these can be interrelated in an automated $N^2$ chart (on page 143 and following) and hence formed into an effective structure.

The TRIAD building system generates a process, or physical system design, that contains only elements that can be traced upwards to the PD. Instead of considering threats and risks after the design, the means of addressing them is incorporated seamlessly into the design. The method is highly focused and effective, especially when employed by those familiar with the particular domain and the threats it contains.

The TRIAD building system is holistic, since it starts from such a high level statement (the system's prime directive), and then elaborates the detail including interactions with and threats from other systems, such that every element contributes to the whole, and no element should be overlooked. Using ISM and/or the automated $N^2$ chart in support, the method is also synthetic (as opposed to reductionist or analytic) and organismic, in that the resulting system structure will consist of interacting subsystems making up a whole (process or product) system.

## THE GENERIC REFERENCE MODEL

Previous methods have emphasized system behavior, but sometimes more complete representations of a system are useful. There is, it seems, an almost unlimited variety of systems. Systems appear at all scales, from the miniscule to the universal. At the most fundamental level, a system appears to represent an island of order in a sea of disorder.

Is it possible, in such circumstances, to uniquely represent any system, regardless of type, scale, purpose, and so on? Would such a representation, were it feasible, serve any useful purpose, or would such a representation necessarily be so generalized as to be meaningless? The generic reference model (GRM) responds to those questions.

The value of a universal representation of any system is scientific in the first instance. Such a model must represent the parts, relationships, limits, dynamics, and so on of any system, and so will explain the phenomenon of emergence, and the ability of some systems to adapt. Such a model could also be used as a reference against which to judge real-world systems for content, completeness, and behavior. Finally, such a model could be used practically in concept and design formulation to conceive a comprehensive solution to an issue, problem or need. In this last use, a generic template would be instantiated, element by element, to achieve a solution.

## CHARACTERIZING A SYSTEM

Some things are clear about systems, at least from a systems theoretic perspective: they are made up from parts, themselves systems, that mutually interact. Without the interaction pathways, there would be no pattern; without the pattern there would be no sense of reduced entropy, real or perceived, associated with a system.

In the broadest sense, systems exist; they have being. This is true for all systems, even transcendental systems; they exist. Most systems, but not all, also do something: they have function, and they fulfill some purpose. The sun is a system with a three-layer structure of concentric spheres: it clearly exists, but it is difficult to see function or purpose in that existence. For those who do see purpose in our Sun, consider instead a similar star in a distant galaxy: does that have purpose, or does it just exist? On the other hand, manmade and human systems have function and purpose. A hospital is a manmade system with evident function and purpose; so too are a farm, a waste disposal unit, and a naval task force.

So, while all systems exist at some level, not all systems "do" anything. It is also evident that some systems are sentient; they think. This is evidenced by behavior that does not necessarily repeat itself. While reflex behavior is always the same, sentient or thoughtful behavior changes with experience and events. So, some systems think.

Already we seem to have three categories of system: being, doing, and thinking systems. Some systems exist. Some systems exist and do. Some systems exist, do, and think. Possibly, some systems exist and think, but do nothing, although examples in that category are hard to envisage; after all, how would you know? Figure 8.23 shows notionally how a system might be perceived in this context: existing in an environment, perhaps doing, and–if thinking–then stimulated by the doing of some other system. This other system is necessary to fulfill the idea of behavior as being response to stimulus. Thinking addresses stimulus, while doing addresses response; thinking and doing are inextricable linked, therefore, as shown.

Figure 8.23 shows that the categories of *being, doing,* and *thinking* cannot be considered in mutual isolation. While they might be represented separately for the convenience of a diagram, only *being* could be considered on its own. *Doing* cannot be conceived in isolation from *being,* and *thinking* cannot be conceived in isolation from either *doing* or *thinking.*[8] That is not to say, however, that *thinking* always results in *doing;* a sentient system may decide not to respond. Bearing in mind the inappropriateness of considering these various categories separately, let us elaborate on being, doing and thinking as though–for the moment–they were separate.

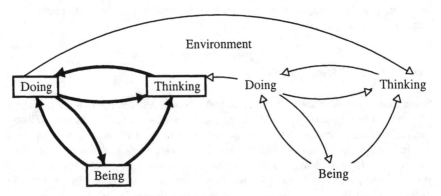

**Figure 8.23** *Being, doing, and thinking.*

---

[8] I am purposely overlooking doubtful supernatural phenomena, such as ghosts, as being outside the realm of this discussion.

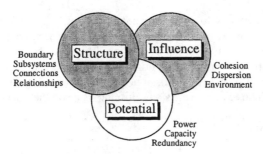

**Figure 8.24** *GRM form model.*

## Being

Being is represented as having form in a systems context; see Figure 8.24. The elements of form are seen as structure, influence, and potential; the energy or capability stored and expressed in form. The logic of the elaboration is to identify form, the availability of form, and the survivability of form, where availability addresses an internal threat to form, and survivability addresses an external threat to form. Since there can be no more than that inside and outside the boundary, the set is complete.

As the figure shows, structure is comprised of boundary, subsystems, connections, and relationships; together these enable description of all structure within[9] the boundary.

Influence supposes that form exists at least for a period, and that in consequence there must be balances between influences that tend to disperse the structure, and those that make the parts cohere. To define the effects of such influences, influence includes the environment in which they operate. Influence does not imply static stability; form may oscillate, reverberate, and distort, but so long as it continues to exist about some mean condition, then the dispersive and cohesive elements must balance.

Potential indicates that form may have capacity, power, and redundancy; which will be no surprise to any engineer. However, the same is true of the form of all open systems that maintain static or dynamic form. Redundancy (generally), the possession of spare replacement parts, or of robust, reliable elements, is common to all living and most manmade systems as a hedge against defect or shortfall, although redundancy can be "consumed" or waste away, leaving the overall system form vulnerable to further internal defects.

## Doing

Doing is represented as performing a function in the GRM: it is comprised of mission management, resource management, and viability management. Together,

---

[9] The GRM in general describes what is inside the boundary of the generic system.

these three are referred to as the management set. The rationale behind the division is as follows. Functions imply purpose, represented here by mission: purpose is expressed as the pursuit of mission. Resources are necessary to sustain that pursuit. Viability, that is, the continuing health and capability of the system, maintains the ability to pursue mission. So, given viability and resources, the system will continue to pursue mission. This, too, is a complete set.

Mission Management

Mission management is elaborated using a causal loop model to promote completeness, closure, and continuity. See Figure 8.25. Essentially, mission management can do no more than:

- Collect information from the operational environment;
- Set/reset objectives based in part on that information;
- Strategize and plan how to achieve those objectives;
- Execute the resultant plan;
- Cooperate with others in the operational environment, if need be.

**Figure 8.25** *GRM: mission management.*

This list applies for any functioning system: from a bat catching a moth, to a person driving a vehicle, to a general managing a war. They will all differ in detail, but fundamentally the list applies. Note that work is done in going around the loop, and that energy will be required–and dissipated–in consequence.

The process of mission management is continual; deciding to act within some operational environment will change that environment, or the perspective of that environment, necessitating the collection of more information about what has changed. So the loop cycles continually. Given sufficient energy, more than one mission may be pursued at the same time; a worm may be on a mission to find food and, at the same time, on a mission to reach the safety of a dark, damp patch of earth. A brigadier may initiate a mission to blow up an enemy fuel depot and, at the same time, a mission to restore regional order.

Resource Management

Resource management is also elaborated using a causal loop model to ensure completeness, closure, and continuity (see Figure 8.26). Essentially, Resource Management can do no more than:

- Acquire resources;
- Store resources;
- Distribute resources;
- Convert and utilize resources;
- Discard excess/waste.

**Figure 8.26** *GRM: resource management.*

As the figure shows, there is an implied external resource environment, where resources may be sought and where waste may be disposed. This may, but need not be, the same as the operational environment referred to in mission management above.

"Discard excess or waste" covers several aspects in the GRM resource management model. Waste disposal is a vital, but often neglected part of resource management. In sieges of medieval castles, for instance, one of the greatest risks to defenders was the accumulation of waste, and the potential for disease that it raised. For Apollo missions, the ability to discard spent rocket booster casings was essential to mission success, but so too was the need to dispel combustion gases; also waste. For animals, waste comes in many forms—excreta, of course, but also the exhalation of carbon dioxide, the gas that plants need to thrive.

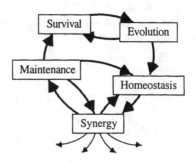

Figure 8.27 *GRM viability model.*

Resources are often converted, perhaps to energy, during the cycle; food and fuel are obvious examples. Energy reserves are stored within most systems, sometimes in condensed form, where space is at a premium. Body fat is an obvious example, together with rocket propellants and chlorophyll.

Viability Management

The viability model in Figure 8.27 has a different derivation. For a system to remain viable, that is, able to exist and remain sound in some environment, then it must:

- Resist external threats (survivability);
- Overcome internal defects and deficiencies (maintenance);
- Adapt to changing situations (evolution);
- Maintain stable internal conditions (homeostasis);
- Have all its internal parts work in harmony to produce external effects (synergy).

So the elaboration logic is subtler than for mission and resource management. In addition to internal threats and external threats (which account for all threats) there are essential concepts of change, resistance to change and synergy. Evolution promotes adaptation in the longer term, while homeostasis resists change in the shorter term. Synergy is the key notion; the other four factors enable and ensure synergy in both the short and the longer term because, without synergy, there is no functional capability, no integrity, and no system.

The three separate models, mission, resource and viability management, may be combined, Figure 8.28, to produce a GRM function model. Viability management is at center, mission management at top, and resource management at bottom. Threat and change have been added to show their different impacts. The various causal lines that would indicate synergy have been omitted for clarity. The

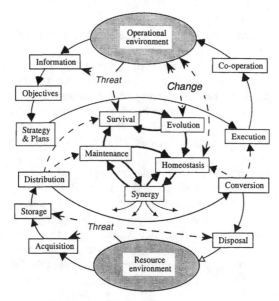

**Figure 8.28** *GRM function model.*

model, as a whole or in areas, can be elaborated further, using similar logic of complete sets. Survival, for instance, elaborates to the complete set: avoidance of detection, self-defense, and damage tolerance. Synergy elaborates to coordination and cooperation. Maintenance becomes defect detection, replacement, and disposal.

## Thinking

Thinking is elaborated into the GRM Behavior Model, which is founded in a classic nature versus nurture paradigm. Behavior is addressed, also classically, as stimulus leading to response. The response may be instinctive, suggesting nature; or considered, suggesting nurture. Response may be nurture overriding nature, or "knee-jerk," that is, nature reacting instinctively.

The broad form of the behavior model is shown in Figure 8.29, in which nature and nurture are shown as two shaded areas. Nature is shown to be the result of evolution. Elaborating nature is the field of the psychologist, the anthropologist, and others. Karl Gustav Jung [9] might have suggested that the natural mind was comprised of some or all of the following:

- The collective unconscious;
- Archetypes;
- Aggression;
- Emotion;

- Instinct;
- Libido;
- Energy;
- Character.

The collective unconscious is, according to Jung, a state of mind shared by humanity; one that we inherit. Together with the unconscious and the conscious mind, which are unique to the individual, they describe the layers in the mind. Within the collective unconscious are archetypes; indicative of definite forms in the human psyche that are present always and everywhere. This not to say that we inherit memories, but that our human brain has evolved such that we all have similar patterns of thought, we all go though similar stages of cerebral development, and we all have unconscious, as well as conscious, mental processes at work.

Jung provided a model of how people the world over behave in ways that are clearly human; regardless of creed and culture. We are largely instinctive creatures, although we may not wish to see ourselves in such a light.

Nurture, on the other hand, is largely learned behavior. As we develop from infancy, we come to possess stored tacit knowledge, models of how the world looks and works, and beliefs about the way things are–or should be. Our individual belief systems may be comprised of some or all of the following:

- Beliefs;
- Stereotypes;
- Values;
- Morals;
- Training;

- Roles;
- Categories;
- Ethics;
- Ideologies.

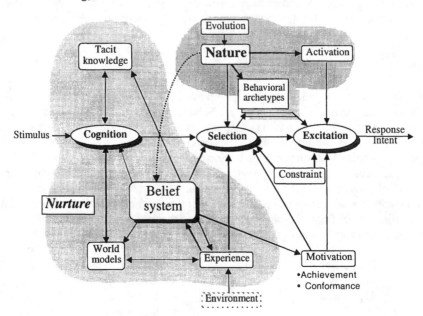

**Figure 8.29** *GRM behavior management model.*

Stimulus/response might come about as follows:

1.  A stimulus is sensed and interpreted on the basis of tacit knowledge, world models, and beliefs (that is, what is expected or believed to exist).
2.  The interpretation of stimulus is processed at (behavior) selection, where reaction is initially by nature, followed by nurture.
    a.  The sequence is important: reactive behavior is fast, and it evolved to save life when under immediate threat.
    b.  Considered behavior probably evolved as part of our social inheritance, where we learned to work and hunt in groups, develop strategies and plans, and make decisions in complex situations.
3.  The processing of behavior selection results in responsive behavior being selected from a range of feasible archetypal behaviors–some learned; some reactive, instinctive, or inherited.
4.  The chosen responsive behavior then excites response intent, with a degree of excitation that is constrained by circumstance and enhanced by motivation and nature.
5.  The response intent activates the remainder of the GRM model elements to realize the physical response.

The behavioral model describes, not only the stimulus-response behavior of individuals, but also that of groups that act on the basis of a shared belief system. For instance, a military group may consist of many individuals who will have received the same training and who may have shared belief-reinforcing experiences. Their training may have been such as to become second nature. It would not be unreasonable to describe the behavior of the group using the GRM behavior model.

**Synthesizing the Whole GRM from the Parts**

The form, function and behavior models can be brought together, as in Figure 8.30, to form a reference model for any system. Form and behavior fill the lower and central panels. Function is separated out to show mission management at the top, with resource management at the left, serving all the other panels, and viability management at right, maintaining viability by serving all the other panels.

Figure 8.31 shows the GRM with the various parts overlapping to signify that, although they may be viewed as separate, they are not really separable. The whole may be considered as a virtual machine by substituting the central panel titles, from bottom to top, with technology, people, and process. Form, behavior, and function are more generic titles, and hence more widely applicable.

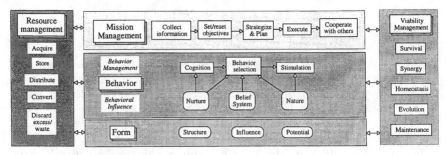

**Figure 8.30** *Generic reference model: unconnected.*

Causal lines are included to show some of the intra-activity within the overall representation. Different system instantiations would emphasize different channels of causality. The following example shows a common channel of causality:

1. Technological sensors detect signals–stimuli–from an operational environment.
2. These stimuli pass to the mission management feature called "collect information."
3. The stimuli are passed to cognition (in behavior) so that they may be recognized and interpreted, using tacit knowledge, world models, and beliefs.
4. The interpretations of stimulus are passed back to "collect information," where a situation picture is built up from successive stimuli.
5. Interpretations are also passed to behavior selection, provoking the

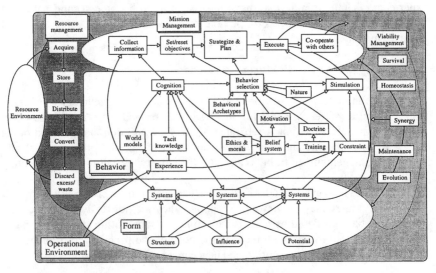

**Figure 8.31** *Generic reference model: connected.*

selection of one or more behaviors based on behavioral archetypes.

6.  The interpreted situation developing in "Collect Information" is passed to "set/reset objectives," and thence to "strategize and plan," both of which are influenced by the behavior already selected in response to the stimulus.

7.  The plan is then executed, possibly directly into the operational environment, but more likely by activating technological systems: communications, transport, weapon, sensor, and so on.

8.  Executing the plan changes the operational environment and, possibly, the resource environment, too. Certainly, resources will have been utilized during the process. Sensors will detect a changing situation and the cycle repeats from 1 above.

This list might apply to an individual, a command and control system, or a business. The model applies equally when considering one person, or a coherent group of individuals, such as a platoon, a hive, or a colony; with or without technology. In instances where there is no ethic, morality, or doctrine, the terms would be ignored. Such terms are usually indicative of a system with people involved.

**Using the Generic Reference Model**

There are various ways in which the GRM may be used. When auditing a system design, the GRM maybe used as a checklist. Here, the elements of the GRM are set out in a table and the auditor goes through the design, seeking the corresponding part or parts in the design. The auditor also develops an $N^2$ chart, with the GRM entities on the leading diagonal. External systems with which the design is intended to interact are shown on the border of the $N^2$ chart. The auditor then enters into the chart the interactions and information flows shown in the design and pursues the interchange and flow logic; both internally and between the designed system and external systems.

Using this approach, the auditor–with the design team–should be able to

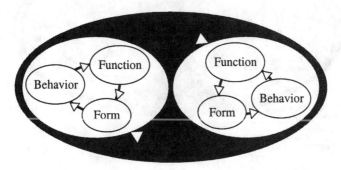

**Figure 8.32** *Interacting GRMs.*

check completeness and "walk through" the operation of the design solution.

A second way to use the GRM is as a design template, instantiating each part of a system-to-be-conceived in correspondence to the GRM. This approach is amenable to dynamic simulation. The GRM describes only the "internals" of any system. Those internals will exist within an environment supporting other systems, with which a particular system of interest will interact. (The future tense is important here; system design is about creating the uncertain future.)

An SOI is shown in Figure 8.32 as a basic GRM, interacting with another system within an environment. (There could be many more systems, all mutually interacting, but two will serve to explain the process.) Each system is shown as comprised of form, function, and behavior. Because each system is shown in an environment, interacting with another system, the diagram is system-theoretically sound.

An elaboration of the figure is shown in Figure 8.33. The mission management element of each system has been elaborated internally, and other systems within the resource environment have also appeared. Further levels of elaboration, following the pattern of Figure 8.31, will then give way to instantiation. If the SOI is, for instance, a military platform in combat with an enemy platform, then the systems of the form model will become technological systems; sensors, communications, processors, displays, weapons, and so on.

Alternatively, if the SOI is a lead manufacturing company, in competition with a rival, or cooperating with a partner, then the technology would be machinery, information systems, and communications; the mission would be the management of production; and the behavior would reflect the perceptions and culture of the company personnel.

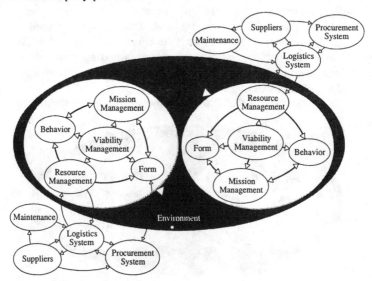

**Figure 8.33** *Interacting GRMs: first level of elaboration.*

## NONLINEAR DYNAMIC SYSTEM SYNTHESIS

The system models and methods presented so far are dynamic only in the sense that they may show lines or channels of causality; change brought about by causal influences is left to the imagination.

In Figure 8.33, for instance, the lines of causality between the two systems are shown, but what effect does each system have on the other? If we call the two systems "red" and "blue" respectively, and assume that the both red and blue are naval ships exchanging artillery fire, then when blue hits red, red is changed by the impact. Red's function, form, and behavior may all be affected. The same is true when red hits blue; blue is similarly changed.

This line of thought leads to some radical conclusions. The effectiveness of, say, the blue ship is determined not just by the blue ship's attack capabilities, but also by the red ship's defenses. Moreover, as combat exchanges proceed, the effectiveness of both ships changes, as each receives damage and utilizes resources. With a damage repair crew in operation, effectiveness may not only fall, it may also revive.

So, widely cherished notions, that it is possible to measure and predict the effectiveness of an operational system design, are evidently optimistic. Instead, effectiveness (and performance, availability, survivability, maintainability, and so on) varies over time according to the situation.

The effectiveness of a system might be viewed (almost) in isolation, if it interacts with another, identical system. Following the naval engagement example, if a type X destroyer interacts with an identical type X destroyer, then the interaction will test both the offensive and defensive capabilities, both of which will change as combat proceeds.

The method of pitting identical systems against each other affords significant advantage. If the systems are identical, then not only are the technological facilities identical, but so too are the people, their behaviors, beliefs, training, and processes. Combat between two such identical type X destroyers, for example, might be expected to result in identical effectiveness, and an evolving effectiveness profile as the engagement proceeds.

Suppose, then, that one factor in only one of two otherwise identical ships were changed. Any resulting change in the effectiveness profile of both ships would be exclusively attributable to the one element of change in the one ship. Using this approach, it is possible to explore the effects, if any, of individual changes, or of groups of changes, in a design. Moreover, it is possible to compare the effects of seemingly unrelated changes. Would a smarter weapon system compensate for a reduced level of training? Would increasing radar transmitter power improve the ship's effectiveness, or might it make the ship more detectable, and hence vulnerable, at longer ranges?

The ideas of effectiveness being determined as much by environment and competition as by an SOI are easy to understand in the stark, detached terms of a hypothetical naval engagement. They apply equally to other system in other environments. Ecologies and economies can be viewed in this light too.

If an insect predator is introduced into an environment to combat an insect pest, then the outcome can be modeled not as two systems, but as many systems interacting. The outcome may be favorable; it may on the other hand be counterintuitive. Introducing a new business into an economy may be viewed similarly.

In each and every case, the dynamic interchanges between systems are too complex to imagine. The systems may change beyond expectations and intuitions during the course of a simulation. There is a need for modeling. The nature of the issues dictates that the models must accommodate nonlinear dynamic effects.

## STELLA™

Various tools are available in the market for such activities. Early tools and seminal methods [10], have paved the way for today's powerful desktop tools. One such is STELLA™, an object-oriented, visual-programming tool. The acronym stands for: systems thinking environment and learning laboratory approach. The tool is intended to do just that: to enable the user to think about systems using dynamic simulation models, and to be able to try out different ideas that would be difficult or unwise to try in reality.

The tool is able to represent systems at three distinct levels: the subsystem, system and containing system levels. Of itself, the tool does not necessarily represent nonlinear behavior, but programming objects may be chosen and linked to represent feedback, discontinuities, limits, and so on.

To simulate system models dynamically, therefore, it is important to develop the structure and architecture of the model outside of, and before using, STELLA. Causal loop models and interacting GRM models make good starting points where they represent real world feedback and reaction. Pitting one system against another, as outlined above, becomes simple. First represent one system in an environment, then replicate it and cross connect the two representations in the manner of Figure 8.33. Establish starting conditions and a scenario.

The resulting STELLA model is sound in that it is consistent with open systems theory, and complete in that it is instantiated from the full GRM. More complex models would follow the same general route. Two naval task forces might engage each other, with aerial, submarine, and surface weapons; a number of ships on each side; and a number of noncombatant support, minesweeper, and other ships. Although more complex, the process is essentially the same.

Simple Examples

STELLA and similar tools are straightforward to understand and to learn, although it is quite possible to use them in a way inconsistent with systems theory. Several of the elements and simple constructs to be found in STELLA are shown in Figure 8.34.

At the top can be seen a reservoir, called level, with an inflow leading to it and an outflow leaving it. Inflow and outflow are rates, as indicated by the tap symbols. Inflow draws from a cloud, indicating some undefined but infinite source; outflow similarly spills into an undefined, but infinite sink. This simplest of all models could represent a bath with the taps open and the plug removed, or a population with births and deaths, or a store with goods coming in and goods going out, and so on.

The second diagram shows a conveyer in place of the reservoir. Conveyers are in continuous motion, as in a factory or at an airport, with items entering, going through and coming off the other end. Conveyers can leak, as shown, to represent, for example, defective items being rejected during manufacture.

The third diagram shows a queue and an oven. An oven, like those in a bakery, can be opened to place items inside, and it is then closed for a period during which the items inside are inaccessible. Since there may still be a flow of items coming into the system, they will form a queue, as indicated, until the oven

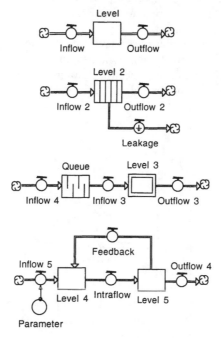

**Figure 8.34** *Basic STELLA constructs.*

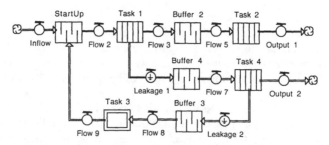

**Figure 8.35** *Process feedback model.*

disgorges and accepts a new charge of entities from the queue. Ovens might be used to represent, for example, a doctor's surgery and waiting room; a vehicle-spraying plant that can accommodate a limited number of items to spray, which then take time to dry; the gestation of a placental mammal, during which time the mother cannot become pregnant; and so on.

The final diagram shows a forward flow through two reservoirs, with feedback from the second to the first reservoir. This might typically be used to represent quality inspection in a factory, where rejected items are returned for improvement. A converter is shown; these are devices for converting values and dimensions.

Additionally, STELLA includes a variety of preset mathematical and statistical functions, graphical input devices to accommodate empirical inputs, and both graphical and tabular outputs.

While it is easy to predict the behavior of the models in Figure 8.34, the model in Figure 8.35 might be more challenging without the aid of dynamic simulation. The model represents the workflow in a business, factory, or plant. Sequential tasks are to be performed, but there is also the possibility of remedial

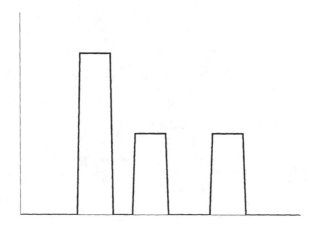

**Figure 8.36** *Task 4 activity from Figure 8.35.*

action to be taken. Task 4 is the inspection task, which either passes items as "good" direct to output 2, or returns them for rework. There is a 25% rework rate, while leakage 1 is 75%, that is, 25% of the flow goes through to output 1. What effect will the inspection and rework have on the pattern of activity?

To observe the effects of feedback, a single pulse of work inflow is initiated. Task 4 activity duration and level are shown in Figure 8.36, which shows work level on the y axis against time on the x axis: there are two rework pulses in addition to the straight through pulse corresponding to those items requiring no rework. This has occurred because task 3 has limited capacity, and is overloaded by the feedback, which it accommodates in two equally sized "bites."

The simulation model in Figure 8.35 is not a full system model; there is no external environment and other interacting systems are not represented. If they were, it might be that the delay caused by rework would invoke some reaction. Nonetheless, such partial models are useful to investigate phenomena and to represent parts of systems. (As in this case, the behavior of parts may change when they are connected to other parts to make a full, interacting system representation.)

Simulations of full systems models such as that of Figure 8.33 can be fashioned from modules representing subsystems; these may be tested individually and then interconnected to create the full system; stage by stage. Some subsystems are typical or archetypal, and can form part of a library of subsystem representations; thereby greatly reducing the time needed to develop new simulation modules. The process is analogous to progressive testing and integration of a complex product system.

## SUMMARY

A variety of tools and methods is introduced: causal loop modeling, $N^2$ charts, interpretive structural modeling, R-Nets, the soft systems methodology, the rigorous soft method, the TRIAD building system, the generic reference model, and STELLA. Together, these tools and methods, and others with similar capabilities, provide the ability to think about systems in a structured, system-scientific, dynamic, and rigorous way. Examples of systems thinking using these methods will follow in subsequent chapters.

## ASSIGNMENTS

1. Using the GRM as a template, describe a systems engineering team, identifying, but not detailing, the kinds of technology the team members might use, the process they would employ, the doctrine they might work to, the relevant beliefs they would hold, and so on. Include another GRM,

the production team, with which the first team interacts, showing the interchanges between the two GRMs/teams.

2. Create an $N^2$ chart of the principal members of a corporate board of directors–presidents, vice-presidents, chief executives, and so on, listing no more than 12. Identify in the interfaces how they relate to each other and what information they would exchange. Include an additional layer around the $N^2$ chart to include external inflows and outflows to the appropriate individuals.

3. You are tasked with organizing the boss's birthday party. Develop a behavior diagram showing the processes, in sequence, that you would undertake; there should be no more than six at the first level. Also identify the data and resources you need to undertake each process and the result of each process.

4. You are a marine officer[10] tasked with seizing and holding an enemy dockyard. Using the TRIAD building system, identify your prime directive, your consequent mission objectives in a logical time sequence, threats to achieving those objectives, and the strategies you propose to overcome them.

---

## REFERENCES

[1] Mandelbrot, Benoit, *The Fractal Nature of Geometry*, New York, NY: Freeman, 1977.

[2] Lorenz, Edward N., "Deterministic Non-periodic Flow," *Journal of the Atmospheric Sciences*, Vol. 20, 1963, pp. 130–41.

[3] Lano, R. J. A., "Operational Concept Formulation," *TRW Series on Software Technology*, New York, NY: North-Holland, 1980.

[4] Hitchins, D. K., *Putting Systems to Work*, Chichester, England: Wiley 1992, pp. 139 – 145.

[5] Warfield, J.N., "Intent Structures," *IEEE Transactions: Systems Man and Cybernetics*, Vol. 3, Part 2, 1973, pp.133-140.

[6] Saaty, T. L., "Operations Research: Some Contributions to Mathematics," *Science*, Vol. 178, (4065), 1972, pp.1061-1070.

[7] Lano, R. J. A., op.cit.

[8] Checkland, P. B., *Systems Thinking, Systems Practice*, Chichester, England: Wiley, 1981.

[9] Campbell, Joseph (ed.), *The Portable Jung*, New York, NY: Penguin Books, 1971.

[10] Roberts, Nancy, et al., *Introduction to Computer Simulation*, Reading, MA: Addison-Wesley, 1983.

---

[10] OK, so you're not a marine officer; watch a few James Bond movies and make it up! It is easier than a computer game, and more fun, too.

# Chapter 9

## Systems Thinking at Work: Case Studies

*How can I tell what I think until I see what I say?*
*E. M. Forster*
Quoted in Barth, Washington College Magazine, Winter, 1992

Systems thinking is thinking about systems in a structured way. The systems of interest are open systems, interacting with other open systems in an environment. In nature, open systems adapt and adjust to their situation by sensing, reacting, and responding to feedback. People and sociotechnical systems similarly adapt and adjust, sometimes through experience, sometimes in anticipation, or by design.

In thinking about open systems, therefore, it is important to identify and represent feedback and reaction, and to allow and enable representations of systems to adjust and adapt. In this respect, systems thinking is fundamentally different from conventional simulation and modeling. In particular, it aims to be organismic, allowing and enabling systems and subsystems representations to interact as they would in an organism.

Consider a typical battle simulation with aircraft attacking a ground installation defended by surface-to-air missiles. A conventional simulation approach works out the probabilities and progressively assesses the damage; eliminates aircraft due to SAM, and eliminates SAM due to aircraft. It is a calculator–elegant, perhaps, but still a calculator.

An organismic simulation, on the other hand, recognizes that both the attackers and defenders would be unlikely to fight to the last man, and would either change tactics or give up as attrition levels rose. In other words, the parties react and respond to a situation; they exhibit behavior, and the outcome is different. The various groups may, because of their mutual interactions, exhibit emergent properties, capabilities, and behaviors that would not be evident in the mechanistic calculator simulation.

Systems thinking models set out to be holistic and synthetic, as well as organismic. The first example, the doctor's surgery, is a case in point. It would be simple and straightforward to treat a family doctor's surgery as a queuing problem; with patients assembling in a waiting room and being called in for consultation. However, there is rather more to it than that. There may be several doctors, forming a so-called panel, or practice, and, as a result of each consultation, there may be several outcomes. Patients may be reassured that there is nothing wrong, they may be prescribed medication, they may be referred to a hospital consultant for further investigation, or they may be sent to the hospital as an emergency admission. Unless they die, patients are likely to find their way back into the doctor's surgery for follow-up action; depending on the issue, they may return in days, weeks, or even years, and they may return continually for chronic conditions.

The overall system is closed in the sense that people and patients may pass through several open systems, may be fed back in several different ways, and the only escape is death. Even that is questionable in modeling terms, since for zero population growth, every death is matched by a birth.

Paradoxically, then, systems thinking typically employs closure in addressing open systems. Causal loop modeling is particularly valuable as a precursor to dynamic modeling because CLMs are formed from closed loops. Simulating the causal interactions of such closed loop models creates powerful models of causal interactions without representing complete open systems: the causal effect *implies* a system, of which the effect is an emergent behavior.

## THE DOCTORS' SURGERY

The issue to be addressed concerns the excessive waiting times being experienced

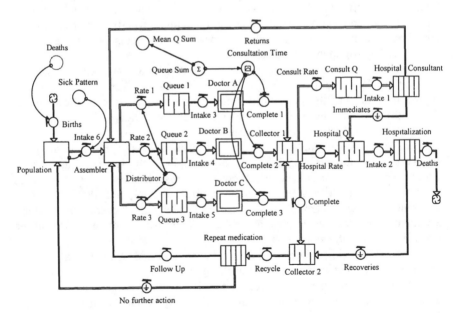

**Figure 9.1** *Doctors' surgery: STELLA model.*

by patients attending a general practice consisting of three doctors plus the usual reception, records, and clinics. There seems to be only two choices facing the managers of the practice: increase the number of doctors, or cut down on the average consultation time; presently 5 minutes per patient.

As this is a straightforward problem, a model can be developed directly in STELLA or some similar tool; that is, without the need for a precursor paper model. The model, Figure 9.1, shows three doctors, A, B, and C, in the center, each with his own queue. Patients are drawn from the population at left, and the pattern of arising can vary daily, weekly, and especially annually with the weather; and with annually recurring diseases such as influenza. The outcome of a consultation with any of the three doctors may be: an appointment to see a hospital specialist consultant, a direct transfer to hospital where necessary, prescribed medication by the doctor, or no further action. These are all represented in the figure, which also shows all of the patients returning sooner or later to the general population, unless they die or have a chronic illness; in the latter case they cycle around the loops periodically.

Unlike a conventional queuing model, all the loops are closed. Even the birth rate is coupled to the death rate. Note also, at center top, that the sum of queues 1, 2, and 3, queue sum, is connected to consultation time. This connection allows the consultation time to be reduced as the queue sum increases, to see if there is a suitable balance where queues to see doctors can be managed by reducing consultation times. Clearly there are limits: too short a consultation time will be unacceptable to both doctor and patient; the model explores this issue.

Simulation results are shown in Figure 9.2, from which it can be seen that, after an initial settling period, the consultation time settled to a value between 3 and 4 minutes. During this time, the mean queue awaiting each doctor–according to the simulation–would be some seven patients. This compares with a mean queue of some 19 patients per doctor, for a fixed consultation time of 5 minutes.

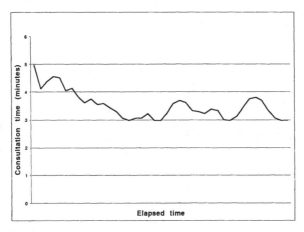

**Figure 9.2** *Surgery consultation time.*

This simulation does not immediately solve the problem: we are developing a learning laboratory. While a mean queue of 19 patients is clearly unacceptable, a mean queue per doctor of seven is still not good. Moreover, a consultation period of between 3 and 4 minutes is hardly generous. Another solution needs to be explored, and there seem to be two options: introduce at least one more doctor into the practice or reduce the number of patients served by the practice.

Figure 9.3 shows the effect of halving the number of patients. A mean consultation time of between four and five minutes results, and the mean queue of patients per doctor reduces to four–a less unsatisfactory state of affairs from the viewpoint of the practice managers. Of course, this means that half of the original patients are not now being served. Perhaps the best solution would be to increase the number of doctors, although there may be practical problems: are there doctors available? Can they be accommodated?

At this stage in addressing the surgery issue, it would be sensible to review the simulation model:

- As it stands at the moment, each doctor takes the next patient in the line from the assembler reservoir; while that does happen, it is also true that some patients prefer to consult one particular doctor, rather than consulting whomever is available.

- The potential for overloading patient records has not been addressed, and may be a concern as patient numbers rise.

- Patients need somewhere to queue, generally in waiting rooms, and the model does not present any limits on available space.

- Patients who are made to wait too long will find alternative doctors; some may even die as a result.

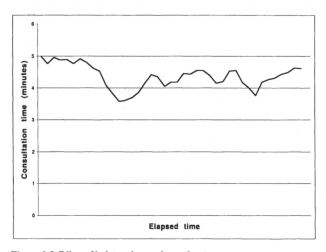

**Figure 9.3** *Effect of halving the numbers of patients.*

- Doctors are frequently called out for emergencies, leaving surgeries on hold.

- Doctors take time to hold clinics for mothers with new babies, for asthma, and so on.

- The model shows only three doctors, in continuous work, where four, five or even six[1] seems to be more sensible.

The model could easily be extended to include all of these factors and more, using the existing closed loop model as a core theme. To extend it sensibly may involve research into the particular practice; its facilities, limitations, and competition. That work, as they say, is left as an exercise for the systems thinker.


## HOSPITAL WAITING TIME

Politicians have a tendency to seize upon one particular measure of a public service and criticize the whole system on the basis of that single measure. With railways, it might be traveling delays, or overcrowding on commuter lines. With schools it might be examination pass rate comparisons between different types of school. With medical services, durations that people have to wait for hospital operations can become a political issue, especially where those health services are publicly funded. Attitudes can become acrimonious when politicians involve themselves in any professional services, but none more so than the life-or-death issues of hospital surgery.

One such issue concerns the use of laparotomy–the so-called "keyhole surgery" technique for operating on internal organs via small holes in the abdominal wall. Surgery by laparotomy has many potential advantages, not least that it reduces both operating time and recovery time; thereby freeing up time in the operating theaters and beds on the surgical wards.

Laparotomy is an acquired technique, however; it takes considerable practice and hand-eye coordination. Some surgeons may find the technique difficult, even impossible, to master. So, when politicians and hospital managers see an instant solution in the extensive introduction of laparotomy, conflict is inevitable. Figure 9.4 shows the situation as a causal loop model.

The model contains a loop at the top that represents the expected advantage of the new technique: laparotomy will reduce operating times. Reducing operating times increases operating theater utilization, which increases the rate of operations. This reduces the patient queues, and hence patient waiting time, especially for those patients whose problem is suited to operation by laparotomy. This encourages and reinforces the wish to gain advantage from laparotomy.

---

[1] Increasing the number of doctors to six, and restoring the number of patients to its original higher value, results in a mean queue length of less than two patients per doctor with a five-minute consultation period – in the simulation.

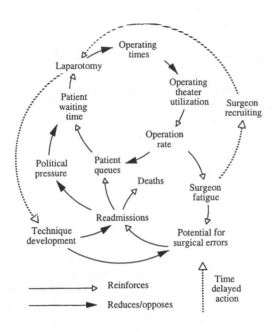

**Figure 9.4** *Hospital waiting times.*

As the figure shows, there is rather more to it. Surgeons take time and practice to become adept with laparotomy. In the meantime, as they develop technique, mistakes are possible. Surgeons become overloaded with the higher rate of operations, fatigue may set in, and mistakes are possible. For both reasons, readmissions may increase, tending to extend, rather than reduce, the patient queues and waiting times.

Politicians may be dismayed by the increase in readmissions, and may lose faith in laparotomy as a technique. As the model shows, however, they have only themselves to blame. Introducing such a novel technique was bound to take time and practice; moreover, they should have foreseen that to increase the rate of operations would necessitate recruiting more surgeons. This is not rocket science; it really is rather obvious–at least, once you have seen Figure 9.4. Given time, laparotomy can become a valuable surgical technique, helping both surgeon and patient. It is not unlikely, however, and as the model shows, that early disappointment as evidenced by readmissions will discourage politicians from encouraging operations by laparotomy.

In this instance, while a dynamic simulation is possible, it is not necessary; the explanation provided in the CLM is sufficient. Note the closed loop form of the model, with the only exit being death. A dynamic simulation would add flesh to the bones of the CLM by estimating times to reduce waiting lists, with and without the technique, how many more surgeons would be needed, and so on.

## SOCIETAL POWER [1]

Western society is evolving; visits around the world confirm that it is evolving differently from other regions. In the West, commercial pressure promotes societal factions by age, persuasion, and sex to sell more fashion, music, stationery, drugs.... In the West, competition is seen as an ideal.

At the same time, the West applauds the individual. The lone individual is an icon, standing against all odds, as in the conventional cowboy Western. Western culture emphasizes, and Western laws defend, individual human rights. There is a cult of the individual, where in other regions there may be cults of the family, the group, the team, or the culture.

A rise in the general standard of education has also coincided with a reduction in those professing a religious faith, particularly among the young. Significantly, there is an emergence of arcane beliefs in:

| | | |
|---|---|---|
| Little green men; | Corn circles; | Astrology; |
| Pyramids; | Black magic; | Ghosts; |
| Crystals; | Reflexology; | Alternative everything; |
| Palmistry; | Spiritualism; | Management science. |

People need *some* belief system, and it need not be rational. With so many alternative belief systems, it is difficult to avoid the implication that a fragmenting belief system is an indicator of a fragmenting society.

Younger people hold many of these beliefs. Psychologists suggest that we all go through various stages of so-called individuation. This is the process of becoming an individual that, according to Jung, is completed by our early thirties. During the process we go through different stages of maturation; each of which may be identified by individual and group behavior. So, the way we dress, the music we like, even the drugs we take, or do not take, mark out stages in our individuation.

Social groups individuate [2]–go through stages of development–just like individuals. Gradually, as maturation leads to individuation, it leads also to interests in power, and to challenges of established beliefs, cannons, and traditions. With individuation come: factions, energy absorption in internecine struggle, resistance to change, uncontrolled change, and, perhaps, societal breakdown.

Figure 9.5 presents a view of the developing theme. At left is a causal loop model showing Western culture with its competition, commercial pressures, forced individuation, and the endemic cult of the individual. It shows also the essential duality with which each of us faces the world: black or white, for or against, left or right, and right or wrong. We see things in stark contrast, even where reality is shades of gray. The growth of factions and fractions that this

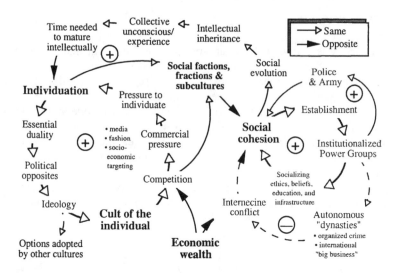

**Figure 9.5** *Causal loop model: Western societal dynamics.*

pressure, coupled with a lack of unified beliefs, encourages, militates against social cohesion.

At the right of the figure is the establishment, with the reins of power seeking to maintain social cohesion, in the face of factions and fractions in society. At top left is an outside loop, suggesting that as society becomes more complex, developing individuals have more stages of development to traverse, and more to learn at each stage. For instance, 50 years ago a son marked his rite of passage to manhood by wearing a suit that was just like father's suit. Today there are distinct fashions for most ages between 10 and 25. The kind of music that 14-year-old girls like is passé by the time they are 15. The drinks that are fashionable for 18 year olds, they would not touch at age 25, and so on.

Also in the figure at bottom right is the growth of so-called autonomous dynasties: these are major international groups that exist and operate outside of local or even national governments. Some are criminal; many are big business.

The whole model is driven by economic wealth, at the bottom. As shown, wealth reduces the need both for competition and for internecine conflict. In other words, the system might be stable during good times, but less so during lean times.

The CLM is far from being definitive; it expresses a viewpoint, that may or may not be valid. One advantage of such a model is that it does present the issue openly, allowing others to disagree. It is not too clear from the CLM just what the outcome might be: is Western society hell bent on self destruction, as Marx believed, or are we worrying unnecessarily?

It is possible to develop a nonlinear dynamic simulation directly from the CLM of Figure 9.5 and to develop some time-dependent graphs showing how

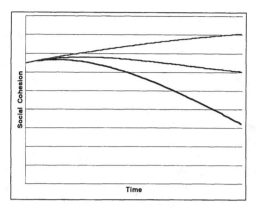

**Figure 9.6** *Social cohesion of a nominal Western society for different economic conditions. Bottom line: poor. Middle line: moderate. Top line: robust.*

social cohesion might vary over time. Figure 9.6 shows three successive simulations, suggesting that social cohesion increases with national economic wealth. This was indicated by the CLM and is confirmed by the nonlinear dynamic simulation.

The simulation allows us to postulate other situations; ones that would be rather difficult to assess using the CLM alone. Suppose, for example, that the people in an economy shared a strong belief. Instead of a nation of factions, suppose there was a one-nation outlook. Would that make a difference during a time of economic difficulty?

Figure 9.7 suggests that shared belief is a more powerful force for social cohesion than a robust economy.

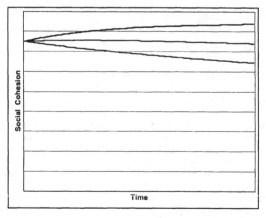

**Figure 9.7** *Social cohesion. Top: poor economy, and a widely shared belief system. Middle: moderate economy, and some shared belief. Bottom: robust economy, but little shared belief.*

It would be wrong to suggest that either the CLM of Figure 9.5, or the two graphs from the equivalent STELLA model, give a definitive answer to a problem. The CLM expresses a view about how a complex social system might be driven. The STELLA model (not shown) requires figures to be entered, representing values for parameters such as social cohesion; it would be impossible to justify such figures.

Nevertheless, the models are able to express ideas with some rigor. The graphs, with their scales deliberately omitted, suggest behavior. The behavior is consistent with what we might expect once we see how the model works. And the model works in the manner in which we suspect that society works.

This example, then, presents a conceptual model, one that is grappling with the representation of an abstract and complex aspect of social behavior. It is system theoretically substantial in that the model is holistic, synthetic and organismic. And it is conjectural, since it represents circumstances and conditions that may never really exist.


## NATIONAL ENERGY STRATEGY


Energy is a hot topic. Energy consumption is forecast to rise by over 200% in Europe in the next 20 years, while at the same time some countries are phasing out their nuclear energy supplies. They may hope to bridge the gap with alternative energy supplies, but these are proving difficult to realize. Addressing such a substantial issue is a case for the rigorous soft method; see page 157.

The following scenario is real: the identity of the industrialized country has been omitted, along with many factors contributing to the issue. Broadly, the issue is one of concern about future energy supplies and global warming, exacerbated by fossil fuel $CO_2$ emissions. The national government has already committed to dismantling its nuclear energy industry within a decade, while at the same time signing up to the 1997 Kyoto protocol, aimed at the international reduction in greenhouse gas emissions.

*"There is uncertainty about the future of nuclear power. Activists are protesting about both the dangers from power stations and the difficulty of waste disposal. Illnesses, particularly infantile leukemia, are being ascribed to radiation from power stations, although scientists are unable to detect any causal relationship.*

*"Government is concerned, both about public opinion and about the massive costs of decommissioning power stations. The alternatives seem uncertain, too. Fossil fuel is problematic, with the cost of coal rising slowly but inexorably. The availability and cost of oil is always a problem, associated as ever with a volatile Middle East.*

*"Government has committed, too, to reducing greenhouse gas emissions, and has signed up to the international Kyoto Accord. The major generator of $CO_2$ is fossil fuel. If $CO_2$ emission control is to be realized, it seems inevitable that fossil fuel burning will have to be reduced, but how, and if there is to be no nuclear industry, what is the future source of energy going to be?*

*"A new factor has arisen: the rail link to the continent. This promises to reduce the cost and time of freight, but it also carries the threat of ingress from undesirable elements.*
*"What to do? There seems to be no sensible way forward" (Anon.)*

Reading through the situation, there seems to be no obvious solution and, in addition to that, precious few symptoms; that is, indications of change from a prior, supposedly satisfactory state. A short list follows:

| Symptom | Description |
| --- | --- |
| 1 | Activists are protesting. Although the topics of protest may not be new, the act of protesting appears to be, and government is concerned over public opinion. |
| 2 | There is a concern that radiation may be causing infantile leukemia, although there is no scientific evidence. |
| 3 | Nuclear power stations are widely seen as dangerous and expensive, especially when decommissioning is considered. |
| 4 | The cost of coal is rising slowly, while the international cost of oil fluctuates wildly and is quite beyond control. |
| 5 | The government has committed, under the Kyoto Accord, to reduce greenhouse gases; mainly $CO_2$. |
| 6 | The new rail connection to the European continent is seen as a two-edged sword. |

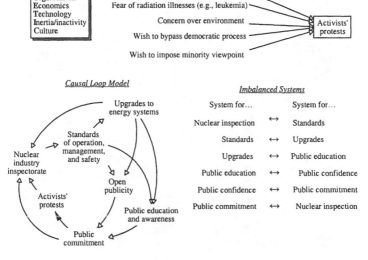

**Figure 9.8** *Addressing symptom 1: activists protests.*

**The Symptoms**

The RSM procedure commences by identifying and elaborating each symptom in turn. Figure 9.8 shows a useful template in use. The symptom is shown, at top right, together with possible causes, at top center. The panel at left shows the expanded acronym POETIC, which stands for: politics, organization, economics, technology, inertia/inactivity, and culture. These factors serve as an aid to memory; they are the most likely causes of symptoms. At bottom left is the causal loop formed from the possible causes and the symptom.

At bottom right are the imbalanced systems, implied by the causal effects. A symptom emerges due to an imbalance or dysfunction between these implicit systems that were previously balanced.

Note the use of pejorative terms in the laundry list of possible causes; these have disappeared in the CLM, which outlines a strategy that would obviate the need for protest. This has introduced elements in the CLM that were not present in the list of possible causes; they are needed to complete the logic flow in the causal loop.

Addressing the first symptom looks at activists, but does not address their concerns; justified or not. The second symptom addresses those concerns; see Figure 9.9. The cause of concern may be a genuine fear of radiation (engendered by the aftermath of Hiroshima and Chernobyl, and in no small degree by media hype), coupled with a lack of objective scientific evidence in the public domain, and the ever-present need for a scapegoat to blame for illnesses such as infantile leukemia; the causes of which are poorly understood.

The resulting CLM proposes an ideal world in which objective scientific evidence is made widely and freely available to the public, and research into the

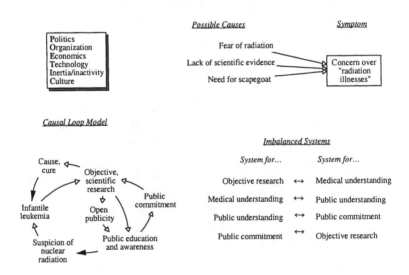

**Figure 9.9** *Addressing symptom 2: concerns over radiation.*

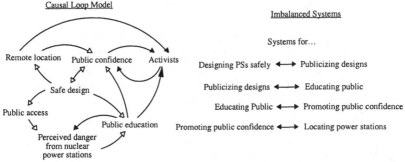

**Figure 9.10** *Addressing symptom 3: perceived dangers from nuclear power stations (PSs).*

real causes of such unfortunate and emotive diseases is intensified. Going around the CLM identifies the implicit systems that are most probably imbalanced for the symptom to have emerged.

The third symptom concerns the perceived dangers from nuclear power stations; see Figure 9.10. Possible causes include perceptions that the power station designs are intrinsically unsafe, that the stations are sited too near centers of population, or that the activists have "got it wrong."

The CLM creates another ideal world by dropping the pejorative adjectives from the laundry list of possible causes. The model assumes that public concerns would be assuaged by education and openness, giving access to all. The model concedes that putting power stations in remote locations would enhance public confidence and appease activists.

High decommissioning costs for nuclear power stations arise because of difficulties in disposing of spent fuel, which can be highly radioactive, and of decontaminating sites.

There are many ways to dispose of waste safely: some are relatively safe; some radical. Ideas vary from burying waste under tons of concrete at the bottom of shafts drilled into remote granite bedrock, to placing it in tectonic plate subduction zones, to locating it at one of the Lagrangian points on the moon's orbit where the balance of the earth's and the moon's gravity creates gravity "wells." Politicians and pundits will argue about the relative safety of these and many other methods of disposal. Disposal is an unresolved issue.

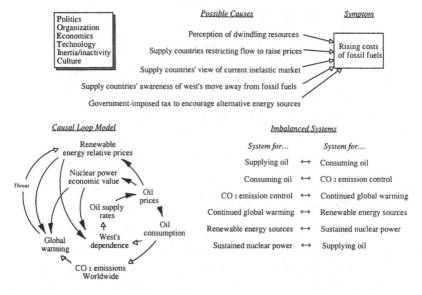

**Figure 9.11** *Addressing symptom 4: fossil fuel issues.*

The rising cost of fossil fuels is an issue; it constitutes the main source of domestic and industrial power, yet its supply, particularly from the Middle East, is uncertain. Moreover, it is heavily implicated in the rise of greenhouse gas emissions, which, according to many scientists are already resulting in global warming.

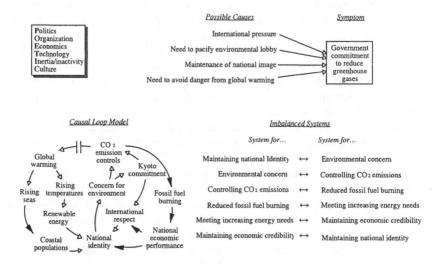

**Figure 9.12** *Addressing fifth symptom: greenhouse gas emission control.*

The issue is developed in Figure 9.11, which suggests a more global view of the developing situation. The CLM introduces both alternative energy sources and nuclear power, as factors in reducing the West's dependence on Middle Eastern oil. The CLM suggests that, as oil prices rise, the relative cost of both nuclear and alternative energy sources (wind, wave, hydroelectric, biomass, and so on), are effectively reduced. One possible fly in that ointment is alternative coastal wave energy, which could be adversely affected by rising sea levels due to global warming.

The set of imbalanced implicit systems is formed by going around the CLM, implying a system as the source of each and every causal effect.

The fifth symptom, which surrounds the Kyoto Accord, is illustrated in Figure 9.12. The national government decided that it was morally right and proper to reduce greenhouse gas emissions. They therefore signed up to the Kyoto Accord on greenhouse gas emissions, and undertook to reduce fossil fuel burning as a primary power source, and to replace it with "clean" alternative energy sources: wind, wave, hydroelectric, and biomass.

The sixth and final symptom is shown in Figure 9.13; the new rail link. The CLM is quite full, showing the effect that the new link might be expected to have on fuel consumption and $CO_2$ emissions. The new link will greatly reduce journey lengths for freight vehicles that can travel over the much faster and cheaper rail link on special carriers. This will reduce oil costs per journey, and hence will reduce commodity prices. As a byproduct, exports and imports are expected to rise and rail transport will increase to carrying more long-distance trucks. Hence more, not less $CO_2$ will be emitted in the longer term.

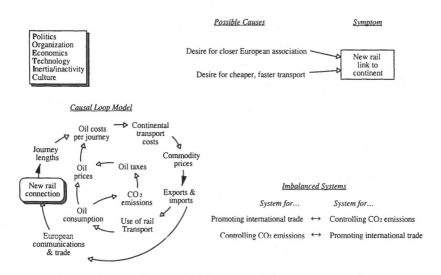

**Figure 9.13** *Addressing symptom 6: the new rail link.*

**Table 9.1**
Consolidated List of Implicit Systems in Imbalance

| | *Implicit System for…* | | *Implicit System for…* |
|---|---|---|---|
| | Nuclear inspection | ↔ | Standards |
| | Standards | ↔ | Upgrades |
| Nuclear Power | Upgrades | ↔ | Public education |
| | Public education | ↔ | Public confidence |
| | Public confidence | ↔ | Public commitment |
| | Public commitment | ↔ | Nuclear inspection |
| | Objective research | ↔ | Medical understanding |
| Radiation | Medical understanding | ↔ | Public understanding |
| & Leukemia | Public understanding | ↔ | Public commitment |
| | Public commitment | ↔ | Objective research |
| | Maintaining national identify | ↔ | Environmental concern |
| | Environmental concern | ↔ | Controlling $CO_2$ emissions |
| Kyoto | Controlling $CO_2$ Emissions | ↔ | Reducing fossil-fuel burning |
| Accord | Reducing fossil-fuel burning | ↔ | Meeting increasing energy needs |
| | Meeting increasing energy needs | ↔ | Maintaining economic credibility |
| | Maintaining economic credibility | ↔ | Maintaining national identify |
| | Supplying oil | ↔ | Consuming oil |
| Fossil Fuel and | Consuming oil | ↔ | Controlling $CO_2$ emissions |
| Global Warming | Controlling $CO_2$ emissions | ↔ | Continued global warming |
| | Continued global warming | ↔ | Renewable energy sources |
| | Renewable energy sources | ↔ | Sustaining nuclear power |
| | Sustaining nuclear power | ↔ | Supplying oil |
| The New | Promoting international trade | ↔ | Controlling $CO_2$ emissions |
| Bridge | Controlling $CO_2$ emissions | ↔ | Promoting international trade |

## Bringing the Imbalanced Implicit Systems Together

Each of the symptoms generates a list of imbalanced implicit systems. These are accumulated and rationalized in Table 9.1. The entries in the table are then transferred to an $N^2$ chart–Table 9.2–without alteration. The table is a printout from a simple tool that has limitations on the size of title that may be used; titles have therefore been abbreviated, but can easily be identified by reference to Table 9.1 above.

As it stands, the unclustered $N^2$ chart reveals little. It would be helpful to see the archetypal patterns indicated in $N^2$ charts on page 143 and following. One way

**Table 9.2**
Unclustered $N^2$ Chart of Imbalanced, Implicit Systems

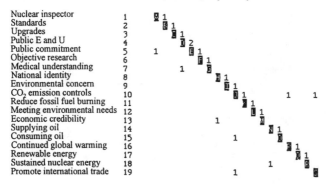

| | |
|---|---|
| Nuclear inspector | 1 |
| Standards | 2 |
| Upgrades | 3 |
| Public E and U | 4 |
| Public commitment | 5 |
| Objective research | 6 |
| Medical understanding | 7 |
| National identity | 8 |
| Environmental concern | 9 |
| $CO_2$ emission controls | 10 |
| Reduce fossil fuel burning | 11 |
| Meeting environmental needs | 12 |
| Economic credibility | 13 |
| Supplying oil | 14 |
| Consuming oil | 15 |
| Continued global warming | 16 |
| Renewable energy | 17 |
| Sustained nuclear energy | 18 |
| Promote international trade | 19 |

to do this is to symmetricalize the $N^2$ chart, as well as to cluster it. Symmetricalizing is a reasonable process in this instance, since an imbalance between any two systems A and B means that there must also be an imbalance between B and A.

The result, in Table 9.3, is revealing. Not only are there clusters around the leading diagonal, and a node formed around system 7–a system for controlling $CO_2$ emissions–but there are also some obvious omissions, marked with question marks on the chart. Omissions can be identified by gaps in the cross of interfaces forming at a node: in the chart there is a gap, marked by a cross, between controlling $CO_2$ emissions and renewable energy sources. The gap is marring the symmetry of the cross.

Another way to identify omissions is by noting the absence of an obvious connection. The block of implicit systems from 13 to 19–see Table 9.3–is disjoint; there are no connections with the other implicit systems, 1 to 12. The other two

**Table 9.3**
Imbalanced Implicit Systems: Clustered $N^2$ Chart

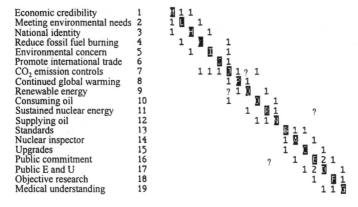

| | |
|---|---|
| Economic credibility | 1 |
| Meeting environmental needs | 2 |
| National identity | 3 |
| Reduce fossil fuel burning | 4 |
| Environmental concern | 5 |
| Promote international trade | 6 |
| $CO_2$ emission controls | 7 |
| Continued global warming | 8 |
| Renewable energy | 9 |
| Consuming oil | 10 |
| Sustained nuclear energy | 11 |
| Supplying oil | 12 |
| Standards | 13 |
| Nuclear inspector | 14 |
| Upgrades | 15 |
| Public commitment | 16 |
| Public E and U | 17 |
| Objective research | 18 |
| Medical understanding | 19 |

**Table 9.4**
Hierarchy-Shifted N² Chart.

| Group | Item | No. | | | | | | | | | | | | | | |
|---|---|---|---|---|---|---|---|---|---|---|---|---|---|---|---|---|
| | Standards | 1 | ◼ 1 1 | | | | | | | | | | | | | |
| | Upgrades | 2 | 1 ◼ 1 | | | | | | | | | | | | | |
| Nuclear | Nuclear inspector | 3 | 1 ◼ 1 | | | | | | | | | | | | | |
| energy | Public E and U | 4 | 1 ◼ 2 1 | | | | | | | | | | | | | |
| | Public commitment | 5 | 1 2 ◼ 1 1 | | | | | | | | | | | | | |
| | Medical understanding | 6 | 1 ◼ 1 | | | | | | | | | | | | | |
| | Objective research | 7 | 1 1 ◼ | | | | | | | | | | | | | |
| | Sustained nuclear energy | 8 | 1 ◼ 1 1 | | | | | | | | | | | | | |
| | Supplying oil | 9 | 1 ◼ 1 | | | | | | | | | | | | | |
| | Renewable energy | 10 | 1 ◼ 1 | | | | | | | | | | | | | |
| Environ- | Consuming oil | 11 | 1 ◼ 1 | | | | | | | | | | | | | |
| ment and | Continued global warming | 12 | 1 ◼ 1 | | | | | | | | | | | | | |
| energy | Promote international trade | 13 | ◼ 1 | | | | | | | | | | | | | |
| | CO₂ emission controls | 14 | 1 1 1 ◼ 1 1 1 | | | | | | | | | | | | | |
| | Reduce FF burning | 15 | 1 ◼ 1 | | | | | | | | | | | | | |
| | Meeting environmental needs | 16 | 1 1 ◼ 1 | | | | | | | | | | | | | |
| | Environmental concern | 17 | 1 ◼ 1 | | | | | | | | | | | | | |
| | Economic credibility | 18 | 1 ◼ 1 | | | | | | | | | | | | | |
| Sovereignty | National identity | 19 | 1 1 ◼ | | | | | | | | | | | | | |

question marks denote a missing interface between public commitment and Sustaining Nuclear Power, which becomes evident in reviewing the overall logic of the chart. Clearly, there is no public commitment at present to sustaining nuclear power, so the question marks in this second instance denote an error or omission[2] in the prior analysis.

At this point, the chart needs to be reclustered, to account for the additional imbalance entries; see Table 9.4. Three groups can be identified; in this instance, a deciding factor in identifying groups is to look for clusters that share the minimum of interfaces. Sustaining nuclear power marks a boundary because it has the only interface between nuclear energy, the top group, and environment and energy, the middle group. At the bottom, environmental concern shares the only interface between the bottom group, sovereignty, and environment and energy.

The clustering process, then, is a mixture of three things: some systems science, to identify that subsystems are loosely coupled to others; some cerebral pattern identification, to detect clusters and nodes; and some common sense, to associate like things using the column of implicit systems at the left. As an example of this last, consider whether sustaining nuclear power belongs more to nuclear energy or to environment and energy. Rationally, it seems to be an even choice, but the chart shows more interfaces with environment and energy, so it sits better there.

**Assessment**

What has been achieved so far? The issue has emerged as a concern over the future of the nation's energy supplies. We have created a kind of rich picture, and

---

[2] Using the N² chart to identify errors and omissions in this way adds considerable power to the RSM process.

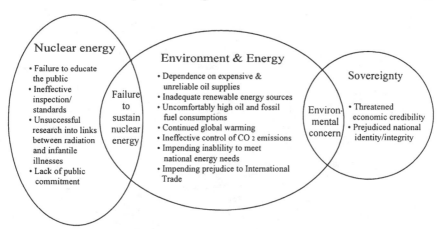

**Figure 9.14** *Systems interaction diagram.*

from it we have identified three themes. (None of these is earth shattering, but then this is a deliberately simplified example to explain the process. More complex problems might have 30 or 40 symptoms, hundreds of implicit systems, and a plethora of red herrings and irrelevancies.)

The $N^2$ chart is essential for the clustering process and for identifying nodes, and the like, but it is lacking in the visual presentation department. The systems interaction diagram (SID) presents the same information in a more digestible form, and–remembering that the $N^2$ chart is derived from imbalanced/ dysfunctional systems–the SID restores the missing pejoratives. So, sustaining nuclear energy becomes failure to sustain nuclear energy; economic credibility becomes threatened economic capability, and so on.

The SID in Figure 9.14 is valuable for understanding, and is particularly useful for briefing. It is backed up by an audit trail leading back to the individual symptoms and their possible causes. The process is synthetic, holistic and organismic and therefore stands as system scientific. It may be proven mathematically (see Appendix B) and can be used–as here–to predict outcomes, so is falsifiable.

The SID can also be represented as a CLM; see Figure 9.15. As a result of the hierarchy shift introduced by the $N^2$ clustering process, this is a high-level model, looking much more at the issues than at the detail from which the model was drawn. There are three loops, each corresponding to the ovals in the SID.

The central loop is at the core of the issue. Oil and coal consumption will increase $CO_2$ emissions, exacerbating global warming. Global warming, with its concomitant rising seas and shifting patterns of rain, wind, and temperature, will prejudice renewable energy supplies, such as hydro-electric, wind power, and biomass (e.g., tree farming). These alternative sources of power are the very ones in which national government has put its faith. Renewable energy supplies are intended to reduce dependence on nuclear energy, too, which is being phased out

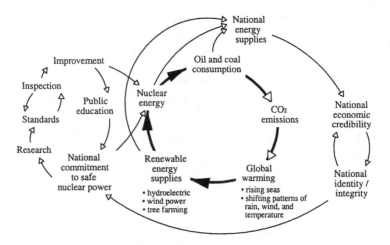

**Figure 9.15** *High-level CLM: power, environment, and sovereignty.*

largely for political reasons, although it is a "green" power source. As such, nuclear power is the one sure way of reducing harmful fossil fuel consumption. So, the central loop is dysfunctional; it cannot be depended upon to function effectively.

One way to address the dysfunction is to reverse the decision to phase out nuclear power. The loop at the left indicates what this implies: gaining public trust and support, together with even tighter inspection and control regimes. In any event, reliable and expanding sources of national energy supplies are essential to sustain the economy, national identity, and national integrity; see right hand loop. If nuclear power is phased out on schedule, the outlook is bleak both for the national economy and for the national contribution to global warming.

(The CLM at Figure 9.15 is a conceptual model and is amenable to simulation modeling. In this example, however, it is simpler to identify a number of strategic options facing government, and to choose between them.)

**Options for Addressing the Issue**

The last part of the RSM process is to establish a sensible solution to the issue; see Figure 8.21 on page 160 for the overall RSM process. Faced with a dilemma, national government has at least four options:

1. Continue/increase using fossil fuels:

     o   Phase out nuclear power; increase $CO_2$ emissions;

     o   Maintain international trade, national economic performance using increasing amounts of fossil fuels;

     o   Renege on the Kyoto Accord.

2.   Retain, sustain, and enhance nuclear energy:

    o   Reduce oil, and reduce greenhouse gas emissions, using mixed energy sources;

    o   Maintain international trade national economic performance;

    o   Educate the public to the advantages and safety of modern/future nuclear power;

    o   Introduce a new generation of nuclear power stations of safer design, remotely sited and under even tighter control;

    o   Ratify the Kyoto Accord.

3.   Phase out nuclear and fossil energy, and control $CO_2$ emissions:

    o   Accept reduced industrial and transport capability;

    o   Accept reducing international trade and national economic performance;

    o   Ratify the Kyoto Accord.

4.   Retain nuclear energy as in 2, but:

    o   Seek alternative sources of energy that do not prejudice the environment or the people;

    o   Inform the public of the reasons and research progress;

    o   Educate the public to the advantages of retaining nuclear power in the near term, and of switching to advanced energy sources when available;

    o   Ratify the Kyoto Accord.

Option 4 has challenged scientists for some time. The Dyson sphere was originally proposed in 1959 by the astronomer Freeman Dyson [3] as a way for an advanced civilization to utilize all of the energy radiated by their sun. It is an artificial sphere the size of a planetary orbit. The sphere would consist of a shell of solar collectors or habitats around the star, that would be able in principle to catch the total energy output from the sun. This would create a huge living space and gather enormous amounts of energy.

Less elegant, but within the current technological capability, would be to locate solar "umbrellas" in earth orbit, to catch solar energy and to focus it on to receptors at ground level. Solar umbrellas may convert solar energy from the visual to the microwave spectrum. There are evident dangers in such focused beams of energy passing vertically downward through the atmosphere, but here at least is one viable source of nuclear[3] energy with no radioactive emissions, and no expensive disposal problems.[4]

---

[3] Solar energy is, of course, remote nuclear fusion energy.
[4] Nuclear fusion here on earth is also an option, but at the time of writing it does not look too promising as a practical solution.

**Table 9.5**

Assessing Energy Options against Original Issue Symptoms and Factors

| Symptoms and Factors | Option 1<br><br>*Continue with fossil fuels* | Option 2<br><br>*Retain and enhance nuclear energy* | Option 3<br><br>*Phase out nuclear and fossil fuels; depend on alternative energy* | Option 4<br><br>*Retain nuclear energy until clean source available; e.g. solar umbrellas, nuclear fusion* |
|---|---|---|---|---|
| Activists are protesting | 4 | 2 | 3 | 1 |
| Concerns about radiation and leukemia | 4 | 2 | 3 | 1 |
| Nuclear energy seen as dangerous | 4 | 2 | 3 | 1 |
| Fossil fuel prices rising and fluctuating | 4 | 1.5 | 3 | 1.5 |
| Government signed up to Kyoto agreement | 4 | 1.5 | 3 | 1.5 |
| New rail connection to the Continent | 4 | 2 | 3 | 1 |
| Factors | | | | |
| Alternative energy inadequate to sustain economy, *and* threatened by global warming | 3 | 2 | 4* | 1 |
| Intent to maintain national sovereignty, identity, and integrity | 3 | 2 | 4* | 1 |
| | 30 | 15 | 26 | 9 |
| | 4th | 2nd | 3rd* | 1st |

*A nonviable option for economic and global stability.

One way to assess the various options is to see if they address the original symptoms. In some cases this would necessitate simulation, but in this instance it is simpler to use rank matrix analysis (RMA); see Table 9.5.

Here, a matrix of options and symptoms is set out and the degree with which each option addresses a symptom is placed in order; for example, 1 is an ordinal number representing "first." The process is repeated for each symptom, the ordinals are then summed and the column with the lowest ordinal sum is deemed the preferred option. The whole table can be subjected to statistical pattern recognition techniques to judge the probability that the ordinal pattern could have arisen purely by chance–the null hypothesis.

Table 9.5 was drawn up assuming that the public can be persuaded that a new generation of nuclear power generators will really be safe. Option 3 may be eliminated as impractical, although that is the strategy being currently pursued. When that strategy was introduced, it was believed that alternative energy would provide for all the nation's needs; that belief has proven optimistic. When the estimated 200% increase in energy demand, forecast for the next 20 to 50 years, is included, it is evident that radical rethinking is necessary.

## Conclusion to RSM Issue Development

RSM is thorough and comprehensive. It produces results that are traceable all the way back to the original issue symptoms; the more symptoms, the more robust the solution is likely to be. The preferred optional solution to the issue may be socially and politically unacceptable even where, as in this case, essential factors are included in the trade-off table. These factors were derived from the SID and from the high-level CLM.

Using RSM, a major advantage accrues when a team works together to produce a result: RSM allows and encourages a melding of minds and expertise; allowing a team to exceed the efforts of individuals. RSM is holistic, as it treats the whole body; it is synthetic, as it builds a solution outward from the symptoms; and it is organismic in that both the process and the solutions are derived from interactive parts creating a whole.

## THE RAILWAY DILEMMA

### Stakeholder Analysis

Interpretive structural modeling is particularly effective when used as a basis for stakeholder analysis. Stakeholders are entities that stand to gain, or–particularly–to *lose*, from the successful implementation of a project or process. Pundits often omit potential losers, which impoverishes the results of their analysis. For example, if there were a plan (there is none!) to introduce a hotdog stand in the entrance to the Great Pyramid, stakeholders would have to include those with an interest in preserving the sanctity, dignity, and numinousness of the ancient tomb, as well as those who stood to make money.

Similarly, if there were a plan to introduce a skyscraper into a remote seaside beauty spot, stakeholders would have to include those who want to preserve the natural beauty of the landscape. If there were a plan to bomb an enemy town, the inhabitants would be stakeholders, since they stand to lose. If, say, archaeological remains were at risk, then heritage might be at risk that could be of value on a worldwide scale.

Stakeholders are often those who "own" issue-containing systems. Thus stakeholders in a new taxi design are taxi companies, taxi drivers, potential passengers, and bus and train companies, where the new taxi could reduce their business. The general public in the area is also a stakeholder, since the new taxi will become part of their environment: according to its characteristics, they could be deemed winners or losers, but in either event they are stakeholders.

To undertake stakeholder analysis, first establish a rich list of stakeholders, and then find the individual *objectives* of each stakeholder. Ideally this process would involve asking individual stakeholders, but that is not always practicable: people are sometimes inaccessible, and they sometimes lie, too.

Where obtaining a clear and plausible set of stakeholder objectives is difficult, it may be necessary to resort to role playing; the process is inferior, but may be the only option.

Next, identify relationships between the objectives of all stakeholders, using interpretive structural modeling (on page 148). Then pinpoint the bases for synergy/dysfunction between all stakeholders.

**The Railway**

A major train company is experiencing revenue problems and, over many years, there has been a lack of investment in rail infrastructure, signaling, stations, and rolling stock.

Stakeholders include the business (and shareholders) as a profit making concern, passengers, signal and rail staff, train drivers and guards and, of course, the station staff who take the brunt of complaints, but who have perhaps the least control over events.

From a short-term business profitability viewpoint, the rail company would want to run as few trains as possible, packing each train full, so that they

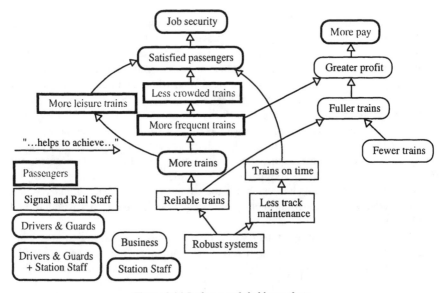

**Figure 9.16** *Railways stakeholder analysis.*

maximize the revenue to running cost ratio. Passengers, on the other hand, prefer more frequent trains, and would like them to be less crowded.

The resulting intent structure is shown in Figure 9.16 for all the stakeholders concerned. While there are several causality strands, there are two fundamentally opposed viewpoints, with the profit-making concerns of the business lining up against the interests of all the other stakeholders: passengers, station staff, drivers, guards, and signal and rail staff.

This, of course, raises the issue of railway ownership. Should railways be a public service, run perhaps by a nonprofit-making company, or subsidized out of local or regional taxation? Throughout the world there are variations in the way that railways are run; there appears to be no overall right answer, only a locally workable solution.

Such stakeholder analyses provide an interesting, holistic view of complex situations; all stakeholders can be given a say, and their points of view and objectives can be fitted into the intent structure. Sometimes, to the surprise of implacable opponents, they may find that their different objectives are complementary, and help to achieve the same ends. At other times, as in the example, there is little hope for reconciliation, but at least the differences are out in the open where they may be addressed.

## COMPUTING INTEGRITY

As a society becomes wealthier, it creates more complex, sophisticated information technology (IT) and information warfare (IW) systems. It also creates more capable viruses, worms, and so on, to attack those systems. This is clearly a matter of major concern. Valuable and important information systems, fundamental to business, commercial, and government operations can be attacked and disabled by individuals with a grudge, or a distorted sense of humor. Of particular concern in the context of IW is the ability of a foreign power or group, of itself no threat, to mount a devastating attack on an economy; perhaps by falsifying or destroying bank records, by transferring large amounts of money, and so on.

The choice of the term "virus" is fortuitous: computer viruses do, indeed, behave in a broadly similar way to biological viruses, with their ability to multiply and spread through the networked "body" of the information and computing system. Is it feasible to carry the analogy further and protect information systems against viral infection in similar ways to those employed in nature?

What choices do we have to protect our IT and IW systems? Should we isolate our systems inside a protective shell to avoid IW attack; open our systems to meet "infections" head on, and so create an auto-immune system with "antibodies;" or both?

The provision of an effective shell inside which to protect and isolate an information system is fraught with difficulties. Of course the policy works if the machinery is completely isolated, but then it is of restricted value. Total isolation would mean that no external media could be introduced: no floppy disks, no tapes, no data links, and so on.

Attempts to introduce such a policy in a police force failed. A new information system was introduced with a firewall and without any floppy drives. A decision was taken that personal processors within the network would not require dedicated antiviral software; reliance was placed instead on the firewall to prevent viral infection: initially, it worked. However, bored policemen on weekend shifts brought in their own floppy drives, and loaded games onto the computers to pass the time away. Soon the whole network was heavily infected with viruses injected along with the games software.

So, should we meet such infections head on, employing all the latest antiviral software in firewalls, networks, and personal computers? Is the advantage not always with the attackers; the ones creating the next virus or worm? This is a dilemma facing information warfare in particular.

One way to address such a vague, yet important, question is to maintain the medical analogy and look at the issues in life cycle terms. If viral infection "brings down" a network of computers, this can be viewed either as a debilitating illness or a death (an interrupted or truncated life cycle). We could attempt to prevent losing our total capability by having more than one such network, or–more sensibly perhaps–by having parallel networks employing different operating systems and different hardware, such that a virus that would debilitate one network would be unlikely to debilitate the others. This is "adding variety" to enhance system cohesion, see The Principle of Connected Variety on page 109. It would, after all, be foolish to concentrate all vital resources in machinery that used

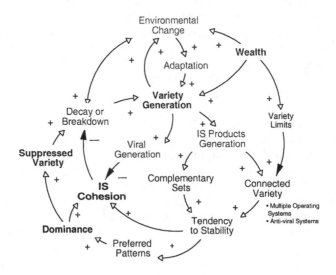

**Figure 9.17** *Information systems' life cycle map.*

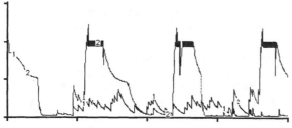

**Figure 9.18** *The impact of variety on IS cohesion.*

only one operating system; that would simply present an attractive target to an enemy.

An information systems (IS) life cycle map, as shown in Figure 9.17, shows how the issue may be viewed. This is based on the system life cycle map of Figure 6.3 on page 113. Wealth has been inserted in place of energy, viral infection instead of dispersive influences, IS products generation instead of interactions and reactions, and IS cohesion instead of systems cohesion. It is possible to simulate the IS life cycle map, and to explore the various interactions between wealth (the amount of money spent on IS implies energy), and the generation of viruses. Note that variety generation produces both viruses and IS products.

The first parameter to be examined is that of variety: in the IS context, this refers to operating systems, applications, and antiviral systems. As the life cycle map suggests, limits to variety are also limits to stability and cohesion.

The corresponding result of the dynamic, nonlinear simulation is shown in Figure 9.18. The x axis represents time and the y axis represents IS cohesion. There are two lines: line 1 employed very little variety; line 2 had a greater variety, but is still modest. The difference in system cohesion between the two conditions is remarkable. With additional variety, IS is much more cohesive. However, the periods when it is cohesive are interspersed with longer periods when its condition is poor. The low-variety alternative, on the other hand, might have poor cohesion and stability, but it is available most of the time.

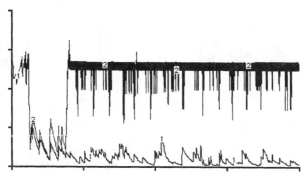

**Figure 9.19** *The impact of wealth on IS cohesion.*

If the amount of wealth entering the system is increased, the picture changes; see Figure 9.19. It makes little difference to the low-variety IS, line 1, but the higher variety IS, line 2, becomes permanently available and stable–after a shaky start. The perturbation at the start is caused by the time it takes, both within the model and in the real world, for all the parts in a higher variety system to come together and form a stable group. Note that this high level of stability is based on a continuing process of injecting new wealth, creating and incorporating new products, and, in consequence, overcoming new viruses. The fine structure within line 2 represents the impact of new viruses being attacked and overcome. One way in which the increased variety could be utilized is to detect and isolate virally infected parts or regions, leaving the rest of the system to operate normally while the infection is eradicated.

It is possible within the simulation to increase both the level of wealth injected, *and* to increase the level of viral attack. The result is shown in Figure 9.20, from which it can be seen that even the wealthier, more sophisticated system with its greater variety, is unable to fend off a determined attack completely. As in the human body, it is possible to be overwhelmed by viral infections. This should not surprise us. There is no such thing as perfect defense; the initiative is always with the attacker–nowhere more so than in the creation of new ways to attack computers. In the end, if increasing the injection of wealth and increasing connected variety fail to overcome the aggressor, then it will be necessary either to change the IS completely for something that is virus proof, or to neutralize the aggressor.

Analysis using the system life cycle theory and the life cycle map may seem abstract. Certainly, it does not produce hard, dependable numbers. It does however, get to the heart of complex system behavior and, as such, is able to tackle intractable issues where other approaches fail. In this case, for instance, it clearly suggests that the best bet in the short to medium term is not to imagine that we can somehow protect ourselves from viruses, but to build more varied and better protected IT and IS. In this instance, at least, it seems that money works. In

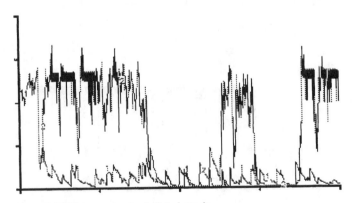

**Figure 9.20** *IS: increasing level of viral attacks.*

the longer term, the answer will be found in IT and IS that cannot be infected. The key weakness in current systems is that the instructions that tell computer hardware what to do–software–are capable of being rewritten in the machine. Software, one of the supposed strengths of such systems, turns out to be a major weakness.

## CRIME MANAGEMENT

### Crime and Punishment

Crime is invariably an issue in any democratic society. Freedom, the inalienable right to do whatever you want provided you do not interfere with the similar rights of others, seems to be open to abuse.

Crime comes in many guises: from noisy neighbors, to multiple killings, to polluting the environment, to...; the list is endless. Some crimes are considered by society to be more heinous than others, and are punished more severely. Lesser crimes may attract lesser punishments, unless the offender persists, in which case he or she will face escalating penalties.

This assumes, of course, that penalties deter a criminal; there is little evidence to support such an assertion. Instead, it seems that those about to perpetrate a criminal act either do not care about potential punishment, or consider that they will never be caught, and–again–do not care. For a few, the greater the risk, the greater the buzz.

A very simple model of crime and punishment is shown in Figure 9.21. There is a general population that includes lawbreakers–some of whom are awaiting trial, while others are in jail. People may move between the various categories. The model explores the notion that locking up persistent offenders for longer times may reduce crime on the streets. This appealing idea is based on the rather

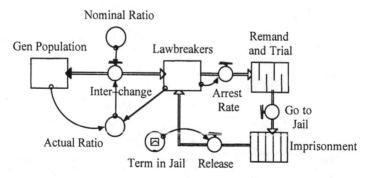

**Figure 9.21** *Crime and punishment model.*

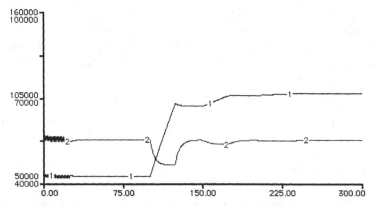

**Figure 9.22** *Increasing the imprisonment term for persistent offenders.*

obvious notion that a repeat offender cannot be offending while in jail. So, the argument goes, lock up repeat offenders for longer times (e.g., "three strikes and you're out").

In the model, released prisoners return to the group of people in society more likely to offend–the so-called "university of crime" phenomenon. While it might be thought that this is cynical, it also seems to be generally true; otherwise there would be no persistent offenders.

Results from the simple model are counterintuitive; see Figure 9.22. When the average jail term is increased from 24 to 48 months, the initial effect is a sharp drop in the number of lawbreakers, line 2, as the persistent lawbreaker is taken off the street for longer. The number of persistent offenders to be housed, line 1, doubles in direct proportion to the doubling of the sentence. The cost of housing prisoners therefore also doubles–at least, since there will probably be an associated demand for the building of new prisons.

Unfortunately, the number of lawbreakers, line 2, then starts to climb back to

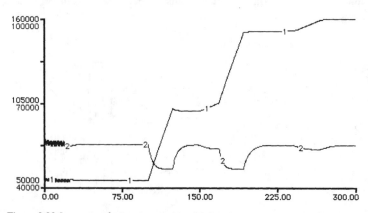

**Figure 9.23** *Increasing the imprisonment term twice.*

its original value. This happens because locking up an established villain leaves a "job opportunity" which is gradually refilled from the ready supply of "wannabes" in the general population. Outcome: the same crime rate, and double the prisoners in jail.

There is a risk that politicians, seeing the initial drop in the number of lawbreakers and hence in the crime rate, will decide that their policy of increasing jail sentences is effective. They may then increase sentences even more, before realizing that the crime rate will always rebound to its original level. Potential results are shown in Figure 9.23, where the imprisonment term was raised initially from 24 to 48 months, as before, and then from 48 months to 72 months. In this way, a society can end up with a large proportion of its population in jail, at great expense, and with no consequential benefit in terms of a reduced crime rate.

## Proactive Policing

If locking up persistent offenders for longer is not the solution, then what is? One consistent political mantra is: "tough on crime; tough on the causes of crime." Politicians vie with each other to emphasize their willingness to be tough, but successive administrations nonetheless face the same crime issues; either unable to agree on what causes crime, or unwilling to act for fear of unpopularity.

The police suggest that, given more manpower, they can catch more criminals. They may be right, but there is more to policing than catching criminals. In fact, the very first metropolitan police force in London, England, was set up by Sir Robert Peel in 1829 to prevent crime, not to catch criminals [4].

Since that time in the 19$^{th}$ century, much has changed. Laws have changed so that it is generally necessary for an individual to have committed a crime before the police can act. While this may be admirable from a personal freedom viewpoint, it means that the damage has to be done first–before police action. That has led to a general air of fear among the young, the infirm, and the old in particular. In many societies, social atomization has occurred, with people barricading themselves into their homes; afraid to go out on the street, especially at night. Social cohesion has decayed, and the quality of life for many citizens is poor.

A policeman's view of the situation might appear as in Figure 9.24: a causal loop model. From this viewpoint, society may be seen as composed of ordered and disordered citizenry, where disordered has a broad definition. An ordered society is one in which people socialize, go about their lawful business, become members of social groups, religious groups, and clubs, go on family holidays, attend sports events, and so on. Essentially, an ordered society is one in which each and every individual is linked and connected to others in a variety of relationships that regulate and stabilize social behavior. See The Social Genotype above, and Social Capital on page 392.

As the figure shows, low-level disorder occurs when people move between areas, so that there is a transient population element, with few relationships and hence limited stability. Disorder also occurs where people live in isolated fear; not joining in with social activities. The central causal loop indicates that crime originates more often from within a disordered society. Antisocial activities increase, encouraged by an atmosphere of criminal behavior. These activities, together with continual media reporting of crime, engender fear among the vulnerable, leading to greater disorder and atomization.

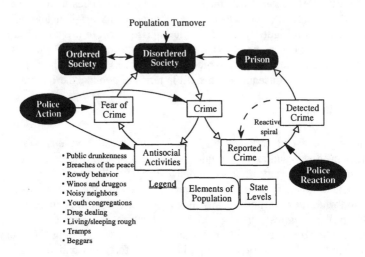

**Figure 9.24** *Police and society.*

Crime leads to reported crime in the figure, and from there to detected crime. Police have observed a so-called reactive spiral, in which greater success in detecting crime leads to a greater willingness on the part of the public to report crime, demanding more police resources. The price of success is high. As the figure goes on to show, detected crimes lead to convictions and prison, from which criminals emerge to rejoin disordered society.

Police concern themselves, very properly, with quality of life [5]. There is a direct connection between crime, disorder, fear, and quality of life, which is evident in the willingness, or not, of neighbors to help each other; see Figure 9.25. The figure also shows one approach to police task allocation: reacting to crime has first call on manpower, with whatever is left being allocated to reduce disorder, to reduce fear (particularly of crime) and to support neighboring and victim helping.

The hope and expectation is that raising the level of neighboring and victim helping will reduce fear among vulnerable citizens, which will reduce both disorder and crime, believed to be an escalation of disorder. As the causal loop model shows, once the process starts to work, fewer officers should be needed to react to crimes, leaving more officers available for controlling disorder and reducing fear.

The police employ technology in an attempt to cope with rising crime trends. Fast cars and radios are their mainstay. However, it has been calculated that the chance of catching a criminal at the scene of a crime by responding [6] to a 911 call is some 3%, rising to about 5% if the journey were made at the speed of light! One of the reasons for this are threats by potentially violent criminals against anyone who phones 911 until the criminal has left the scene; typically, 911 calls may not be received by the police until some 20 minutes after a crime has occurred. Whatever the reason, the police as currently deployed, even with their technology, find it difficult to catch criminals in the act.

So, the police are inhibited from acting before the crime, and are generally too late to catch criminals immediately after the crime. What, then, is the case with subsequent crime clear-up rates? Some figures from the U.K. give food for thought. In a typical county area, reported crime clear-up rates were about 20%. Estimates suggest that only some 25% of crimes were reported, making the net crime clear-up rate only some 5% (that is, 20% of 25%).

If these various factors are put together, the implications are evident. We do not live in a de facto free society. Instead, we live in a society where crime goes largely unsolved and unpunished, and where many law-abiding citizens go in fear, or do not go at all. Politicians are either unwilling or unable to rectify the situation and restore the quality of life that many older people nostalgically recall from their younger days.

How can the police act in such circumstances to restore the quality of life for the law-abiding citizen? There is no single answer; no magic bullet. However, it is possible to conceive of a sensible approach that would improve things, yet exist within the current legal framework, in a free society.

If criminals cannot be caught, then perhaps they can be deterred. One way to deter them would be to increase visible policing, by having more officers on foot, in uniform. Experiments have shown the direct effects on crime from such patrols to be limited. However, patrols need to be widespread, forming a stabilizing "web" over communities, towns, and cities, with uniformed officers on active patrol at intervals of, say, 100 meters in metropolitan areas, able to assist each other in seconds should the need arise. Moreover, aimless wandering would give way to an intelligence-led patrol, with officers knowing their territory, knowing

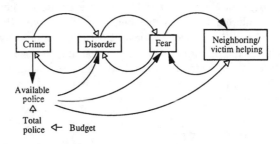

**Figure 9.25** *Crime and fear.*

their citizens, knowing what is normal behavior, and knowing "who is doing what to whom" in their area.

Intelligence-led patrolling is proactive, whereas reacting by car to 911 calls is reactive policing. Patrols on foot enable the officer to become familiar with everyone on the patrol route: the good, the bad, and the ugly. The visible officer reassures vulnerable people, while potential criminals are deterred, or elect to operate elsewhere; if the web is wide enough, there is no "elsewhere." In addition to reactive and proactive policing there is, potentially, a third kind of policing: problem-solving [7] policing. Here, the police cooperate with social services, youth groups, probations services, and many others to solve the kinds of problems, such as social deprivation, that lead to disorder, corruption, and crime.

The three levels of policing may be compared to the three stages in United Nations (UN) Peace Operations:

| *UN Peace Operations* | *Civilian Police Operations* |
| --- | --- |
| Peacebuilding | Problem-solving policing |
| Peacekeeping | Proactive policing |
| Peacemaking/peace enforcement | Reactive policing |

The UN peace operations terms at left reasonably describes the principal roles of the police, at the right. There is an implied sequence, too, from bottom to top, which is the same in each case. Reactive policing restores order, where disorder describes anything that is disturbingly different from the norm. Proactive policing maintains the state of order, preventing outbreaks of disorder by being "on the spot" to nip any disturbance in the bud. Finally, problem-solving policing addresses and resolves root conflicts and potential sources of discord and disorder.

The whole policing approach[5] might be arranged as shown in the systems diagram in Figure 9.26. At the center of activities is a computer-assisted dispatcher, who allocates officers to tasks in real' time. Some officers are on proactive patrol/peacekeeping duties, level 2; some on problem solving/peace-building duties, level 3, and some are reacting to events, peacemaking, level 1.

The dispatcher is a co-ordinator rather than a controller. Officers have a so-called menu of prearranged tasks to perform, in any order, as their term of duty unfolds. An officer might be on patrol when an event occurs nearby, to which he or she, may be alerted by the dispatcher. The officer will divert to the event; switching seamlessly from level 2 to level 1 activity. On the way, the intelligence officer may brief the attending officer. In attending the event, the officer may question witnesses, and gain crime intelligence in the process that will be passed on to crime intelligence, which may have already warned the attending officer of any threats he or she might face at the event scene.

---

[5] The approach outlined before was developed in co-operation with Ian Beckett, a respected senior policeman recently retired from the post of deputy chief constable with one of the counties in England.

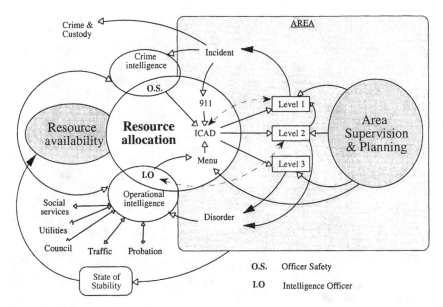

**Figure 9.26** *Policing systems diagram.*

Note how the systems diagram is symmetrical with respect to a horizontal line through the center, with crime above the line, and so-called operational policing below the line. Police resources are able to move smoothly from one half to the other and back on a minute-by-minute, hour-by-hour basis, as the situation dictates. This maximizes the effectiveness of police resources.

**Police Test Bed**

The systems diagram may provide the basis of a nonlinear dynamic simulation; see Figure 9.27. In this instance, the purpose of the model is not to solve a problem. Instead it forms a training tool for police officers–senior as well as junior–to try their hand at allocating officers in the face of variable and mounting crime rates. The officers they have available are indicated at lower left.

Officers are notionally allocated to levels 1, 2, or 3 using the chained sliders, top left. (Chained sliders will not allow the sum of officers allocated to exceed that available.) Actual numbers on the job are shown at center top, where PC indicates police officer. At top right, the trainee manager may select whether the patrol officers are to be on foot, in vehicles, or waiting in the station. Below are buttons to select either a full-time or a part-time intelligence (India) officer.

Dials at bottom right advise the operator of the developing situation in terms of crime. The state of stability is a measure of social entropy, with high entropy/disorder equating to low stability. Detection is the rate at which officers are detecting crimes and processing the offender. Crime is the rate at which crimes

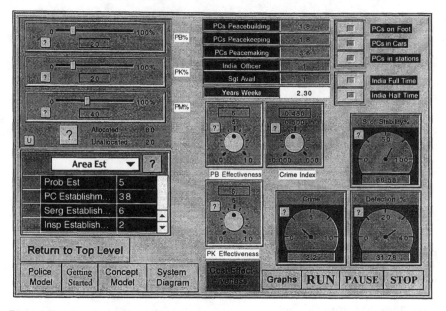

**Figure 9.27** *Police management trainer control panel.*

are occurring. The dials have warning levels: if crime rises above a preset level, the dial changes color and flashes to warn the operator that action is needed.

The control panel fronts a societal simulation, not unlike that of Figure 9.24, which employs genuine crime and population statistics, gathered over many years. There are also simulation modules that represent the likely effects of peacemaking and peacebuilding, using the experience of police officers as a guide.

The resulting test bed allows officers to try many different ways of restoring and maintaining order over a 20-year period, during which there is an underlying gradual rise in crime, which can be controlled only by peacebuilding; level 3. That is a slow, resource-consuming activity, however, and resources are needed in the shorter term for level 1 and 2 activities.

Also running during each operation of the test bed is a cost-effectiveness model. This calculates effectiveness as the sum of crime detection performance at level 1 and stability enhancement at levels 2 and 3, divided by the cost of the officers deployed on these various activities. This helps managers to assess the various strategies that different operators may devise to achieve order and stability.

The general approach is cautious. It would be unfortunate to introduce a new and untried system of policing, only to find that it did not work or, worse, that it exacerbated current difficulties. The supporting simulation is holistic, synthetic, and organismic, but it represents some possible future where there is no clear right or wrong. The best that can be hoped for is that such test beds inspire professional policemen to try new approaches in simulation before reality, and that in so doing they gain a better understanding.

## A Different Approach to Policing

Policing in a democracy is a balancing act. The simulation shows that increasing the number of police on duty at any time would increase crime detection rates, reduce crime rates, and enhance social stability and social cohesion. Higher police numbers would need to be maintained to secure the new status quo. The simulation also indicates the likely cost of such increases, which would easily double or triple the numbers of police in some areas. Objections can be seen on two fronts: "overpolicing" and cost. Some sectors of society would object to greater numbers of police as inhibiting, rather than guaranteeing, freedom. There would also be objections to the increased cost, which would surely fall on the shoulders of the citizenry as increased taxation.

Figure 9.28, a causal rather than a systems model, explores the likely costs of increasing the number of proactive police patrolling the streets, picking up intelligence, and so on. Instead of just looking at the cost of the additional police, however, the figure also considers the costs of prosecuting lawbreakers, and the cost of maintaining inmates in a penitentiary. If increasing the number of police deters lawbreakers and hence reduces the numbers in jail, then perhaps the reductions could help to offset the cost of the additional police. (This notion allows that society would consider and allow such accounting practices, where reductions in one area of social cost could be offset against rises in another area; this would not be considered normal accounting practice. Accounting practice is highly reductionist.)

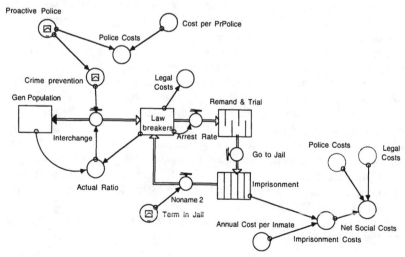

**Figure 9.28** *Exploring the social cost.*

The counterintuitive results are shown in Figure 9.29, which should be compared with Figure 9.21 and Figure 9.22. Figure 9.29 assumes that the policy of increasing prison terms for persistent offenders has been implemented, but in addition there has been a progressive increase in proactive police officers on intelligence-led patrol. The left-hand chart in Figure 9.29 shows lawbreakers at line 1 and prison inmate numbers at line 2. After rising sharply because of the increased jail-term policy, the number of inmates drops, due to the increasing numbers of officers on patrol deterring would-be lawbreakers. The numbers of both lawbreakers and those incarcerated drops eventually below the numbers at the start, even though the policy of increased jail terms is maintained.

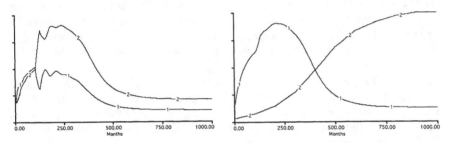

**Figure 9.29** *Social costs of increasing proactive police numbers. Left chart: line 1, lawbreakers; line 2, numbers imprisoned. Right chart: Line 1, net social cost; line 2, numbers of proactive police officers.*

The chart at right shows the net social cost at line 1, and the numbers of proactive police at line 2. Although the numbers of proactive police have trebled, the net social costs have been reduced dramatically. This is, of course, due to the reduction in the costs of prosecuting alleged lawbreakers, many of whom do not go to jail, and the reduced costs of housing fewer inmates; these factors more than compensate for the increased cost of policing.

There are, of course, many other social costs that have not been addressed. The cost to society of having up to about 400 per 100,000 of the population in prison is more than just the cost of maintaining the prisoners. While in jail, they have no opportunity to contribute to society, and, no opportunity to work, be paid, and pay taxes. Meanwhile, their families are left to fend for themselves, their teenagers get into trouble with the law, and a repeating cycle of crime and violence establishes itself in a social genotype. Perhaps the establishment and maintenance of social order is a subject that should be addressed as an overall system. At present, the police, the courts and the legal system, prisons, probation, and so on, are often viewed politically as mutually independent, rather than as parts of one system for maintaining social order.

Perhaps, after all, democratic societies get the level of policing they deserve, or at least are prepared to pay for.

## DEFENSE ACQUISITION

### Cold War Inheritance

Defense procurement is an interesting, if sensitive, subject, surrounded by convention, regulation, security, and even espionage. The following treatment eschews such ideas, and "thinks systemwise" about the process by which nations procure from industries that manufacture the defense components and the systems.

Classic Cold War procurement strategy was based largely on intelligence. Given sufficient information about the enemy it was possible, in principle, to: identify potential shortcomings in our own capability, and specify a solution that would plug the gap and restore own capability to be on par with, or even better than, that of the enemy. This so-called operational requirement was translated into an equipment or system requirement for industry to build.

It did not work well. Defense intelligence was invariably incomplete; it was considerably wide of the mark, too, as the collapse of the Soviet Union showed. Moreover, the process of developing operational requirements and imposing them on industry was, and is, challenging. Combat during the Cold War was rare; that was, after all, why it was a cold war. Defense procurement expertise was provided by military operators who were themselves operating 15 to 20-year-old equipment designs; were increasingly dependent on exercises for surrogate fighting experience; had little contemporary technology understanding; and were therefore poorly placed to conceive new tactics, using new technology in future conflict.

Complex operational and equipment characteristics were, and are, not

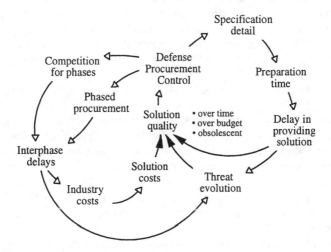

**Figure 9.30** *Ineffective procurement control.*

amenable to straightforward, consistent, precise, complete text description by industrially naïve authors. Defense industry engineers who read such specifications lacked military backgrounds and experiences, so the defense context in which textual specifications needed to be understood was missing. As a result, requirement specifications grew in length and deepened in specificity, taking longer and longer to prepare in ever more detail. During this time the need evolved, resulting in "carefully specified obsolescence;" that is, solutions were out of date before delivery.

The situation is presented in Figure 9.30. Western governments also introduced phasing, so that major defense projects comprised several discrete phases; this enabled customers to vet progress at project feasibility and definition stages in particular.

Competition was introduced to improve control of escalating costs; unfortunately, each phase of a project was subject to competition, and had to be undertaken by a different contractor. The competition for phases took time, sometimes years, during which time industry either had to maintain project teams at great cost, or disband them and lose the knowledge and expertise they had built up. Defense projects would typically take over 20 years from concept to first delivery, and the weapon systems might then last in service an additional 20 years or more. Military personnel could well find themselves in the front line, operating systems that were designed over 40 years previously.

**A Changing World**

Defense procurement since World War II had been characterized by applications of successive layers of government control over the defense industry. It is reasonable to suppose that tighter control of requirement specifications, budgets, contracts, schedules, milestones, payments against progress, and so on, would result in predictable project outcomes. Reality has been counterintuitive: ever tighter control has led invariably to escalating cost and timeframes, suggesting that defense procurement is a "nonlinear dynamic" procurement system, where cause and effect are hard to unravel.

The importance of a robust defense industry to a nation's economic and political stability should not be underestimated; see Figure 9.31. Producing effective weapon systems requires advanced industrial technologies, which raise and maintain a nation's technological capability. Possessing effective weapons makes a nation an attractive international partner in cooperative defense pacts. As the figure shows, a solid defense industry therefore contributes to national pride and confidence, economic progress and political stability. In effect, it qualifies a nation as an international player. There is all the more reason, then, for nations to exercise ever-closer control over defense procurement–even though control can be seen not to work.

Global changes threaten the defense status quo. Societies are fragmenting along old fracture lines. Commercial industries are being revolutionized by

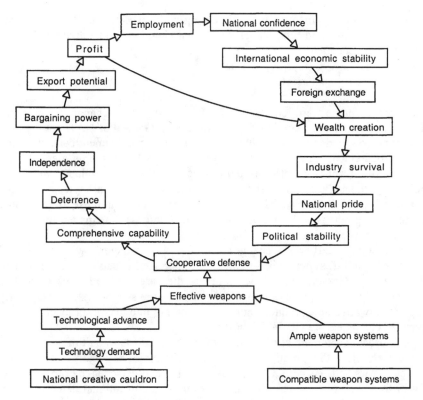

**Figure 9.31** *Importance of a defense industry.*

Japanese-inspired agile lean volume supply systems to produce high-quality, low-cost, consumer products: automobiles, personal computers, domestic electronics, white goods, and many more.

Whereas the defense industry was at the forefront of technology, now commercial industry is the major supplier and consumer of high-performance, high-reliability automobile, electronic, and processor goods.

Cold War predictability has given way to high levels of uncertainty in defense. International policing appears to be a significant future military role; usually as part of some international force, perhaps with former enemies as new allies. Nuclear proliferation continues, with India and Pakistan squaring up, and China in the wings.

Recognizing the rapidly changing world, the U.S. defense industry was obliged to introduce commercial lean practices to reduce costs and procurement timeframes. U.S. defense acquisition reform objectives include: emulating the phenomenal success of commercial volume supply, and reducing the U.S. defense tax burden–sound politics, and economics at any time. Reform tactics include the creation of super-aerospace companies that are able to afford their own defense research and development (R&D), dispensing with restrictive military standards and specifications, and introducing a so-called single process initiative (SPI) to streamline procurement.

## U.K. Defense Procurement

European countries were left standing by the speed of U.S. change. In the U.K., defense minister Lord George Robertson (Secretary General of NATO at the time of writing) responded in 1996/1997 by introducing a procurement concept called smart procurement. The plan was to introduce lean volume supply methods into the creation of military platforms and products. While most products would still be largely specially designed and developed in the traditional style, one of the cornerstones of the new policy was the procurement and use of available commercial products and systems in operational defense systems and platforms. This was intended to cut the procurement cycle time.

The commercial world changes very rapidly, as anyone with a personal computer or a hi-fi system will be aware. Commercial products have a half life on the order of 18 months to two years before they are outdated. This is vital to commercial viability, which requires that lean volume supply systems constantly supply innovative new products; otherwise, they fall into disuse. Products such as personal computers are viewed as consumables.

To procure commercial-off-the-shelf (COTS) products as part of a solution to a defense project, then, requires that procurement cycles be ideally less than two to four years; otherwise, COTS products will be superseded before delivery. Every three years or so, hardware and software in an operational defense system need to be updated or replaced. Experience with introducing new operating systems and updating software programs suggests that, while this may cause glitches, it also enables systems to be updated frequently and to continually improve in capability and performance. The idea of continual update is, however, alien to both defense procurers and to military logisticians and maintainers. It is also difficult to reconcile with former defense procurement cycles of 20 years or more.

On this basis, smart procurement should aim to reduce platform procurement times from about 21 years to two to four years. Is such a reduction feasible? Clearly not, by holding with the outdated paradigm of phased procurement; with delays to allow competition for each phase. Projections suggest that streamlining and eliminating some phases might reduce the procurement cycle from 21 to 14 years; projections are educated guesses, however. Radical change appears to be necessary if smart procurement's original—and laudable—aims were to be even approached.

Two possibilities come to mind:

1.  Eliminate phases altogether; simply provide industry with an objective and leave it to produce the result. This requires:

- Government trust and "hands off" during design, development, and proving.

2. Switch to the alternate way to buy a solution: go out into the marketplace and see what is already available, as one might do with a new automobile, computer, industrial drilling machine, office complex, and so on.

- This is a completely different approach, familiar in other fields of endeavor.

Eliminating phases in bespoke procurement[6] equates to customers providing a requirement at the start and trusting the contractor to produce the goods to order some years later.

Procurers are mandated to safeguard public money, however. Releasing tranches[7] of government money against the evidence of visible progress, exercises control over industrial projects. Doling out money in tranches will cost more in the long run, because the repeated demonstration of progress seriously interrupts and delays industrial work flow, but the drive to demonstrate "being in control" is, it seems, paramount. The concept of trusting contractors, even when a firm price has been agreed beforehand, is apparently incompatible with the mandate.

In any event, without taking other measures, eliminating phases is unlikely to reduce the procurement cycle to less than 10 to 11 years–still a long way from the two-to-four-year target.

The alternate approach suggested above, which we will call total systems acquisition (TSA), is to buy defense products and systems from a national or international market, as we buy cars and washing machines. A robust market implies international sources. If governments do not buy from their indigenous defense industry, how are those industries to survive? Once freed to compete in an open defense market, defense industries will become much leaner and meaner through competition. Once freed, they can form agile, lean volume supply associations, nationally or internationally.

How could defense customers and users integrate and maintain different systems purchased from an open international market? Systems would be designed to accommodate differently sourced products, just as computers accommodate plug and play, variously sourced motherboards, and so on. Conventionally procured platforms would be equipped with suites of COTS avionics systems, COTS weapons, or COTS data communications, for instance; each of these may interact with other, conventionally procured systems, to which they would be only loosely coupled. In the final analysis, the whole platform

---

[6] The term "bespoke" is taken from the world of tailoring, where a made-to-measure suit is "bespoke" as opposed to a ready-made suit, which is "off-the-peg."

[7] The term tranche is used to describe a portion of an overall amount in financial terms.

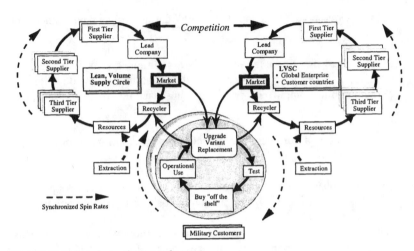

**Figure 9.32** *Defense lean volume supply system.*

might be procured as a COTS solution to a defense need. In this event, the platform procurement cycle could indeed fall to four years or less.

What if some of these COTS products were subject to a continual, commercial upgrade to both software and hardware? This would place special responsibility on the operational organization to act as a "consumer" of "consumable products," with all that implies in terms of logistics, maintenance, and keeping operational systems consistent–so that operators can switch instantly from platform to platform as needed.

If future smart procurement is to take advantage of the potential for agile lean volume systems to supply faster, better, and cheaper, then operational users of defense equipment will be seen as purchasers of defense consumables. Operational user consumption rates and patterns will be matched to lean volume supply patterns of provision: see Figure 9.32.

The figure shows three circles, or causal loops. The upper two are competing lean volume supply systems, formed into circles by the process of recovering and recycling obsolescent systems previously supplied to military customers, seen in the bottom causal loop. Each supply circle comprises a lead company, with several tiers of suppliers in support, offering goods in a marketplace. Military customers shop in the market for competing COTS products; some as complete new systems, others as upgrades and variants to existing systems.

Examination of the figure shows that the spin rates of the three circles have to be synchronized; otherwise, either the supply circles will not supply sufficient products to maintain profitability, or the military will not buy enough of the supply circle output. To facilitate spin synchronicity, the lean volume supply systems would need to supply upgrades, variants, and replacement products that were so innovative and attractive that the military customer could not afford *not* to buy them.

Total systems acquisition incorporates: competing supply systems, a market, multiple military customers, and multiple commercial customers (not shown). Defense is generally a smaller customer (e.g., for electronic goods) than the commercial market, so commercial customers will be necessary to establish and maintain a steady throughput of products from each lean volume supply system. This will increase the wealth of the suppliers to the point that each can finance its own R&D, thereby reducing the defense tax burden further.

Lean volume supply circles (LVSCs) need not be constituted from one large aerospace company. Instead, each tier may be made up of different companies. The lead company, instead of buying supplier companies, may "seduce" them by offering a business deal they cannot refuse. The lead company organizes supplier companies and provides communication systems, at its own expense, such that the whole chain, or loop, becomes a single, connected system. Nor need the supplier companies be exclusively national organizations; offshore companies with appropriate technology, skills, and cost structures could be suitable, too. In this way, the makeup of lean volume supply nations within TSA might reflect major defense alliances, binding the alliances economically as well as militarily and politically.

Although employing different tactics, TSA would achieve the same objectives as the U.S. defense acquisition reform program. Optimizing the complete TSA system would: build national wealth and reduce the national defense tax burden, enhance standings in both world politics and economics, supply armed forces with the latest technology, make that latest technology affordable.

National security would be an issue, and sensitive products would likely be developed in secure premises. For many products, however, the sensitive features could be contained in electronic memories or programs; products can be manufactured using dummy data and programs, with sensitive material being supplied only when the product is in operational use.

TSA can be applied to individual defense products, to complete fleets, or to anything in between. It is possible to start small and work up, taking just a few COTS products at a time, or it is equally possible to go "big bang." Third world countries have little choice. Their defense procurers tour world markets in the shape of exhibitions and firepower demonstrations, see what they like, test it, and then buy it.

Western defense suppliers may find themselves supplying third world countries with systems that are ostensibly more capable than those supplied to their own armed forces. This frequently happens where, for instance, a new sensor has been developed and fitted to a platform (tank, ship, plane) that is for sale to approved third world buyers.

## COTS Procurement Behavior Diagram

Figure 9.33 presents an instance of a behavior diagram for COTS procurement. The behavior diagram illustrates how, in principle, a complete bureaucratic structure could be set in place to conceive defense capabilities and to procure some, or all of that capability from the commercial sector or, for that matter, internationally.

There are, as might be expected, some sticking points that the behavior diagram illustrates. It may not be possible to identify suitable components in the marketplace, as proposed in the diagram; they may not have been developed commercially because of technology limitations, development cost, security, and so on. Given a robust commercial defense market, however, the behavior diagram identifies a potential defense capability procurement strategy.

The example is illustrative, too, because it represents only one of several behaviors, or strategies, for procuring defense equipment. There is no absolute right or wrong in such diagrams. A behavior diagram may represent a socially or politically unacceptable practice or policy; it does so logically, rationally, and openly, however.

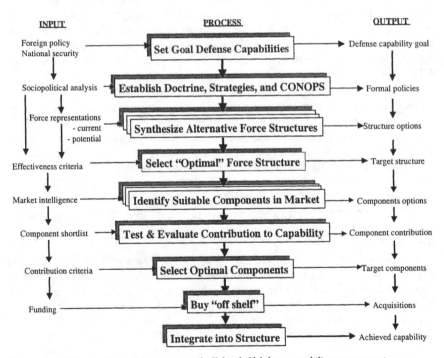

**Figure 9.33** *Behavior diagram for commercial off-the-shelf defense capability procurement.*

## BUILDING PYRAMIDS

Precisely how the ancient Egyptians built the Old Kingdom pyramids is not known. What is known is that the largest pyramid, the Great Pyramid of Giza, took some 20 to 23 years; it is still the largest stone building in the world, and the only extant wonder of the ancient world.

Without knowing precise constructional techniques, it is possible to work out the sequential logic; see Figure 9.34. Some things are obvious from the remaining evidence. The site was chosen with care: it was on a plateau near the river; the principal means of transport. There was an ample supply of high-quality limestone nearby on the plateau, from which the pyramid is largely constructed. The remains of a large causeway can be seen, up which provisions would have been hauled, together with granite from Aswan and white limestone from the Tura Mountains; used to encase the finished pyramid. The causeway had to be built early on so that heavy granite beams, needed in construction, could be hauled up.

The pyramid complex consisted of much more than the pyramid, although that still dominates the scene. There was a valley temple at the foot of the causeway to receive the dead king's funeral party arriving by river. Against the side of the pyramid, at the top of the causeway, was a second, mortuary temple, where the king would be worshiped during and after interment. There were smaller queens' pyramids to house the king's wives. Near the queens' pyramids was a small *ka* pyramid, associated with the cult of the king. There were solar boats–disassembled real boats buried near the pyramid–presumably to enable the dead king to sail on the Nile River, or to sail across the heavens by day and by

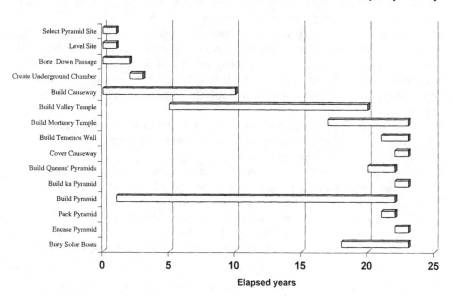

**Figure 9.34** *Pyramid complex: Gantt chart.*

**Figure 9.35** *The Giza Pyramids.*

night. A Temenos wall surrounded the pyramid, creating a numinous space. The causeway was eventually covered over, to serve as a dark passageway for the funeral party and priests to go between the valley temple and the pyramid.

Figure 9.35 shows the stones of the Great Pyramid at the right, with Khafre's pyramid partly revealed in the center: the top of this pyramid retains vestiges of the outer casing stone with which all the pyramids were originally finished. In the distance can be seen the smaller Menkaure's pyramid, and a tiny pyramid right on the horizon that is one of Menkaure's three queens' pyramids.

The valley temple of the Great Pyramid complex could not be completed while provisions and stones were being delivered to the dockside and hauled up the causeway. The mortuary temple could not be built until the pyramid had risen to a sufficient height, since the temple leaned against the pyramid. The causeway covering could not be completed until the mortuary temple was nearing completion.

The Temenos wall, which surrounded an area paved with marble slabs around the pyramid, could not be completed until construction gangs had finished, for fear of obstruction and damage to wall and paving.

The pyramid could not be packed[8] until built, nor cased until packed. And, the smaller queen's and *ka* pyramids would probably have waited until the bulk of the building was over, and manpower became available.

So, the sequence of events can be plotted, together with estimates of how long the various parts of the complex might have taken to build, knowing the overall timeframe. The resulting Gantt chart is not evidential; it is an educated guess.

---

[8] Pyramids were built in steps. To create the true pyramid shape, the steps were then filled in with packing stones, before being finished off with smoothed limestone (casing stones). Figure 9.35 shows the Great Pyramid packing stones.

It is also possible to speculate within sensible limits; see Figure 9.36. The central flow shows the early part of the conception process for the Great Pyramid: the introduction of new features (each pyramid seemed to incorporate experiences gained from building its predecessors), a supposed modeling process to explore new features prior to building in earnest, and the location of materials needed in the construction.

At left can be seen the various factors needed to initiate the central processes, and at right are the outputs from each central process. There is no hard evidence that the architects developed models before building the Great Pyramid, so the behavior diagram is based on reasonable supposition. Is this the way the ancient Egyptians thought, planned, and worked? Was their systems thinking like ours today?

It is tempting to think so, but for one problem. Using all our present day technology, it would be difficult, if not impossible, for present-day builders to replicate the Great Pyramid in 20 years. We rely on our technology to such a degree that it is difficult for us to comprehend how such an enormous undertaking could possibly be achieved without it. If we can reasonably view the ancient Egyptians as systems engineers, then we may still have a lot to learn about systems engineering.

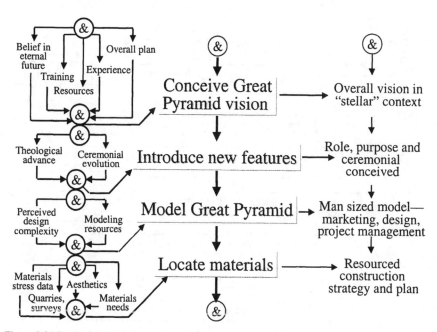

**Figure 9.36** *Pyramid design behavior diagram.*

## ARCHITECTONICS–THE STUDY OF ARCHITECTURE

Although various kinds of systems can be brought together in different configurations, there is, it seems, no science of architecture–or architectonics. We have seen earlier that systems may be classified by their architecture, but how does architecture contribute to a system's emergent properties, capabilities, and behaviors?

Figure 9.37 shows one view of the purpose of architecture. From this viewpoint, it is an active entity within a system, central to system design and operation, and would need to be designed itself, rather than allowed to "happen."

Layered architecture is widely used in many systems, and it is amenable to dynamic simulation. Figure 9.38 shows a notional diagram of building security, in which up to 10 layers can be distinguished. A layer is considered as an obstacle that a would-be intruder has to overcome. In the figure there is an external fence, with mounted sensors and an intended entrance. Beyond the entrance is a space; a so-called sanitized zone, where no unauthorized persons should find themselves.

In the diagram, weapons guard the sanitized space; more extreme security systems may couple the weapons to sensors, so that the weapons fire automatically when anyone or anything is detected in the sanitized zone. Beyond the zone is a wall with a door. Beyond the door is another space. A door in the

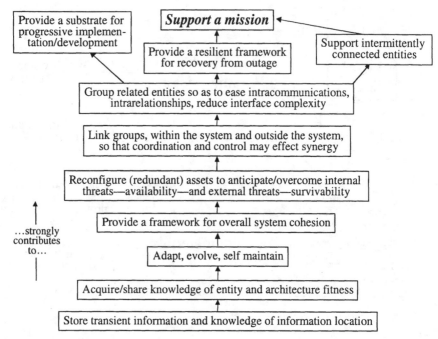

**Figure 9.37** *The role of architecture?*

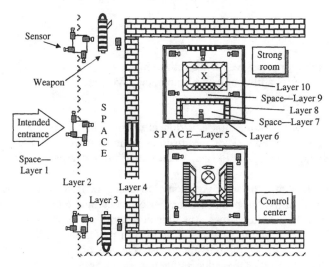

**Figure 9.38** *Building security as a layered architecture.*

wall defining the central space leads to a strong room with a u-shaped counter. Beyond the counter is another space, and finally there is a safe within which is the object to be secured.

Security seems sound. However there is a vulnerable point:

- There is a control center, where a security guard sits in front of screens that show situation and activity.

- The guard is able to grant or deny access through the remote locking and unlocking of doors and barriers.

- The guard's facility is a potentially vulnerable point: power supplies may be sabotaged, rendering the guard effectively blind and with no controls; sensors and controls may be intercepted, presenting the guard with a false picture, or taking over controls.

- The security database may be hacked, providing false credentials to infiltrators, and so on.

- The security guard may be subverted, bribed, family threatened, and so on.

In the final analysis, security often boils down to people rather than technology, but the technology helps.

Simple analysis of multilayer defenses reveals how we might produce effective, economical designs in modern systems; defensive and nondefensive. The following equation uses the concept of neutralization and assumes that each layer has an identical probability of neutralizing an intruder passing through the layer:

$$P = 1 - (1\text{-}p)^N \qquad\qquad (9.1)$$

where:  P is the expected overall neutralization;
        p is the probability of neutralization per layer; and
        N is the number of layers.

The results of plotting this equation are shown in Figure 9.39. The graph illustrates the so-called law of diminishing returns: increasing the number of layers draws the lines closer to each other, until adding another is of no apparent benefit.

Figure 9.40 shows the same equation plotted in 3-D, from which it is easier to see that equivalent overall neutralization performance may be achieved at several points, indicated by white arrows. The greater neutralization per layer reduces the number of layers needed to give the same overall performance. More layers also generally mean more complexity and cost. On the other hand, reducing the number of layers increases the vulnerability to losing a layer.

The assumption in the equation, that each layer has the same probability of neutralization, limits further analysis. A layered architecture simulation model, shown in Figure 9.41, consists of four identical layers, and more may be added as needed. Each layer consists of a reservoir at left, a conveyer representing the layer, and a second reservoir to record the number of attacks that were neutralized. The conveyer "leaks" successful attackers to the following layer, where the process is repeated. Each layer neutralizes a proportion of the attackers, but successful attackers penetrate to the next layer.

The test bed simulates 100 separate attacks, and shows which layer neutralizes each attack. The sum reservoirs count the neutralizations per layer,

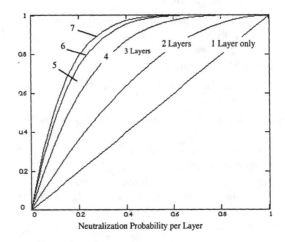

**Figure 9.39** *Expected multilayer neutralization.*

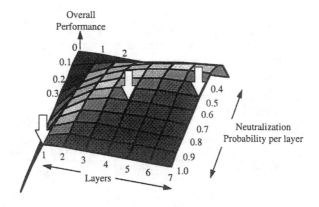

**Figure 9.40** *3-D view of neutralization equation.*

giving the answers as a proportion or percentage of the 100 attacks. The attack pattern, the times taken to traverse each layer, and the probability of leakage can be selected and varied for each layer.

As shown in the figure, the layers behave independently of each other. In defending, say, a medieval castle, that might not have been the case. Such castles, with their sanitized zones, moats, walls, baileys, keeps, and so on, had upwards of 10 to 12 discrete defensive layers.

If the outer wall were being scaled, then those defending the wall could be left to fend off the attackers. On the other hand, they could be reinforced using reserves from within the castle, or by redeploying those allotted to defend an inner

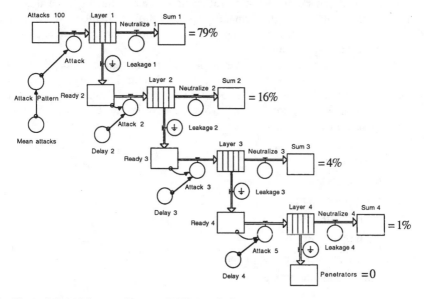

**Figure 9.41** *Multilayer architecture STELLA test bed.*

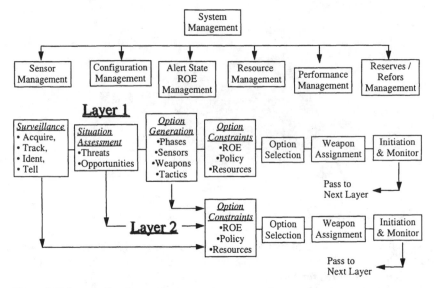

**Figure 9.42** *Layer leakage management.*

wall or bailey. To reinforce successfully during the heat and noise of battle required communications, discipline, and organization. Some castle defenders were able to redeploy men to good effect. Others waited until the outer wall was breached and then withdrew behind the first inner wall, sometimes leaving the overwhelmed outer wall defenders to their fate.

The test bed can be enhanced to reflect intelligent behavior, such that the inadequate upstream layer performance causes reinforcement of subsequent, downstream layers. As might be expected, tests show that this enhances the overall neutralization probability.

Figure 9.42, taken from early Strategic Defense Initiative concepts, shows schematically how multilayer defense command and control architecture might be developed to take advantage of layer-to-layer co-ordination. Running from left to right are the management controls for two successive layers. Each layer forms a process pipeline. In sequence, they are:

- Surveillance;
- Situation assessment;
- Option generation;
- Option constraints;
- Option selection;
- Weapon, or response, assignment;
- Initiation and monitoring.

The last four functions are repeated in following layers so that an intruder who survives the first layer is handed over to subsequent layers. Note how the command and control, or management, organization reflects the physical layering.

Running down the figure are the elements of management that relate to each and every layer:

- Sensor management–the optimum allocation of sensors to enable each layer, substituting good sensors for failed, poor, or damaged sensors, and so on.

- Configuration management–the rearrangement of sensors, defenses, layers, and so on, to accommodate shifts in threat, accommodate damage, use reinforcements, and so on.

- Alert state and rules of engagement (ROE) management–the ability to raise or lower the stakes; an essential feature of effective management, particularly in politically or socially sensitive situations.

- Resource management–to reallocate resources in real time according to the demands of the enemy, casualties, and so on.

- Performance management–monitoring overall performance, identifying strengths and weaknesses, redesigning architectures, organizations, retraining, and so on, and evolving improved performance.

- Reserves and reinforcement (refors) management–drawing upon back-up forces.

The figure sets the scene for architecture design. Configuration management, for instance, will require information about the progress of the threat, damage to sensors, loss of defensive weapons capability, and so on, and will require the controls to effect changes to overall system configuration. Similarly, other management functions will require sensors, processing, and controls.

There may be an opportunity to "double-up," that is, for some management functions to share sensors and controls, and therefore perhaps to share information and communications pathways. That would create a vulnerable point, so sensors, pathways, and controls either have to be independent, or replicated and independently routed, to avoid single-point failure or damage. Ideally, by knowing what each management function requires, the architecture should be able to reconfigure itself and heal itself if damaged, so that the mission may be pursued.

Architecture, then, is determined not only by mission and system functions needed to achieve that mission, but also by threat, damage, and failure; any or all of which may necessitate reconfiguration, damage repair, jury rigging, and so on. The design of fault tolerant, damage tolerant, and self-healing architectures is challenging, and requires modeling and simulation support.

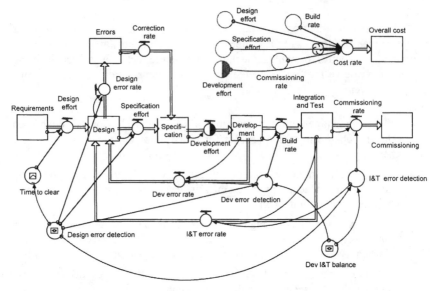

**Figure 9.43** *An elementary systems engineering cost model.*

## SYSTEMS ENGINEERING: THE COSTS OF REWORK

Rework is a concern in many manufacturing and systems engineering projects. Reworking may be triggered by a variety of causes: an ineffective process resulting in a substandard part; the inadvertent inclusion of defective parts, such that the product fails to pass tests; and, particularly, errors in the requirement that have not been detected until the process is under way. This last cause became of special interest when a tool vendor made the seemingly outrageous claim that, no matter how long was spent on eradicating errors in the requirements provided by the customer, the overall project length would not increase.

Prima facie, this is ridiculous. True, some customers supply extensive requirements documentation, written by several/many people, and, true, some of these requirements may be laced with inadvertent errors, contradictions, and omissions. But, surely there must come a point at which the process of finding and correcting these errors at the very start of a project becomes counter-productive, and it is more expedient to get on with the rest of the work, and sort the remaining errors out as and when you find them.

Experience shows that finding and eradicating errors in a large and complex statement of requirements is time consuming. Moreover, although some errors are obvious and can be sorted out easily, others are more intractable. As more errors are found, it takes longer and longer to find the more obscure errors. Experience also suggests that it may be impossible in practice to eradicate 100% of errors.

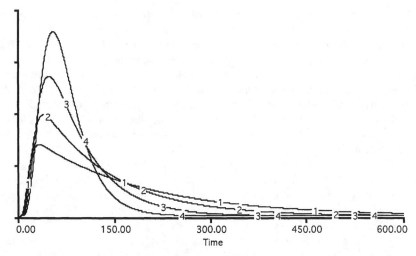

**Figure 9.44** *Systems engineering requirements commissioning.*

To disprove the proposition, a rudimentary STELLA model[9] was constructed (see Figure 9.43). The model shows a simple, straightforward series of processes: requirements, design, specification, development, integration and test, and, finally, commissioning. At the start is a process for detecting and correcting errors. There are also two feedback paths, one from development, and the second from integration and test. These represent rework caused by errors found in development, and errors found in the integration process, respectively. The model also had a simple accounting system, at top right, which accounted for the manpower and materials costs; the latter arising only once development is underway.

Controls are included, to vary the proportion of errors found during the initial design process, and to vary the proportion of those errors that escape the design net, and are then discovered either in development or in integration and test. Importantly, it is also possible to vary the amount of time it takes to clear each defect.

The model was set up so that 50% of defects would be discovered in one unit of time, 60% in two units, 70% in four units, 80% in eight units, 90% in sixteen units, and the theoretical 100% in 32 units of time. This square law was felt to be harsh, and therefore a sound test of the tool vendor's proposition.

The simulation executed by putting 1,000 requirements in the left-hand reservoir and seeing how long they took to clear into the commissioning reservoir, having gone around the rework loops–perhaps more than once.

---

[9] The model as shown is not a full systems model since it shows neither the source of requirements nor the end point of commissioning. It may be synthetic and organismic, but it is not holistic. However, in this application, the simulation model is sufficient to investigate a phenomenon, which is its sole purpose.

The results of the simulation are shown in Figure 9.44. Line 1 corresponds to an initial error detection of 70%, line 2 to 80%, line 3 to 90%, and line 4 to a theoretical 100%. Following line 4 initially shows that 100% error eradication at the start results in the commissioning of all 1,000 requirements by 240 time units; by far the shortest time. However, line 4 was the slowest to climb near the origin, in keeping with the additional time needed for error eradication. Lines 3, 2, and 1 were quicker at the start, indicating that a few requirements got through more quickly, but all took longer to complete. Indeed, line 1, the 70% line did not complete within the 600 time units shown, and extending the duration showed that line 1 reached zero at about 1,250 time units.

Figure 9.45 shows the equivalent results for the systems engineering cost model. Line 4, the 100% error eradication line, climbs quickly toward the horizontal, indicating that no more costs are being accumulated; that is, the project is complete. The other lines take correspondingly longer to reach greater overall costs. Line 1, the 70% line, clearly has some way to go.

So, the tool vendor's seemingly outrageous claim was justified: within reason, no matter how long it took to eradicate errors at the start, the overall project duration and cost would not increase; indeed, the simulation suggested they would fall. The reasons are evident: rework dominates the duration, and hence the cost, of such projects.

It has long been known, and repeatedly stated by experts for decades, that the costs of correcting errors after the start of any major project are prohibitive, and that errors should always be eliminated as fully as is possible at, or before, the start. Despite this, some project managers still find the urge to get going irresistible. (Some customers exhibit the same impatience, being eager to see tangible progress.) Perhaps project managers should develop their own test beds,

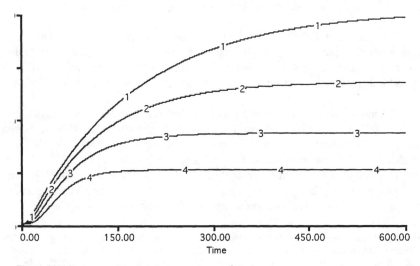

**Figure 9.45** *Systems engineering costs versus error detection.*

like the systems engineering simulation of Figure 9.43. Only then, perhaps, will they believe. Meanwhile, projects will continue to run over time and over budget....

## SUMMARY

The many and various tools and methods introduced in Chapter 8 have been illustrated in examples of systems thinking. Some examples have used only one tool, while others have employed several tools; strung together to form a method. Most methods are "handraulic;" that is, capable of being worked by hand without need of tool or machine support. Tools such as STELLA can be purchased for very little money, and they come in a variety of guises from a variety of suppliers.

The results of using such methods and tools is more rigorous systems thinking, and a sound, system-scientific approach to addressing problems and issues.

## ASSIGNMENTS

1. An intelligent building is proposed for introduction into a run-down business area of a major city. Opinions are divided: some say that it will bring much needed work to the local area, both in its building and in the creation of new and permanent jobs; others suggest that the new building will be a major eyesore and will seriously damage the picturesque skyline, which brings in a lot of sightseeing tourists. Identify six significant stakeholders-surrogates in the proposed new project, and undertake a stakeholder analysis.

2. The Roman Empire rose to magnificence during the first millennium, but it contained the seeds of its own eventual decline. Create a life cycle map showing the rise, establishment, decline, and eventual fall of the Roman Empire, giving reasons for state transitions.

3. Expand the doctors' surgery model shown in Figure 9.1 on page 182, to include up to 12 doctors, and make time allowances for clinics, emergencies, training, holidays, and staff sickness. Include a medical records department and reception; with staff. Without changing the number of patients to be received, identify how many full-time doctors are needed to provide an effective service, the number of records to be maintained, and the overall salary bill for the complete practice. Make sensible allowances for salaries and any other numbers you may need.

4. A major retailer of household goods, fashion, food, and so on, has gone into decline. Its fashions are no longer seen as fashionable; its household goods no longer have their previous high quality. Its food is still good,

but sales are down because of the high prices. Locate this organization on the life cycle map and, using the principles of life cycle theory, propose ways in which the organization can return to its former glory.

5. An electronics company is in some difficulty. The company's products appear to be sound and are well priced, but they are not selling well. Morale is low. There have been several initiatives to try and kick start the company, but none has been sustained. Many of the company's products in use with customers have incipient faults, which will cost the company dearly to repair. The various parts of the company–marketing, design, manufacturing, and customer service–seem to be working independently. Use RSM as a means to explore the issues and develop a potential solution to the company's difficulties.

6. The systems engineering model in Figure 9.43 is an analog model; that is, it does not work with discrete packets representing individual requirements. Expand the systems engineering model of Figure 9.43 as a paper model, making it discrete by introducing queues, conveyers, and ovens as you see appropriate. What difference would you expect your changes to introduce in the way the model operates, and in the graphical results?

7. Zero tolerance is a philosophy suggesting that larger crimes follow on from smaller crimes, and that if people are stopped from perpetrating the smaller crimes, then society may become crime free. A police force elects to introduce a policy of zero tolerance, such that even the slightest misdemeanor is prosecuted. Using the CLM technique, represent the effects of this new policy on: the rate of prosecutions; the number of police; the pressure on facilities such as cells, courts, lawyers, and so on; policing costs; and reactions by generally law-abiding citizens. Form and express a considered view on the efficacy of zero tolerance: a) as a short-term expedient, and b) as a permanent police practice.

8. Using Figure 9.44 and Figure 9.45 as source data, estimate the process entropy for 80%, 90% and 100% initial error detection in the systems engineering model of Figure 9.43, using the method outlined in Appendix A. From this work, develop a graphical relationship between entropy and rework.

9. The architecture of castles and prisons differs: one is trying to keep intruders out, while the other is trying to keep prisoners in. In each case, the first major obstacle to be scaled is the most daunting: the outer wall for the castle, and the inside fence for the prison. Develop a model, after the style of Figure 9.38, and use it to explore the effects of different layer leakage patterns. Should the outer castle layer be the hardest to breech, or should it perhaps be the inner and final layer? Justify your conclusions, taking account of psychological impacts on would-be intruders/escapees.

# REFERENCES

[1] Hitchins, D. K., "Growth and Collapse of Societal Power," *Systemist*, Vol. 12, special edition, 1999.

[2] Rice, Michael, *Egypt's Making*, London, England: Guild Publishing, 1990.

[3] Dyson, F., "Search for Artificial Stellar Sources of Infrared Radiation," *Science*, Vol. 131, 1960, pp. 1167-1668.

[4] Peak, Kenneth J., and Ronald W. Glensor, *Community Policing and Problem Solving*, Upper Saddle River, NJ: Prentice Hall, 1996.

[5] Ibid.

[6] Kansas City, Missouri Police Department, *Response Time Analysis*, Vol. 2, Washington, D.C.: U.S. Government Printing Office, 1978.

[7] Peak, Kenneth J., op. cit.

# PART III

# SYSTEMS

# ENGINEERING

# Chapter 10

## System Concept and Design

*'Tis a thing impossible to frame*
*Conceptions equal to the soul's desires;*
*And the most difficult of tasks to keep*
*Heights which the soul is competent to gain.*
William Wordsworth, 1770 - 1850

### SYSTEM SOLUTIONS

System concepts and designs find potential solutions to problems: identifying a viable solution concept is the first step in "systems engineering" a solution.

A problem exists within a conceptual problem space; a solution exists within a conceptual solution space. Within both those conceptual spaces, systems may be interacting with each other in their respective environments. A solution may result from rearranging the systems in the problem space, from changing interactions, from changing environments, or from deletions or additions; for example, of new or altered systems.

A system solution, then, exists within the context of other systems in other environments, with which it interacts. A system design exhibits purpose, or is purposive, that is, purpose can be ascribed to it. When a putative system design is "inserted" into a solution space, it purpose will present as effects on other interacting systems within the solution space. The solution to the original problem is to be seen in the sum of the effects, and any changes they may invoke in all the interacting systems; including the putative system design.

The following brief transport issue exemplifies this abstract process:

> *A transport problem exists between two industrialized towns. Roads are single lane highways, and there is a single-track railway. A solution is proposed in the form of a new canal, able to move heavy loads and freight. A canal system is conceived, taking water off a major river, passing through a lock system, and returning water to the river some miles away. The transport system now consists, conceptually, of three modes of transport: road, rail and canal. Heavy, nonurgent transport will switch–over time–from both the railway and the road to the canal; easing congestion on both.*
>
> *With reduced congestion, the roads in particular will become more attractive for local private cars, through traffic, and holidaymakers, so traffic levels will rise again–but*

*perhaps not to their original level. Railway commuters may temporarily switch to the roads while traffic levels are lower, but will revert to the trains; bringing them back toward their former levels, too. The canal, running through picturesque countryside, will attract pleasure boats and holidaymakers, especially in the summer; businesses will grow up along the river to serve the new trade. Heavy freight using the canal will be impeded during summer months, unless the canal has lock gates able to take at least three canal boats at once.*

The problem space encompasses the existing transport issues. The passage concerns a concept for an enhanced transport system between the two towns; in the process, it suggests design features in the proposed new canal. Note how: the road and rail systems change once the new canal system is introduced; the new canal system is affected by the changes in the utilization of the physically unchanged road and rail systems; and the environment of the new canal system also changes, with new businesses and new users. The solution space encompasses all of the change. And the design of the canal has not even been considered yet–this is conceptualizing only.

It is important to visualize a putative new system in its eventual operating context, to gauge its impact on other systems in the solution space, to gauge their reactions and interactions, and to anticipate reactions and reverberations against the new system, which may invoke particular design characteristics–as in the example of the increased canal lock capacity above.

**Creating Solution Concepts**

Conceptualizing is a creative, cerebral activity. It is also vitally important: from the concept will spring the design, the development, the product, the business, the industry.... Get the concept wrong, and the ramifications could be enduring.

A conceptual solution to a problem describes a way of solving the problem. You know you have a conceptual solution only when you can describe at least one other way of solving the problem. The process of conceptualizing a system solution involves:

- Maintaining the highest level of abstraction;
- Making as few assumptions as possible;
- Challenging any and all presumptions;
- Identifying obstacles to solution;
- Identifying alternative ways to overcome obstacles;
- Creating alternative solution "maps";
- Modeling alternative solution concepts dynamically; exploring: counterintuitive behavior, reactions from other systems, resource demands, and likely costs;

- Selecting the "best" conceptual solution, where best may mean any or all of: the simplest, cheapest, best quality, lowest risk, most appealing, most exciting, most needed, and so on.

Consider, as an example, the conceptual problem of preventing airborne intruders from entering national airspace. One conceptual solution might involve building and deploying a fleet of airborne interceptors, based on the coastline in the supposed direction from which intruders might appear.

Already there have been several assumptions; some evident, others tacit:

- What constitutes national airspace?
  - o Some airborne intruders might be equipped with stand-off weapons, capable of being launched several hundred miles from shore, beyond national airspace.
  - o Would it be legitimate to intercept the intruder aircraft, or only the stand-off weapon once it entered national airspace?
- Need the airborne intruders be aircraft? They could be unmanned aerial vehicles (UAVs), sea-skimming missiles, or even powered hang gliders carrying special forces.
- Presuming the direction of the threat is clearly a risk; a determined enemy might not oblige.
- There may be a tacit assumption that the interceptors are manned fighter aircraft, which need not be the case.

Making such assumptions has artificially constrained the solution space. The first task would be to postulate the full range of potential air intruders. Next would come feasible ways of neutralizing each and every element in the range of threats. Then would come optional conceptual solution maps, incorporating feasible ways of neutralizing all of the threats into competing solution concepts, but also including other, interacting systems. These might include, for example, scheduled passenger flights, hovercraft ferry services, light aircraft activity, coastguards, and so on. The optional solutions would be simulated dynamically to explore their respective implications in terms of feasibility, capability, practicability, risk, cost, timeframes, and so on.

At this stage, nothing has been designed. Getting the concept right first is paramount. A conceptual map showing one approach to neutralizing the threat is shown in Figure 10.1. The figure shows part of a concept map, using causal loops as the presentation medium. The main loop follows a conventional logic: search, detect, identify, allocate, engage, neutralize, and back to search; each element implies a system, for example, "a system for identifying." Also shown are the related activities of other parties: this is network centric combat [1], with a recognized air picture (RAP) being formed cooperatively using sensor information from all parties cooperating through a communications network. ("Recognized" implies that targets have been identified as friend, foe, or neutral.)

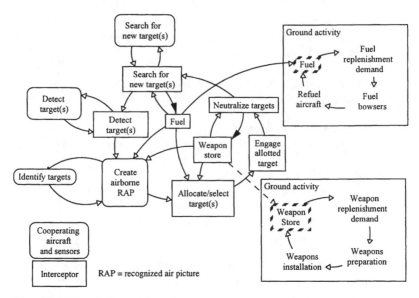

**Figure 10.1** *Conceptual map: interceptor.*

These other parties constitute systems that complement the interceptor system; that is, they are complementary systems. Looked at alternatively, all of these complementary systems, including the subject interceptor, form part of a higher level "system for intercepting intruders."

The figure presents certain assumptions; notably that the interceptor is replenished on the ground, that is, it is a reusable vehicle, not a fire-once missile. However, there is no indication that the interceptor is manned. The conceptual map is consistent with a solution in which an airborne command post (ACP) directs a variety of different, unmanned interceptors with the pilots–if any –operating the interceptors remotely from control stations, perhaps–but not necessarily–within the ACP.

Although the term "ground" has been used for replenishment, implying terra firma, might it also refer to naval carrier decks? Overall, this conceptual map shows only part of a solution concept, but it is useful in provoking ideas and in challenging assumptions. Because of the dynamic nature of the representation, the map provides the bare bones of a concept of operations: a CONOPS.

### Developing Concepts: The TRIAD Building System

The TRIAD Building System was introduced on page 161. It will be employed below to develop a concept for a new business or enterprise: this is a particularly simple example, intended to show how the technique works.

The first step is to develop a prime directive (PD), the ultimate statement of purpose for the new business. An initial PD might look as follows:

"To flourish in the domestic electronic market."

The PD is intentionally bare, to avoid preconceptions and assumptions. It also contains only one action verb; in this case "to flourish." A good PD operates at the highest level of abstraction; it identifies the sphere of endeavor, it is solution transparent, and, of course, it is the ultimate statement of purpose. All solution concepts and design concepts should be traceable exclusively to their PD.

The second step is semantic analysis of the PD. This is a technique of careful elaboration of each word in the PD to establish its scope and implication. The simplest approach is to create two columns: one with the words of the PD; the second with the elaborated version:

| | |
|---|---|
| To flourish... | To operate, survive, and make a profit... |
| ...in the domestic... | ...(trading) in everyday household, life, and family... |
| ...electronic... | ...systems that employ components such as processors, integrated circuits, and transistors in their design... |
| ...market. | ...in the area of economic activity where buyers and sellers operate, and where laws of supply and demand operate. |

*The implied means: a business or enterprise that manufactures and/or trades.*

The semantic analysis (SA) not only elaborates, it also identifies and defines the objectives that must be achieved to fulfill the PD:

- Survive;
- Operate a business;
- Make a profit; and
- Trade (in domestic electronic systems...);
- in an open market.

Following the process of the TRIAD building system, threats to achieving each objective are identified, and strategies are formulated both to overcome the threat and to achieve the objective. Table 10.1 reveals a developing concept for the creation and operation of the proposed enterprise.

Because of high local manufacturing costs, and local cartels working against new entrants to the market, the concept becomes one of manufacturing abroad and selling, not to the open market, but to the dominant manufacturing companies in the domestic electronic cartel–so-called original equipment manufacture (OEM) –where the cartel would relabel the enterprise's products, and sell them as their own. Internet sales and direct sales based on a postal catalog scheme will increase sales opportunities, although these both take time to grow sales volume.

**Table 10.1**
Domestic Electronic Enterprise - First TRIAD

| Objective | Threat to Achieving Objective | Strategy to Achieve Objective/Overcome Threat |
|---|---|---|
| Survive | Quality competition<br>Poor liquidity<br>Financial loss | Conceive, create, and sell niche products<br>Sound financial backing and control<br>Controlled growth rate, with operating costs linked to operating revenues, and a proportion of operating profit reinvested in R&D |
| Operate a business | Inadequate plant and facilities<br>Noncompliance with business regulations<br>Lack of suitable staff | Set up R&D in local area, but manufacturing in high-quality, low-cost overseas factories |
| Make a profit | Excessive manufacturing costs<br>Expensive bought-in resources<br>High machinery and process costs<br>High labor costs<br>Poor sales | Set up R&D in local area, but manufacturing in high quality, low cost overseas factories |
| Trade in an open market | Inadequate marketing, publicity, public relations, and sales<br>Cartels, inhibiting new traders | Initially, sell products to cartels to relabel as their own<br>Set up an Internet and a catalog-based, direct sales capability |

There are, of course, many other strategies that might be adopted, and other ways of addressing the threats. Indeed, there may be other threats to consider, and some of the threats in the table may be more imagined than real. In any event, there is sufficient information in the table to create a concept map and to model the concept dynamically. The model would show the development of the enterprise over time, its success in dealing with threats, its problems in dealing at a distance with remote manufacturing facilities, its likely sales and revenue, and so on.

The next stage in the TRIAD building system is to use the strategies to generate either activities or functions: activities lead to process models, while functions lead to architectures. In this case, either would work, but we shall choose functions to simplify the example. Each function is intended to realize a strategy.

Table 10.2 provides a set of disconnected functional entities; they require connection to become a purposeful structure. The method of choice here would be to use either common sense or, since many real-world problems are too complex to see the forest for the trees, to use interpretive structural modeling to develop first an attribute enhancement structure, and second an $N^2$ chart.

**Table 10.2**
TRIAD Building System - Generating Function from Strategy

| *Strategy to Achieve Objective/Overcome Threat* | *Function to Implement Strategy* |
|---|---|
| Conceive, create, and sell niche products | Marketing research |
| Sound financial backing and control | R&D |
| Controlled growth rate, with operating costs linked to operating revenues, and a proportion of operating profit reinvested in R&D | Sales |
| | Financial control |
| | Financial backing (beyond cartel influence) |
| | Accounting |
| Set up R&D in local area, but manufacturing in high-quality, low-cost overseas factories | R&D |
| | Production design and test |
| | International business analysis |
| | Business development |
| | Manufacturer evaluation |
| | Overseas manufacturing facilities |
| | Manufacturer liaison & control |
| | Quality management |
| Initially, sell products to cartels to re-badge as their own | OEM marketing and sales |

The first part of the ISM process is Saaty's pair-wise comparison of the various functions, asking the question: "does function A strongly contribute to function B." (See Interpretive Structural Modeling, on page 148.)

The resultant matrix is shown in Table 10.3, where "1" corresponds to "yes" and "0" corresponds to "no." The matrix translates into Figure 10.2, with the

**Table 10.3**
Enterprise Attribute Enhancement Matrix

```
Sales             1    1 0 0 1 1 1 1 1 1 1 1 1 1 1
Backing           2    0 1 0 0 1 0 1 1 1 1 1 1 1 1
Market Research   3    0 0 1 0 0 1 1 1 1 1 1 1 1 1
Accounting        4    0 0 0 1 1 0 1 1 1 1 1 1 1 1
Financial Cntrl   5    0 0 0 0 1 0 1 1 1 1 1 1 1 1
Int Bus Analysis  6    0 0 0 0 0 1 0 1 0 1 1 1 1 1
R&D               7    0 0 0 0 0 0 1 1 1 1 1 1 1 1
Business Dev      8    0 0 0 0 0 0 0 1 0 1 1 1 1 1
Prodct D &T       9    0 0 0 0 0 0 0 0 1 0 1 1 1 1
Man'cter Eval    10    0 0 0 0 0 0 0 0 0 1 1 1 1 1
Man Liaison      11    0 0 0 0 0 0 0 0 0 0 1 1 1 1
O'seas Factories 12    0 0 0 0 0 0 0 0 0 0 1 1 1 1
QM               13    0 0 0 0 0 0 0 0 0 0 0 0 1 1
OEM M & S        14    0 0 0 0 0 0 0 0 0 0 0 0 0 1
```

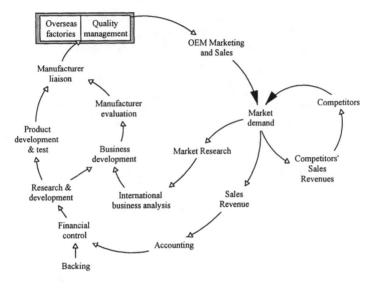

**Figure 10.2** *Enterprise attribute enhancement structure: conceptual map.*

addition of a market connection, and competing suppliers in the marketplace. This is a conceptual map, not unlike that of Figure 10.1, but formed in this case by a more rigorous route, with risk management embedded in the process. Like Figure 10.1, Figure 10.2 presents the bare bones of a CONOPS.

The enterprise attribute enhancement matrix may be converted to an $N^2$ chart and clustered; reducing configuration entropy and revealing potential systems.

The resulting $N^2$ chart is shown at Table 10.4. There are four major groupings in the organization: finance, product potential, product development, and manufacturing and sales. Within each of these groupings there are key functions, identified from the $N^2$ chart: financial control, international business analysis, business development; and manufacturer liaison and control. These key functions

**Table 10.4**
Enterprise Clustered $N^2$ Chart

| Group | Function | # | 1 | 2 | 3 | 4 | 5 | 6 | 7 | 8 | 9 | 10 | 11 | 12 | 13 | 14 |
|---|---|---|---|---|---|---|---|---|---|---|---|---|---|---|---|---|
| Finance | Backing | 1 | ■ | 1 |  |  |  |  |  |  |  |  |  |  |  |  |
|  | Accounting | 2 |  | ■ | 1 | 1 |  |  |  |  |  |  |  |  |  |  |
|  | Financial Cntrl | 3 | 1 | 1 | ■ |  |  | 1 |  |  |  |  |  |  |  |  |
|  | Sales | 4 |  | 1 |  | ■ | 1 |  |  |  |  |  |  |  |  |  |
| Product Potential | Market Research | 5 |  |  |  |  | ■ | 1 |  |  |  |  |  |  |  |  |
|  | Int Bus Analysis | 6 |  |  |  | 1 | 1 | ■ |  | 1 |  |  |  |  |  |  |
| New Product Development | R&D | 7 | 1 |  |  |  |  |  | ■ | 1 | 1 |  |  |  |  |  |
|  | Business Dev | 8 |  |  |  |  |  | 1 | 1 | ■ | 1 |  |  |  |  |  |
|  | Prodct D&T | 9 |  |  |  |  |  |  | 1 |  | ■ | 1 |  |  |  |  |
| Manufacture and Sales | Man'cter Eval | 10 |  |  |  |  |  |  |  | 1 |  | ■ | 1 |  |  |  |
|  | Man Liaison | 11 |  |  |  |  |  |  |  |  | 1 | 1 | ■ | 1 |  |  |
|  | O'seas Factories | 12 |  |  |  |  |  |  |  |  |  |  | 1 | ■ | 1 | 1 |
|  | QM | 13 |  |  |  |  |  |  |  |  |  |  |  | 1 | ■ | 1 |
|  | OEM M & S | 14 |  |  |  |  |  |  |  |  |  |  |  |  | 1 | ■ |

reflect the areas of risk in the venture: insolvency; spending time and money developing a product that will not sell; choosing a poor overseas manufacturing capability; and losing effective control of the overseas factory such that it does not produce the required product with the right quality and cost, and in the due timeframe.

At this point, we have a conceptual structure for the enterprise; see Figure 10.3. To turn this into a design, we would need to look at many more aspects. We have also yet to discuss capacity, throughput, staff numbers and skills, machinery, material sources, personnel, training, and so on. One way to do this would be to invoke and instantiate the Generic Reference Model, on page 163, which will address all those issues and many more besides, and then to model the whole structure in its competitive environment to establish numbers, growth rates, recruiting and training policy, pay structures, and so on.

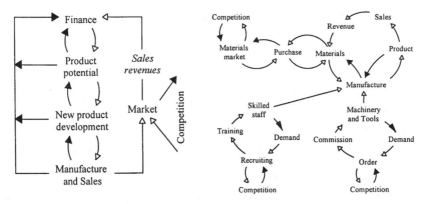

**Figure 10.3** *The enterprise: conceptual systems models. Left-hand diagram: financial view. Right-hand diagram: competing interactions.*

Evidently, the design is much more than the structure or architecture, just as the "design" of a human being is much more than the skeleton. The design of the enterprise will incorporate many other units and facilities; personnel/human resources; an administration, security, and much more; internal communications, plant and machinery, buildings and grounds, and so on. When a functional design has been undertaken, functional features have to be mapped onto a physical layout: a neat functional design may become disordered when the facilities are inadequate, functions have to be split between buildings or floors, allied functions become physically separated, and so on.

The final design will, hopefully, see the many and varied functions, facilities, and subsystems shoehorned into their physical accommodation in such a way that the whole operates effectively. Moreover, the contribution made by each function will be appropriate to the overall enterprise's requisite properties, capabilities, and behaviors. Design, as with the human body, includes the organs and the flesh, as well as the skeleton; and the properties, capabilities, and behaviors of each organ must be appropriate, or else the whole will be unbalanced and less effective than it should be.

Note: at no point during the processes of developing concepts above was there any Cartesian reduction. Throughout, the processes were holistic and synthetic, and the resulting concept maps and structures were, and are, organismic.

## DEVELOPING SYSTEM CONCEPTS: THE SEVEN-STEP CONTINUUM

Codifying the process of concept and design development provides a helpful guide to prevent the conceptualizer and designer from "losing the plot." The following "seven-step continuum" [2] is one such codification, so called because, although there are seven steps, they should not be seen as reductionist. Instead, each step continues smoothly on from the previous step, and the conceptualizer follows the route like crossing a river using stepping stones.

As shown in Figures 10.4 and 10.5, the seven steps are:

- Understand the Issues. As the title suggests, the problem space is explored, and concept maps are created, using the elements of the rigorous soft method.
- Establish the Need. The term "need" is used to distinguish it from "requirement." Conceiving what is needed, rather than what is wanted, is necessary to maintain integrity of process.
- Develop Process and Structure. This is the first appearance of functional architecture, with the minimization of configuration entropy to optimize effectiveness and efficiency.
- Estimate Capacities. Put "flesh" on the bones of architecture. This is the first point at which quantitative information is used to establish "how big," "how fast," "how much," and so on. At this stage the functional design starts to take shape.
- Develop Performance. This includes allocating tasks/roles between man and machine, and identifying cost/performance profiles.
- Develop Effectiveness. This introduces measures to enhance availability and survivability, while maintaining performance.
- Assess Potential Solutions. This measures value, as well as cost effectiveness, harmonizes solution system with siblings, estimates life cycle costs, and explores risks.

The first three steps in the seven-step continuum will be used in the following example: The Far Side.

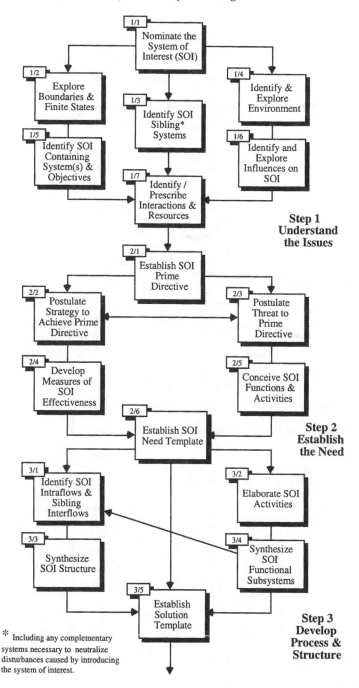

**Figure 10.4** *The seven-step continuum: steps 1 to 3.*

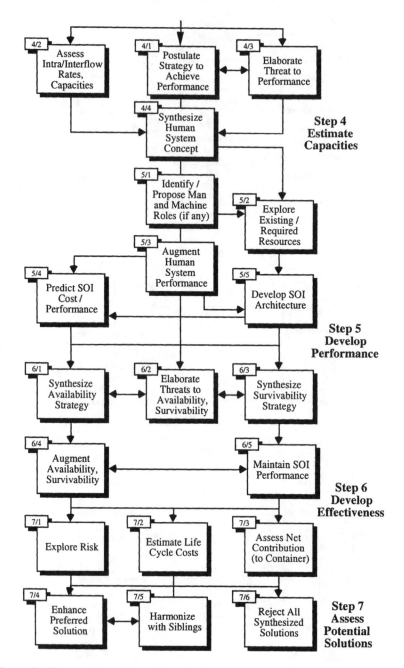

**Figure 10.5** *The seven-step continuum: Steps 4 to 7.*

**The Far Side: Case Study[1]**

The task is to create a design concept for a manned base on the far side of the moon, to be used as a staging post for manned missions to Mars, the sun, the asteroids, and the outer planets.

The base will also support a deep-space radio telescope, placed there because of the extremely low level of manmade radio interference, since the far side of the moon is permanently averted from the earth.

Step 1/1[2] Nominate the system of interest:

This is the Lunar Deep Space Center (LDSC). The name addresses both functions to be performed by the establishment.

Step 1/2 Explore boundaries and finite states:

- Physical boundary encompasses facilities on the lunar landscape;
- Functional boundaries include related earth facilities;
- LDSC will have to be largely self-sufficient, use the physical boundary in this case.
- Finite states include:
  o Operational; that is, fully manned and able to support all missions;
  o Standby; that is, skeleton crews only, but all facilities in working order;
  o Emergency; that is, neither operational nor standby and with some life-threatening situation.

Step 1/3 Identify SOI sibling systems:

- Earth mission control and communications;
- Deep space Missions; that is, the vehicles and their occupants;
- Earth telescopes that may associate with the lunar radio telescope; for instance in an interferometer array;
- Resupply vehicles;
- Earth resource systems, including personnel, foods, gases, tools, construction, and so on.

Step 1/4 Identify and explore the environment:

---

[1] Readers who do not wish to go through the conceptualizing process in detail may go direct to Figure 10.9 on page 264.
[2] See Figure 10.4 for step-numbering scheme.

- The airless, basalt surface of the moon, pockmarked with countless impact craters of all sizes and varieties, indicating a risk from meteorites;
- High-energy cosmic particles regularly and in showers;
- Covered in particle dust;
- Monthly (27.32 days) temperature cycling from very hot direct sunlight;
- No atmospheric UV filtering;
- Starlit darkness;
- Cold descending to only a few tens of degrees Kelvin;
- Gravity 0.1645 that of earth.

Check standard texts for further data.

Step 1/5 Identify SOI containing systems and their objectives:

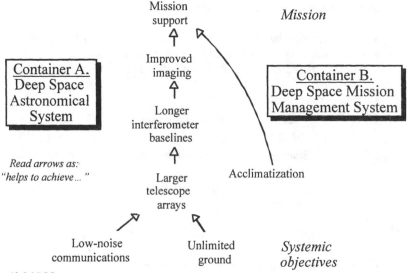

**Figure 10.6** *LDSC mission.*

The intent structure of Figure 10.6 shows the two containing systems, their objectives, and the overall mission of them both.

Step 1/6 Identify and explore influences on SOI:

The two attribute enhancement structures of Figure 10.7 show that an uneven workload will result in the need for high staff turnover, and that the dangers of the situation, coupled with the cost of replacing staff, will militate in favor of self-sufficiency for the LDSC staff.

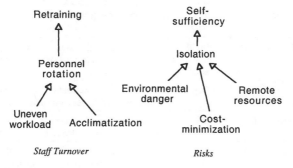

**Figure 10.7** *Influences on SOI.*

Step 1/7 Identify and prescribe interactions and resources:

*Interactions*

- Regular resupply of personnel and logistics;
- Missions to and from deep space;
- Astronomers arriving and leaving;
- Radio communications with earth mission control;
- Radio communications with interferometer;
- Unwanted interactions between the various forms of radio communication; and so on.

*Resources*

- Life gases;
- Food and water—from lunar poles?
- Building tools and materials;
- Libraries;
- Furniture;
- Astronomical and mission support equipment;
- Personnel to support these missions;
- Maintenance personnel;
- Administrative personnel;
- Waste, either for recycling or for disposal; and so on.

Step 2/1 Establish the prime directive and semantic analysis:

To provide...

...safe...

...continuing...

...expert...

...support to...

...deep space...

...missions and to deep space...

...astronomy from a...

...far-side lunar base.

To present as a service...

...minimal risk...

...enduring, survivable...

...comprehensive and high quality...

...facilities and services to...

...missions reaching beyond earth's gravitational influence...

...and astronomy reaching to the farthest depths of space, from a...

...managed resource center on the unseen side of the earth's moon.

N.B. "Deep space" means quite different things to mission control and to astronomers.

| *Step 2/3: Threats to achieving the PD* | *Step 2/2: Strategies to overcome those threats (creative concepts requiring knowledge of threats and environment)* |
|---|---|
| Meteorites and cosmic rays | Build underground by drilling sideways from inside crater rims. Use basalt rock as protection and in place of domes or bubbles to retain atmosphere. |
| Poor communications with mission ground control | Deploy a moon satellite communication system that always has dual sight of both earth transmitter-receivers and LDSC. |
| Lack of trained and willing personnel | Establish a continual training program, and select only the fittest personnel, psychologically and physically. Pay very well! |
| Inadequate resources at LDSC | Provide LDSC with systems to create/mine/manufacture resources, rather than the resources themselves, e.g., smelters, machine tools, borers, and so on. |
| Personnel unable to cope with environment in LDSC | Select only psychologically screened personnel; provide a varied environment within the LDSC, with different "themes" in different parts of the complex; mix the sexes. |
| Inadequate Deep Space Astronomy | Use a large crater as a dish; steer beams electronically, to facilitate time sharing and optimal utilization; ensure radio communications with mission ground control does not interfere; install the best-quality data/image extractors as close to the dish as possible; harden above-ground elements. |
| LDSC operating expense | Use solar power; use lunar resources for power generation, mineral resources, insulation materials, and so on; minimize the management overhead. |

**Figure 10.8** *The sublunar LDSC concept: Air-locked tunnels lead off the central shaft, which acts as a plenum chamber and assembly area.*

*Steps 2/3 and 2/2*
Continued from p. 260.

| LDSC construction expense | Build into lunar surface, rather than creating "bubbles" on the surface; construct from lunar materials where practicable. |
| Inadequate medical facilities for returning missions and LDSC staff | Provide comprehensive medical facilities with compression chambers, artificial gravity machines, psychological state inducement drugs, and so on. |

The emerging LDSC concept may be unexpected. Because of the threats from meteorites and cosmic rays, the LDSC is proposed to be underground; see Figure 10.8. An old crater is cleaned out and rimmed to provide a central shaft, off which radiate a number of tunnels; one is shown in the graphic. Each passage has an air lock set well in from the shaft to avoid impact damage. The airlocks close automatically if the central dome, which acts as a plenum chamber, should experience sudden pressure drop.

Step 2/4 Develop SOI measures of effectiveness:

| *Performance* | *Availability* | *Survivability* |
|---|---|---|
| • Comprehensive support to deep space missions. | • Reliable facilities and services. | • Minimal damage from the environment. |
| • Well-managed, timely lift-offs. | • Minimal downtime. | • Ability to tolerate occasional hits to communications, radio telescope, and so on. |
| • Satisfied, rehabilitated returning mission crews. | • Minimal emergencies. | |
| • Minimal disturbances among LDSC staff on/off duty. | • Minimal resupply costs. | |
| • Unique LDSC research results. | • Minimal upkeep costs. | |
| • Superior quality astronomical imaging. | • | |
| • Enhanced interferometers. | | |
| • Increased astronomical range and variety of image types. | | |
| • Maximum utilization of resources. | | |

Step 2/5 Conceive SOI functions and activities:

| *Strategies to Overcome Threats* | *LDSC functions to achieve strategies* |
|---|---|
| Build underground by drilling sideways from inside crater rims. Use basalt rock as protection and in place of domes or bubbles to retain atmosphere | Construction |

*Continued on p. 262.*

*Step 2/5 continued from p.261*

| | |
|---|---|
| Deploy a Moon satellite communication system that always has dual sight of both Earth transmitter-receivers and LDSC. | Satellite communications management |
| Establish a continual training program and select only the fittest personnel, psychologically and physically. Pay very well! | Earth-based training<br>LDSC continuation training |
| Provide LDSC with systems to create/mine/manufacture resources, rather than the resources themselves; for example, smelters, machine tools, borers, and so on. | Small-scale mining, smelting, refining, working and manufacture from raw materials |
| Select only psychologically screened personnel. Provide a varied environment within the LDSC, with different "themes" in different parts of the complex. Mix the sexes. | Psychological profiling and monitoring |
| Use a large crater as a dish. Steer beams electronically, to facilitate time sharing and optimal utilization. Ensure radio communications with Mission Ground Control does not interfere. Install best quality data/image extractors as close to dish as possible. Harden above ground elements. | Dish construction<br><br>Electronic beam control<br><br>Data extraction<br><br>Image processing |
| Use solar power. Use lunar resources for power generation, mineral resources, insulation materials, and so on. Minimize the management overhead. | Solar power generation |
| Build into lunar surface, rather than create "bubbles" on the surface. Construct from lunar materials where practicable. | Construction<br><br>Development of local materials |
| Provide comprehensive medical facilities with compression chambers, artificial gravity machines, psychological state inducement drugs, and so on. | Perform medical checks<br><br>Manage astronaut rehabilitation/acclimatization |

## Step 3/2.1 Use GRM to generate functional components:

*Internal Architecture Generation Table*

| Mission management | | Viability management | | Resource management | |
|---|---|---|---|---|---|
| Management of... | | Management of... | | Management of... | |
| GRM | SOI | GRM | SOI | GRM | SOI |
| ...information | Communication center<br><br>Image processing center | ...synergy | LDSC management | ...acquisition | CPRM*<br><br>Earth resupply;<br><br>training;<br><br>personnel |
| ...objectives | CPRM | ...survival | Central warning and control | ...storage | Gen stores<br><br>Astro stores |

*Continued on p.263*

*Step 3/2.1 continued from p. 262*

| ...strategy and plans | Operations Astro Laboratory | ...evolution | On-site inspectorate | ...distribution | Administration Power distribution Infrastructure |
| --- | --- | --- | --- | --- | --- |
| ...execution | Operations Astro Laboratory | ...homeostasis | Administration, medical | ...conversion | OJT** Atmosphere conditioning |
| ...cooperation | Operations Astro Laboratory | ...maintenance | On-site maintenance | ...disposal | CPRM; Personnel; Earth resupply |

\* CPRM is contingency planning and resource management.
\*\* OJT is on the job training.

## Step 3/2.2 Use GRM to generate behavioral components

| | *Internal Behavior Management* | | | | |
| --- | --- | --- | --- | --- | --- |
| *Cognition management* | | *Selection management* | | *Stimulation management* | |
| *Management of...* | | *Management of...* | | *Management of...* | |
| GRM | SOI | GRM | SOI | GRM | SOI |
| ...tacit knowledge | Experts; OJT | ...nature | Psychological monitoring | ...motivation | LDSC management |
| ...world models | Lunar; Stellar; LDSC maps | ...experience | "Systems thinking" environment | ...activation | LDSC management |
| | | ...constraint | Power materials | | |

## Step 3/2.3 Elaborate SOI activities:

| Astronomical communications | Satellite communications | Radio communications | Training (earth based) | Training (on the job) |
| --- | --- | --- | --- | --- |
| Materials | Medical | Psychological management | Image processing | Astronomy control |
| Solar power generation and storage Power distribution | Astronaut rehabilitation | Contingency planning and resource management (CPRM) | Operations | Astronomical laboratories |

*Continued on p.264*

*Step 3/2.3 continued from p. 263*

| LDSC Management | Central warning and controls (CW&C) | Inspectorate | Administration | Maintenance |
|---|---|---|---|---|
| Earth resupply | Personnel/human resources | General stores<br><br>Astronomy stores | Infrastructure: sewerage, waste disposal, communications, message handling, and so on | Atmospheric conditioning: heating, cooling, ventilation, purification, humidity, and so on |

The table accumulates all of the functions previously generated; some activities have been incorporated under other headings. For instance, psychological monitoring is incorporated under "medical."

Step 3/4 Synthesize functional subsystems:

The accumulated functions are entered into an $N^2$ chart. Interactions between the functions are also entered, the chart is clustered to minimize configuration entropy, and the diagram of Figure 10.9 emerges.

There are four subsystems: base command; operations; logistics; and resource management.

Step 3/1 Identify sibling interflows:

The final part of Step 3 is to establish the solution template; Step 3/5. A major part of that would be Figure 10.10, which shows the LDSC with its subsystems

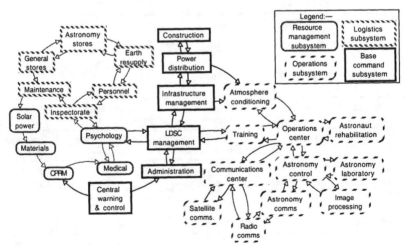

**Figure 10.9** *LDSC functional subsystems.*

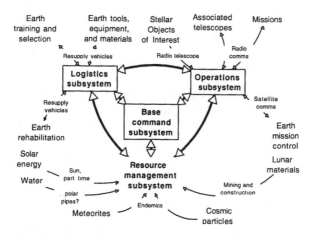

**Figure 10.10** *Connecting the LDSC to its sibling systems via complementary systems.*

interconnected to, and interacting with, its sibling systems and complementary systems (e.g., satellite communications).

There is much more to this particular case study, but enough has been presented within the space available to show the value and effectiveness of the seven-step continuum in developing a robust system solution concept.

As Figure 10.10 shows, the solution concept is organismic; that is, the whole LDSC is built from interacting, mutually supportive subsystems; synthetic, (i.e., built up without any reduction), and holistic, in that it has encompassed the whole of the issue.

Completing Steps 4 to 7 would result in a large amount of specific information about the many subsystems and interconnections from which the LDSC has been conceived.

## SYSTEM DESIGN

A solution concept describes a way or means by which a solution may be achieved. A system design puts flesh on the bones of the solution concept, creating a description of the concept sufficient to enable its being realized. For a system design to be scientifically justifiable, it must be:

- Synthetic, holistic, and organismic;
- Capable of representation in such a way that its potential ability to solve the problem can be proved or disproved.

The first bullet suggests that a system design should be seen in the context of other, sibling systems, operating within one or more containing systems; and that the design should reveal subsystems interacting with each other to create the emergent properties, capabilities, and behaviors necessary to solve the problem.

The second bullet suggests that a design should be capable of dynamic simulation in a representative, interactive environment. It would be insufficient to simulate the dynamics of causal loops; for a system design, the whole system would have to be represented. For example, if resource management were not represented, then causal links would not be sustained; the relationships between resources and causal links that characterize purposeful system behavior must be evident.

Inadequacy in design representation is illustrated in Figure 10.11. The figure shows major subsystems in the human body, including the skin that is arguably the largest organ in the body. In general, organs are not represented; they exist within the subsystems, as the heart exists within the cardiovascular system, and the thymus exists within the immune system.

The form of representation in the figure is reductionist: there is no indication of how the subsystems interact with each other. There is no indication of form: how the skeleton creates a framework on which to mount the nerves, arteries, and veins; how the ribs create a bellows to draw air into the lungs and expel $CO_2$; how the skull forms a local exoskeleton to protect the brain; and so on. Nor is there indication of how the organs are activated with oxygenated blood and, at the same time, bathed in protective lymph fluid. And so on. So, while the figure catalogs the subsystems of the body, it falls far short of design.

The second example, illustrating the design dichotomy, may be presented in

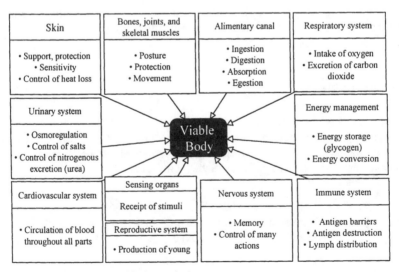

**Figure 10.11** *Subsystems of the human body.*

the form of a system design for a family home; see Figure 10.12. The figure shows major subsystems at left, in direct analogy to the body's subsystems in Figure 10.11. This is not how we think of the design of a home, however, which might be more like the layout at the right. Here, each room is identified by its form and purpose, and it is evident how access between rooms is achieved, too. If we look in systems terms at, say, the electricity distribution system, not surprisingly it will be evident in each and every room at wall outlets, lighting, and switches. Similarly, air conditioning would be evident in every room. Every room has building fabric surrounding it above and below. Several rooms have water supplies, drainage and sanitation, and so on.

To express a system design is to identify all of the subsystems *and* the way in which they are physically configured. When we present the design of a house, while we may identify the subsystems, we view the design with the various subsystems intimately intertwined and interactive. We focus, not on the technological subsystems, but on the activities, comforts, and environment of the human occupants; the family. This becomes apparent when we define the purpose of a family home as: "to nurture, support and protect the members of a family such that they can develop and flourish individually and socially."

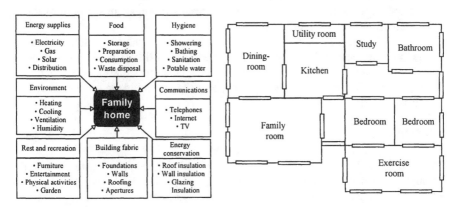

**Figure 10.12** *Family home design dichotomy.*

The process of design, then, appears to be one in which purpose and viability dictate both the subsystems necessary to the design, and the configuration of those subsystems. System design is also influenced by other systems, as well as by containing system(s), and by environment. A high-speed aircraft has a form suited to flying through air. A family home in the tropics may not need doubleglazing. A family home in a remote part of the Andes may not be able to sustain its purpose because of inadequate surrounding facilities.

## DESIGN FOR OPERATION

Design may be developed with different emphases. Systems necessarily constructed from unreliable technology, for instance, may emphasize redundancy and maintenance in the design. For many systems, however, the emphasis is on operational performance.

### Scenario-driven Design

One approach to design for operation is to design a solution to a problem identified by a scenario, or situation. Scenarios are often associated with supposed threats, or threatening situations. A naval scenario might envisage a battle fleet entering foreign waters where an ingenious enemy is believed to have laid down sophisticated mines, and to be operating submarines and fast patrol boats fitted with sea-skimming missiles. The scenario would play out the likely engagement between the two opposing forces.

In a different scenario, the same naval battle fleet might find itself facing an enemy battle fleet with a range of ships having different characteristics, sensor ranges, weapon ranges and types, and so on. The battle fleet might be operating to a different naval doctrine, and under different rules of engagement; both of these factors contribute to different outcomes.

None of these events need ever have happened. Indeed, it would be unlikely in the extreme that a real-world encounter unfolds just like a particular scenario. Instead, the scenario describes a hypothetical, dynamic situation, environment, and opponent that might be encountered, and that it would be advantageous to vanquish.

### CONOPS

In working through a scenario, using either simulation models, or simply talking through situations and events, players make decisions at points along the sequence of events. These decisions, generally about strategies or tactics, affect the outcome. If a particular approach becomes successful, or is prescribed by higher authority, then it may invoke a concept of operations. As the name implies, CONOPS is a description of how operations are to be conducted or, more simply, how things are intended to work. Constraints by superior authority may affect the viability of a particular CONOPS.

The term CONOPS is seldom used in business, where "business strategy" or "corporate strategy" might be more usual. The parallels between business and war are striking at times, and some businessmen adopt *Bushido*, the way of the warrior, in their dealings [3]. Examples from military conflict may serve, therefore, as an analogy [4] for business and enterprise.[3]

To explore the idea, consider a hypothetical scenario:

> *A hostile group has acquired a new antiship weapon. It is a sea-skimming missile that is able to react to anti-missile activities by submerging, and continuing (albeit more slowly) to its target under water. Intelligence advises that the new weapon can be launched from any small boat or from land, but not from a submerged submarine. The weapon's range is over 100 km, flying at 50 to 100 meters above sea level; it is highly maneuverable making up to 20g turns to avoid interception, before returning to its original course. It appears to be impervious to electronic or infrared jamming, and is a fire-and-forget device, that is, the launcher does not need to guide the missile during flight. Intelligence's best guess is that the missile uses inertial guidance with an optically guided terminal phase.*
>
> *A scenario is envisaged in the eastern Mediterranean where an allied naval battle fleet is attacked by a number of small ships that dart out from small islands, launch their weapons, and run for cover.*

Using the information supplied in the text, a problem space may be established, with players, threats, limitations, environment, and so on. There are several fundamental approaches to solving, resolving, or dissolving the issues raised by the new weapon. According to Ackoff [5]: to resolve a problem is to find an outcome that is good enough, one that satisfices; to solve a problem is to select an approach that yields the best of all possible outcomes, one that optimizes; to dissolve a problem is to change the nature of the problem, or its environment, such that the problem disappears.

- Ackoff proposes that to resolve is a qualitative, rather than quantitative approach, relying on past experience and current trial, and is rooted in common sense and subjective judgment.

- To solve a problem, on the other hand, is a research approach based on scientific methods, techniques, and tools, such as mathematical models and simulations.

- On the third (!) hand, to dissolve a problem is to idealize, rather than to satisfice or to optimize, and to change the situation so that the problem cannot or does not arise.

In practice, dissolving would mean not addressing the scenario as given: for instance, one way of dissolving the problem would be to change the environment, perhaps by keeping the allied naval battle fleet out of the eastern Mediterranean. Alternatively, it might be possible to prevent the new weapons from being used by: attacking the bases of the enemy ships; attacking the launch ships prior to launch; preventing the new weapons from reaching the would-be users;

---

[3] The term enterprise is sometimes used as though it is synonymous with business. An enterprise contains within its meaning the context of new, adventurous, and risky, requiring confidence and initiative. Business, on the other hand, implies already being established.

interrupting the manufacture of the new weapons; arranging for the new weapons to contain faulty parts; appeasing the attackers; and so on.

To resolve the problem posed by the scenario might involve tactics, deception (e.g., smoke to defeat the missile's optical tracker), more damage repair crews, enhanced close-in weapon systems (CIWS), and so on. Coupled with such defensive measures might be the use of specifically armed UAV patrols to detect and neutralize enemy missiles and ships.

To solve the problem might involve a simulation of the allied battle fleet ships and enemy ships in the eastern Mediterranean; each equipped with sensors, propulsion, armor, and weapons. Engagements would be "run" under a variety of different conditions, and with the addition of different defensive, offensive, and deceptive facilities to identify solutions to the problem. (An example of this method follows later.)

Suppose instead that the scenario referred to business:

> An overseas business competitor has acquired a new video-recording hub. The hub is a device that enables different video and audio devices to be integrated into a single system. The new hub incorporates automatic interface devices (AIDs): an AID is a revolutionary invention that senses the signals on one side of a connector, and automatically configures the other side of the connector to match. With this facility, the new hub is a direct threat to our business of providing integrated commercial security systems, as it can couple systems made by any manufacturer instantly. Quite how the new system works is not clear, but the AIDs seem to be based on a new chipset produced by a small company in the Far East.
>
> A situation is envisaged in the domestic market where this competitor, armed with the new video/audio hub, targets our business. The competitor is seen as setting up local business consultancies; offering free security system analysis and design, with a view to installing the new hub and associated facilities. And that is only the beginning; the new AID can be used across a wide range of domestic, commercial, and defense products....

Comparison of the two scenarios shows that, at a fundamental level, they are very similar: both are amenable to resolution, solution, and dissolution. For some reason, it seems simpler to tackle the first scenario, perhaps because the threat is so much easier to visualize in life and death terms.

Perhaps, then, the answer to developing business, enterprise, and corporate strategies is to practice initially on simple but stark military analogies, and then to repeat the process using the less dramatic, but nonetheless important, business situation.

## INDUSTRIAL DESIGN PARADIGMS

There can be many different objectives of system designs in industry. Some of the more common include:

- Design for Manufacture. Particularly where manufacture will use automatic assembly tools, it may be necessary to select components in the design that can be tool manipulated. Prefabrication of parts, subassemblies, and so on, may form an integral feature of design for manufacture, as will test access, so that parts can be tested before assembly, or at stages during assembly.

- Design to Cost. Cost may sometimes be referred to as an independent variable, meaning that not exceeding cost is paramount. Design to cost may be associated with reverse engineering, where the concept for some product is reverse engineered to reveal its essential components and subassemblies, together with their relative contributions to the product. By using this approach, it may be possible to select and size the various parts of a system, such that the overall cost is kept within an arbitrary budget.

- Design for Maintainability. This is common for operational equipments, where defects and wear necessitate maintenance. Systems may be designed in modular fashion, and test facilities built in such that: a) maintainers are alerted when a defect occurs, and b) the module that needs replacement is automatically identified. An alternative, but compatible, paradigm requires the design of some product to be compatible with others used in the same operational arena, so that the new product may share spare parts, tools, test equipment, and so on, with existing user products.

- Design for Upgrade. This is a useful concept in a technologically dynamic climate. Systems may comprise loosely coupled parts, such that replacing one part with more advanced technology has no material effect on other parts. If this process is planned in advance, it may be called pre-planned product improvement ($P^3I$).

- Design for Minimal Life Cycle Cost. Since life cycle costs include the costs of failures, spares, maintenance, and replacement, it may be worthwhile to make the initial design more reliable and durable. However, this may also make the initial design more expensive. There may be a minimum sum cost, where the higher cost of the initial product is more than offset by the reduced costs of support during operations.

- Design for Replacement. Some systems acquire large amounts of information during their life cycle. Design for replacement ensures that the information can be transferred easily to replacement systems. A typical example of this might be in air traffic management, or a corporate management system, where valuable information accumulated over many years cannot afford to be lost at switchover.

- Design for Recycling. This is important particularly for consumer products, where the materials from which these products are made may be recovered and reused, thereby reducing the drain on natural resources. Design for recycling would see the various materials used in the design being easily and automatically separable by type (ferrous metal, plastic, refrigerant, and so on). Automobiles, washing machines, television sets, mobile phones, and many more consumable products are being designed for recycling.

Note that in no example is there mention of achieving or enhancing the operational performance of the product. Each and every design objective presumes that product performance is achievable, and that other aspects of the product life cycle must be addressed. Following topics look at some industrial design paradigms in more detail.

### Design for Availability

Availability is one of the key attributes of an operational system. Mathematically, availability may be defined as:

$$\text{Availability} = \text{Mean time between failures}/(\text{mean time between failures} + \text{mean down time}) \quad (10.1)$$

$$A = MTBF/(MTBF + MDT) \times 100\% \quad (10.2)$$

In other words, availability is the proportion of time that an operational system is available for operation; usually quoted as a percentage.

From the formula, availability approaches 100% if MTBF >> MDT. There are different ways in which availability may be maximized, if that is the aim, including:

- Increase the system MTBF:
    o Clearly, this can be achieved by making the parts, their interactions and their cooperation more reliable, dependable, and consistent.
    o It can also be achieved by: introducing redundancy, such that defects in one part do not disable the whole; neutralizing intrusive effects, such as viruses or physical damage, either by incorporating an immune system, or by detecting and isolating intrusive elements; or both.
- Reduce the downtime, MDT:
    o Where this is necessitated by replenishment, then the time taken for replenishment may be minimized by eliminating waiting times, and by maximizing the speed of take-up.
    o Downtime may also be invoked by the need to recover from stress, strain and damage–psychological, as well as physical.
- Self healing. Where a system can self-heal, downtime may be reduced to zero, and the system MTBF goes toward infinity.

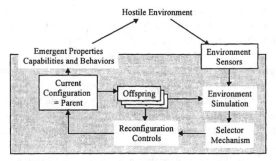

**Figure 10.13** *Autoadaptive systems concept.*

Biological systems are self healing in various degrees, and they generally incorporate immune systems. So, too, do social systems and some sociotechnical systems. See The Social Genotype and Figure 7.1 on page 127. Biological systems continually replace cells that have a limited life; sometimes of only days or weeks. New cells are generated to the same pattern, and replace the older cells as and when they age, in a process of continuous self healing. Like some naval ships, biological systems also have sophisticated systems for repairing damage, and hence for restoring survivability.

In the longer term, as biological systems demonstrate admirably, maintaining availability necessitates evolution, which is—in effect—adaptation to a changing environment. Figure 10.13 shows a general schema for an autoadapting system. An autoadapting system, shaded, exists in a hostile environment; where hostile might imply hot, cold, wet, dry, competitive, predatory, and so on. The system senses the environment, and creates an internal simulation of the environment as the system observes it, complete with threats, opportunities, and competition for resources.

The simulation also contains a representation of itself, interacting with others in the environment. The representation of self (current configuration = parent in the figure), generates a variety of simulated offspring; each slightly different from the parent and from each other.

Each of the simulated offspring is inserted into the environment simulation, to test its behavior, performance, survival, and so on. While the environment is steady and unchanging, none of the offspring will be better adapted to the environment than the parent. Should the environment change, however, it is likely that one or other of the simulated offspring will prove to be better adapted to the changed situation. In that case, the preferred option's characteristics will be used to reconfigure the system, such that the offspring becomes the new parent, and the process will resume generating and testing offspring. Meanwhile, the emergent properties, capabilities, and behaviors of the auto-adapting system will have changed as it adapts to the environment.

A moment's reflection will suggest that this process of adaptation is commonplace. It is the way that we evolve new designs for many systems: cars, tanks, planes, ships, schools, shopping malls, and so on, where the general practice is to create a new system that is based on prior systems, but perhaps with new features, functions or forms chosen from a range of possibilities. The unusual feature in Figure 10.13 is that the capabilities for evolution are contained within the system, as opposed to separate from, and acting upon, the system. The figure is, in effect, a metaphor for evolutionary systems engineering.

Although not shown explicitly in the figure, the autoadaptive system could evolve its autoimmune system, too. Like biological systems, computer systems could, in principle, protect themselves and the data that they contain by evolving their own immune systems; able to detect and neutralize new viruses and worms.

For platforms, availability necessitates both crew and machine: there is little point in having an aircraft serviceable on the ground (SOG), but with no crew available–and vice versa. In this context, it is more sensible to view availability, not in terms of individual platforms, but in terms of service availability, where service might involve having a set number of aircraft airborne and in operation at any and all times.

Techniques such as in-flight refueling can be employed to keep aircraft in operation for extended periods, although they render the platform vulnerable during the replenishment process. Such techniques can be extended to all operational systems in principle. For example, to maintain a set number of visible police in an area, the latter may eat, drink, and be replaced by operational handover while on patrol. The number of police needed overall to maintain the level of operational service would be reduced accordingly, although individual patrol officers may not be too keen on the idea as a standard practice.

These techniques suggest their own CONOPS, which in turn leads to the design of processes to acquire and deploy sufficient resources to maintain operational systems on station. Once the CONOPS is firmly established and detailed, the consequent system design is relatively straightforward.

**Design for Maintenance**

Another design viewpoint emphasizes maintenance as a key design aspect. The GRM showed maintenance as key to system Viability Management (on page 168.) Maintenance was defined as the detection, location, excision, replacement and disposal of faulty or defective parts. In the GRM, maintenance can refer as much to the human element as to technology, so that a defective person–one who is not performing according to role–would be a candidate for maintenance, although we might colloquially refer to the process as dismissal, firing, sacking, sending on permanent holiday, or retraining.

Maintenance is fundamental to system viability; in systems terms, maintenance maintains the system's emergent properties, capabilities, and behaviors. In this context, it does nothing to change those characteristics; that is the role of evolution. Servicing is the process of checking for correct operation, cleaning, replenishing lubricants, and so on, rather than maintenance per se.

Maintenance is the servant of operational availability; see Part I, Chapter 13, starting on page 374 and following. Maintenance can be undertaken:

- When parts of a system become defective or fail. "On-condition" maintenance is done. Popularly used with electronic devices, it is done where prediction of failure is difficult to impossible.

- Before, and in anticipation of defect or failure. This "precautionary" maintenance is generally associated with mechanical devices that show a tendency to wear out. Servicing, too, is often precautionary.

- Only when the system's emergent properties, capabilities, and behaviors change sufficiently as to render it unsuitable for operations. In this case, the system may absorb failures and defects in redundancy, or may accept that some inessential functions are not being performed satisfactorily.

## Accessibility

Since maintenance involves excision, access to the item to be excised is essential. The design of systems to enable access is important to maintenance. Webs and networks of electrical cables and hydraulic pipes may hamper access for maintenance on platforms, as too may the lack of removable covers. Designing, say, a complete airliner for ease of maintenance access is challenging: access panels weaken the skin, in which resides much of the strength and flexibility of aircraft structures.

## Modularity

Modern systems may be constructed from interlocking modules, designed with interfaces such that parts of a system may be removed and replaced without significant impact on the other parts. Modularity is valuable, for the obvious relative ease of removing a smaller part rather than a larger whole, but also because the part to be removed is likely to cost less than the whole. Spares may be provisioned and held as modules rather than as whole systems, reducing overall support costs.

However, there are hidden costs in modularity: the creation of modules invokes internal interfaces and connectors within a system, thereby increasing its overall defect rate. Moreover, because the modules are, preferably, replaceable without post-installation adjustments, modules must be constructed to close tolerances, and in general must be linear and loosely coupled. In general, linear systems occupy greater volume and utilize more power than would the equivalent, nonlinear (e.g., biological) version.[4]

---

[4] If you doubt this statement, consider if you would make a technological mayfly that was able to fly, see, smell, reproduce, and so on, yet was no larger, nor used more energy, than its biological equivalent.

Test Access

Test access goes hand-in-hand with modularity. To determine which module is defective or faulty, a logical process of testing should be able to isolate the particular, offending module. In general, this means that the tests must be made at points of inflow and of outflow to and from each module, generally while it is operating in situ. Tests may be compared with a serviceable item to detect and locate the faulty module. The rationale for this process usually depends on a layered architecture, with process after process in sequence: following the sequence from start to finish shows at which stage the process breaks down, and hence reveals the defective module.

In principle, test access may be provided between modules that are parts of systems buried deep within a complex, many-layered structure. This allows fault diagnosis, but also presents options in remedies. Knowing what is faulty may enable the diagnostician to leave the situation alone until a more favorable opportunity arises, close down the offending system and replace it remotely with another, or undertake the difficult task of replacement in the certain knowledge of the fault.

Some designs have become so smart in this area, that the tests of modules are undertaken automatically, while the system is in operation. Test results from various points in the system may be used to light up individual fiber optics, such that not only is the occurrence of a defect signaled to the operator, but the fiber optics light up in a pattern; spelling out the nature of the defect.

## DESIGNING OPEN, INTERACTIVE SYSTEMS

Designing an open, interactive system requires an open, interactive, and outward-looking approach. First, the system of interest is envisaged as existing and operating in its environment. For the SOI to exist and to persist, it must be continually supplied with resources; this implies a resource environment that contains other complementary systems that supply resources in response to need or demand. These resource systems may also face demands from other systems, raising the potential for resource competition.

Resources come in two principal categories: resources to sustain the SOI; and resources to be "ingested," processed, and ejected as outflow or product. The two categories are not distinct. Food passing through the human gut is ingested and processed; nutrients are extracted and absorbed into the body through the gut wall.

A manufacturing plant, on the other hand, has resources of men, machines, materials, and money that are necessary to sustain and operate the plant. Parts and subassemblies bought from other companies may be integrated to form a higher-level product, which is sold as product/outflow without any of the throughput being used directly to maintain the plant. In this case, the link is revenue from sales that contributes the money needed to purchase more resources and sustain the plant.

As a third example, consider a ground attack aircraft fitted with both air-to-ground weapons and short-range air-to-air missiles. The latter are intended for self defense, so they contribute to survivability–part of platform viability in the GRM; that is, sustaining the aircraft. The air-to-ground weapons, on the other hand, are used to execute the mission, and are associated with mission management, rather than viability. They do not sustain the platform, but they are a resource, they do constitute an "outflow or product," and they do have an effect.

Figure 10.14 also shows operational environments. The outflow from the SOI affects other systems in the operational environment, which react negatively, at least initially, in accordance with Le Chatelier's Principle, or Newton's Third Law: the introduction of a new system, or a changed effect, causes the network of open systems to seek a new equilibrium. Similarly, the inflows to the SOI disturb the status quo initially, and a new equilibrium will be sought. The local "area" or "volume" of systems will reverberate.

The key to open system design, then, is to create closed loops that pass through the SOI in much the same way as flux lines pass through a bar magnet. This is quite unlike the design processes in vogue for creating static systems. By designing in closed loops, it is possible to allow the design to "float" toward its median state under the influence of feedback and feed forward, rather than to dictate beforehand the values of an uncertain design parameter.

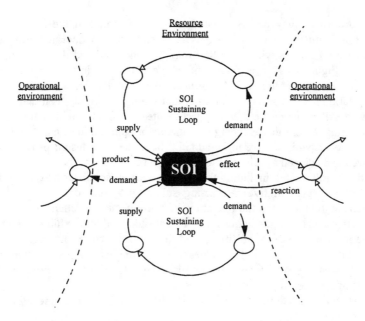

**Figure 10.14** *System design context.*

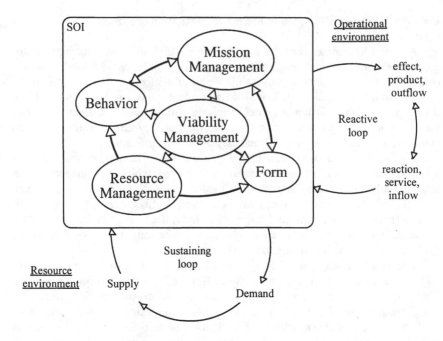

**Figure 10.15** *Elaborating the SOI design in context.*

Figure 10.14 may be instantiated to create a model of the SOI, open and interacting with those other systems it is expected to operate with, and be sustained by. Features within the SOI may be conceived and designed in several ways; see Figure 10.15: the figure shows the GRM within the SOI "envelope," with each part of the GRM connected internally and externally. This representation may be an instantiation of an existing design that is to be enhanced, or of a so-called "green-field," or unprecedented design. It could also be a design of an existing system that is facing new threats or opportunities, and to which new processes or capabilities are to be added. The archetypal design process may be employed and adapted to suit any system design task.

The overall model, including the context, will be nonlinear and dynamic, and will contain representative interacting subsystems, linked to each other; the resource environment and the operational environment. The characteristics of the SOI's subsystems and interactions may be adjusted so that it exhibits the requisite emergent properties, capabilities, and behaviors, while interacting in this representative environment. The values to which the various parts within the SOI were adjusted then represent the nominal specifications for those parts. An example of this design procedure follows.

First, however, how can optimization be achieved in this open, interacting system context?

## DESIGN CAPABILITY RATCHETING

One approach to enhancing the designed capability of an open, interacting system is shown in Figure 10.16. At left, a process of cumulative selection sees a parent organization generate a random set of potential offspring organizations, to be tested for effectiveness in a business performance model. The best performer is selected and becomes the parent in turn. The procedure continues until no further improvements are possible.

The second example, at right, concerns a so-called many-on-many combat between battle flights of own fighters and enemy fighters. First, own fighter design is optimized, using the enemy fighter as a fixed reference, then the tables are turned, and the enemy is optimized using the (enhanced) own fighter as a fixed reference. Finally, own fighter design is optimized again, but this time using the enhanced enemy fighter as a fixed reference. Own fighter will evolve as a team player, with enhanced features for network centric combat *(q.v.)*.

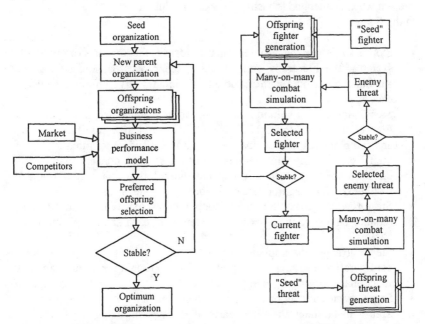

**Figure 10.16.** *Design by cumulative selection. Left-hand diagram: simple cumulative selection of a business organization. Right-hand diagram: systolic[5] cumulative selection, or ratcheting, of a fighter design to optimize it for many-on-many air-to-air combat.*

The objective of this process is to anticipate the phenomenon of design "leapfrogging" that is observed between competitors, where each observes the

---

[5] I call the process "systolic" after the two-beat behavior of the heart.

other's latest designs, and tries to "go one better." Using systolic cumulative selection, it is conceivable that the competitor's next design can be foreseen and exceeded, allowing one's own design solution to jump two steps forward.

This second process, systolic cumulative selection, carries with it obvious risks. Executed without care, it could result in system designs that took little account of the real world or of the laws of physics. The process has to be regulated within practical bounds, but is nonetheless very powerful.


## OPERATIONAL DESIGN APPROACH

It is helpful to codify the process of development from problem space to solution space – see Figure 10.17. Using such a procedural template greatly speeds up the process, while preventing the conceptualizer from jumping to premature conclusions; it also ensures the creation of an audit trail that the conceptualizer may present when challenged to justify the system solution.

In the figure:

- Step 1 explores the problem space, identifying the environment, the systems, the hierarchy and structure, the processes, and so on, that surround and contain the problem, deficiency, dysfunction, shortfall, excess, and so on.
- Step 2 develops a solution concept map as exemplified in Figure 10.1, Figure 10.2, and Figure 10.3. The CLMs are closed, signifying closure and completeness in encompassing the problem and its solution.
- Step 3 creates a dynamic simulation of the solution concept map, should one be necessary, to determine that the solution concept, if realized, would exhibit the appropriate behavior to solve the problem.
- Step 4 works from the causal loop models of Step 2, identifying a (connected) function, process, or system with each and every link in the CLM.
- Step 5 arranges the implicit functions, processes, or systems to minimize configuration entropy; creating clusters and groupings that are candidate functional subsystems, as in Table 10.4.
- Step 6 creates a functional systems solution model.
- Step 7 instantiates the other parts of the system, using the GRM as a template. Behavior and form are added to the functions from Step 6, together with viability and resource management.
- Step 8 synthesizes a full systems solution model from the contributions of Steps 6 and 7.
- Step 9 develops environment and interacting systems models; Step 2 may have identified several complementary systems, necessary to the solution.

- Step 10 links the systems solution model to models of the environment and interacting systems models from Step 9.
- Step 11 introduces systems solution optimizing criteria; for example, value for money, cost effectiveness, maximum all-up mass, "specific shape and center of mass, casualty exchange ratio, and so on.
- Step 12 operates within the system simulation, changing the emergent properties, capabilities, and behaviors of the various subsystems and their interactions so as to achieve the requisite overall system emergent properties, capabilities, and behaviors, and in particular to satisfy the

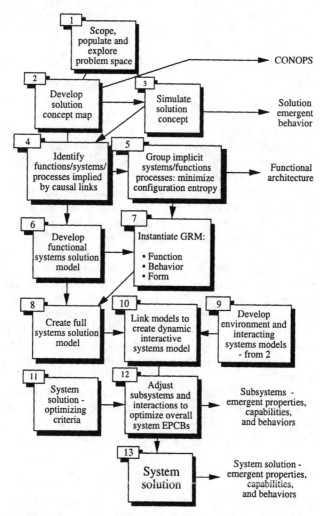

**Figure 10.17** *Codifying the route from problem space to systems solution space.*

appropriate optimization criterion.
- Step 13 produces the resultant optimal systems solution concept.

Although some of the steps are combined, the system design example below follows the code.

## Evaluating and Optimizing Military Capability

Note: The following example of systems design employs a military example. The method used is general, however, and could have been used to evaluate and optimize a business organization, a hospital, a systems engineering project team, or just about anything where there is dynamic competition or interaction, and so on. The approach is also scientific according to Ackoff's definition, since it seeks to solve a problem by finding an optimum solution.

It is current practice in the U.S. and other countries to develop and field so-called military capabilities. Since a military formation is made up from a number and variety of different components, which may be set down almost anywhere in the world against almost any kind of opposition, questions inevitably arise. How can a force capability be evaluated and predicted? How can the optimum combination of elements (the force mix) be identified? How can the right technologies to support, enable, and underpin the capability be identified? What size should the force and its components be, and so on.

On the face of it, these are intractable questions. How can one estimate the capability of a force without specifying the terrain and the enemy it will be facing in some detail? For a start, the capability of any force owes as much to its training and morale, as to its weaponry. And what about the synergy between the force elements? Network-centric ideas have emerged from an understanding that the ability of the various force elements to communicate enables the whole force to be more capable. To use a dated term, intracommunications are a force multiplier because they enhance synergistic operations between the various force elements. The force is an open system, with its complementary parts working synergistically; using the network to create system-emergent properties, capabilities, and behaviors. Looked at in this way, a force capability is not so difficult to envisage, and its potential optimization is amenable to systems methods and ideas.

A key issue in any such optimization is the human element: decision making processes are human; crews in the air and controllers in airborne or ground-based control centers make decisions about rules of engagement, tactics, target allocation, weapon selection, kill assessments, and many more.

### The New Mobile Land Force Capability: Case Study

A new kind of force is envisaged, and it is important to optimize its capability at the force design stage, so that the capability may be created and sustained at a sensible cost. Experienced military strategists have already determined the basic outline of the proposed force, although none has experienced the envisaged warfare scenario. (This is a speculative example only, to illustrate methods of systems design.)

The force is to be a highly mobile land force capable of operating over rough and open terrain. The force will include sensors, weapons, command and control, maintenance, damage repair, and so on. Sensors will include radar, electronic surveillance systems, and UAVs with TV, electronic sensor, and air-to-ground weapons capability. Weapons will include both soft and hard kill: soft kill weapons will include radio and radar jammers; hard kill weapons will include surface-to-surface weapons, surface-to-air weapons, close-in antiaircraft missiles and guns, and so on. Transport will include soft-skinned vehicles, armored personnel carriers (APCs), and light tanks. The whole is a mobile land force not dissimilar in concept to a naval task force, with its potential for fast, coordinated movement, and integrated sensors, weapons and command and control.

## CONOPS

There is an outline concept of operations for the proposed new force. Unlike previous land forces of this type, the new mobile land force (MLF) will fight on the move, with the whole force acting as a single entity. A new network will provide not only voice and data communications, but relative navigation and identification (CNI) facilities, too. The force will move under Formation Control, a new system for determining and maintaining the relative positions of the many and various vehicles on the move.

A typical formation in open country might see armored vehicles leading in a V-formation, with APCs in the center and soft-skinned vehicles toward the rear. Air defense vehicles would operate at the periphery, to provide area cover over the whole formation-on-the-move, and–using close-in weapons–to neutralize surface skimming, or terrain following, missile threats. Formation control is also able to concentrate the vehicles to reduce the target area, or to separate the vehicles to reduce the threat from area weapons.

Whatever the formation, the MLF will sense on the move, create a recognized air and surface picture (RASP) with added communications and satellite intelligence, and share it between vehicles in real time. With all battle sensors, controls, and weapons mobile, an enemy will also be engaged on the move where appropriate.

The MLF may subdivide without losing coherence. Sections with high-speed, all-terrain vehicles may be deployed ahead of the main body for reconnaissance or raiding purposes. Sections may be deployed tactically to take advantage of a static or less mobile opponent. During all of these and similar activities, the MLF will remain under formation control as a single entity.

Instantiation

The challenge is to optimize the overall force capability, given that the enemy is not well known in advance, but could vary from a third world guerilla force to a well-trained force equipped with some sophisticated technology.

The starting point is to represent the force using the GRM, and in particular to employ the dual interacting GRM shown in Figure 8.33 on page 174. Since the precise enemy is not known, it is sensible in the first instance to pit the mobile force against an identical force; that is, itself. Our own force will be called Blue; the enemy force will be called Red. This has several advantages:

- It is unnecessary to find information about a supposed future enemy, much of which would be unavailable in any event. An unspecified enemy could be better, or worse in many respects. The only practicable course is to presume parity–at least as a starting point.

- Defensive as well as the offensive capabilities of Blue will be checked.

- Some aspects of belief and culture might be uncertain, for the mobile force as well as an enemy: such aspects can be considered as affecting both Blue and Red equally and therefore potentially balancing out.

- Using identical Red and Blue forces offers psychological closure. This idea is expressed most succinctly in the Zen saying: "Fundamentally, the marksman aims at himself [6]."

- Red force acts as a dynamic, interactive reference against which to optimize Blue capability/performance/effectiveness. The roles can then be reversed and Red can be optimized against Blue as a reference, thereby pulling each force model up by the bootstraps; that is, ratcheting capability. Since Red and Blue are the one force, this offers a powerful way to design an improved capability.

The proposed force can then be instantiated using the GRM as a template; see Figure 10.18. As a whole, the figure appears complex, but on closer inspection, the various parts of the GRM are evident, and each of those is relatively simple.

Instantiation is a careful process requiring each element in the proposed force to be described and delineated in correspondence to the GRM. Features corresponding to the form model, such as sensors, transport, and weapon systems, may be described in terms of their capabilities. A radar transmitter might be described in terms of its effective radiated power, and receiver sensitivity. A weapon might be described in terms of its launch range, accuracy, and lethality, and so on. The purpose of such description is to be able to calculate such factors as detection range, weapon-firing range, and weapons effects during simulation runs.

Describing some aspects of a force may not be practicable. While the doctrine might be described, what is more useful is the effect of doctrine on behavior

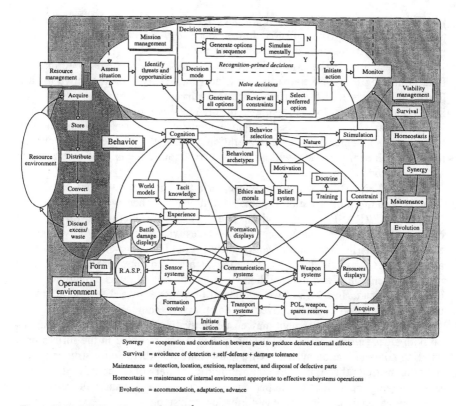

Synergy      = cooperation and coordination between parts to produce desired external effects
Survival     = avoidance of detection + self-defense + damage tolerance
Maintenance  = detection, location, excision, replacement, and disposal of defective parts
Homeostasis  = maintenance of internal environment appropriate to effective subsystems operations
Evolution    = accommodation, adaptation, advance

**Figure 10.18** *GRM for Blue or Red force.*[6]

selection and hence on mission management. Similarly, describing ethics and morals is of limited value: again, their impact on behavior selection is of key interest.

Mission management in this context is recognizable as the command and control process; see Decision Making on page 71. Both naïve and recognition-primed decisionmaking (RPD) may be incorporated into the mission management process; see Figure 10.18, together with behavioral logic determining under what circumstances each method is employed. For instance, it may be that naïve decisionmaking is used by a $C^2$ team until there is an immediate crisis, at which time a commander, operating in RPD mode, may impose his experience on the proceedings.

---

[6] Professor Mike Moulding of the Royal Military College, Shrivenham, U.K., pointed out that, for sociotechnical systems, the GRM could be reconfigured by "shoehorning" technology, people, and process into the form, behavior, and mission management layers, respectively.

The instantiated GRM now represents the MLF technology, the people and the process for both Blue and Red forces, since they are identical. Operation of either force may be examined by "walking through" the model.

First, establish a scenario and an environment. A mobile enemy formation, Red, is moving in a desert area toward Blue which is itself on the move toward Red. Blue detects Red first, using electronic surveillance measures (ESM). The intelligence appears on a display; the RASP. An operator observes the RASP–assess situation in mission management–and interprets it (cognition in behavior management). This interpretation primes the $C^2$ personnel, whose response–behavior selection–will be a mix of reaction (especially at first contact) and considered judgment.

As Red and Blue formations close, primary radar shows Red on Blue's displays, and more information is gathered until a comprehensive situation display is presented (RASP; form model). Blue's $C^2$ team identifies threats and opportunities (mission management,) and decides what it intends to do within the confines of doctrine and rules of engagement set out by higher authority. Blue has a limited set of options. It may: engage the enemy head on; approach to the limit of engagement range and hold; withdraw; or ignore the enemy as though not seen, or not significant. In the first two options, it may move toward favorable ground as a precaution. Red's similar moves will be observed and interpreted by Blue.

At some point, engagement may ensue, activating hard and soft kill weapons on both sides (initiate action in mission management, repeated in the form model entering communication system). Oral and electronic orders and instructions will be transmitted through the communications network, coordinating and synchronizing the actions of the personnel, sensors, weapons and vehicles as they sense, move, navigate, fight, change formation, and so on.

Blue and Red will incur damage, there will be a need for running repairs, the maintenance staff will be involved in mending faulty and broken equipment, casualties will be tended and evacuated, and fresh supplies of fuel, weapons, equipment and vehicle spares, and soldiers will be brought in.

Such walks through are useful in understanding how the model works (and how the real world might work), and in seeing how the various parts might interact. It is, however, qualitative. To progress further, quantitative simulation is needed.

Nonlinear Dynamic Simulation Modeling

*Developing and Running the Simulation*

The move from GRM to nonlinear dynamic simulation is straightforward and methodical. First the Blue force is created in STELLA, or some other similar tool, using a straightforward one-for-one correspondence between the GRM of Figure 10.18 and the STELLA model, wherever possible. Each of the three central modules is created separately and tested, before interconnecting and testing them, using standard test signals. Other modules are created for maintenance and logistics, and for capital and running costs (including the costs of incurred damage). Synergy is represented in the interconnections between the parts of the model that emulate those of the real world. Scenario models are also constructed, including weather, terrain, location, speed of movement, and relative separation between Blue and Red.

Once the basic model of Blue is constructed, its outputs are connected via delays (representing separation distance between Red and Blue) to its inputs. So radar transmissions are routed via representation of the environment and weather, and reflected to radar receivers, again through environment and weather. Inverse fourth power law considerations are applied. Jammers are connected similarly, except that their output is fed back to ESM receivers, using inverse square law rules. Weapons that are fired or launched return to hit and cause damage. Damage is assessed, using damage and lethality models.

In this manner the model is progressively set up and developed until the parameters used in the model match those of Blue system's technology and $C^2$ process. Behavior requires a different approach. The result of good training is assumed to be evident in the rapid, accurate recognition of situations and targets, and in a shorter response delay. In a drawn-out exchange, it would also present itself as persistence[7] and an unwillingness to give up.

Doctrine is represented in different rules of engagement which can be chosen prior to a simulation run, or which can be set to load automatically. For instance, if the Blue force were to find that it was receiving much more damage than it was inflicting, an automatic rule of engagement might be activated to withdraw, regroup and repair damage. Such rules are behavioral archetypes. The model accumulates knowledge of relative damage rates via experience in the behavior model: the model is able to change its behavior in the light of experience, a capability that is often regarded as a sign of intelligence.

Once the Blue simulation model is operating dynamically, with all parts either active or represented, then Blue is duplicated, the duplicate is retitled as Red, and the two models, Blue and Red, are interconnected, using the various models of interactivity constraints as before. There should be no change in Blue system behavior at this point, since it has gone from being connected to itself to being connected to a replica of itself; see Figure 10.19.

Running the simulation should result in identical results for both Blue and Red. Many of the effects may be random, however, notably, damage effects, and some weather and terrain effects. Differences can accumulate, so that the behavior of Blue and Red can be quite different on successive simulation runs. Running the simulations repeatedly for up to, say, 1,000 times results in a distribution of outcomes, which may be analyzed statistically to give a spread of probabilities.

---

[7] Maintenance of the aim is one of the principles of war.

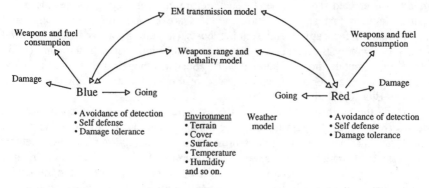

**Figure 10.19** *Symmetrical Red Blue interchange model.*

Conversely, the various random effects can be given seed values such that, although the numbers generated are random, the same sequence of random numbers is generated run to run. By giving Red and Blue the same seed values, the two simulations can be brought into line. This second approach may not be as accurate, but it is much faster and allows more options to be simulated and assessed. Deductions made using this second, faster approach can then be backed up–or not–using the full, statistical approach. Sensitivity analysis can also be applied, varying one or more parameters over a wide range to see the effects on overall behavior of varying one parameter, or combinations of parameters.

*Results, Using the Nonlinear Dynamic Simulation Model*

The first results derive from testing the model by setting up a simulated engagement. The scenario may be set to consider two identical forces closing in on each other in an arid desert. As Red and Blue close, each detects the other, closes to within weapon launch range, and fires, again and again.

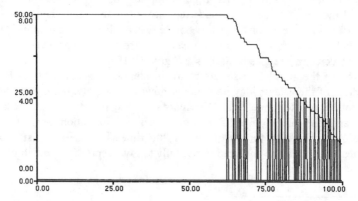

**Figure 10.20** *Weapon firing pattern.*

The consequent firing pattern is shown in Figure 10.20, and is taken directly from the STELLA model. The x axis shows elapsed time. The y axis shows the number of weapons in hand, available to be launched, as a continuous line; as the engagement proceeds, the number of weapons still available is reduced. The y axis also shows individual weapons being launched, as a series of pulses. Each pulse lasts for one fourth of a time unit with amplitude 4, so each pulse has an area of unity, indicating one weapon launch.

Not evident from the graph is the change in the decision making mode. Until time 75, weapons are launched as a result of the naïve decision making mode. From 75 to 100 time units, weapons are launched in the recognition-primed decision making mode. The changeover occurred because the two forces came close to each other and decision time was reduced below a time threshold that precluded structured plan formulation. Instead, weapons were launched at every launch opportunity.

A rather different assessment can be made using the nonlinear dynamic simulation test bed. It is possible to represent individual items of military hardware as being either bespoke, meaning specially made, or COTS, indicating that it is commercially available. Blue can be equipped, for instance, with a COTS weapon that has exactly the same performance as the bespoke weapon fitted to Red. Blue's COTS weapon will, however, be cheaper.

Running the simulation with a COTS weapon for Blue results in Figure 10.21. As before, the x axis represents time, as the two forces, Blue and Red, close in on each other. One y axis represents cost effectiveness, which is scaled to have a maximum value of 100%. Cost effectiveness is calculated as follows: (Capital cost − combat cost)/Capital cost x 100%, where the costs refer to Blue or Red as appropriate. The second y axis represents the cost exchange ratio, where the ratio would be one to one (1:1) for equal Blue and Red costs.

Red cost effectiveness starts out marginally higher than Blue cost

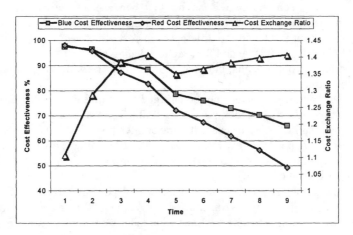

**Figure 10.21** *Cost effectiveness and cost exchange ratios.*

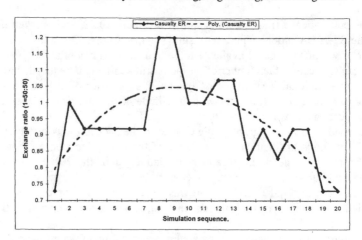

**Figure 10.22** *Varying Blue radar transmitted power.*

effectiveness in Figure 10.21; the situation reverses once the shooting starts, with Blue a clear winner at the end. Since the performance of the two weapons is identical, the reasons for any differences must be concerned with money. While Blue's COTS missiles are cheaper, COTS maintenance is based on throwing away the whole weapon when faulty. The bespoke Red missile, on the other hand, is modular and fault repair is based on module replacement, where modules are relatively cheap. Blue must stock complete missiles as spares while Red can stock modules as spares.

At the start of the simulation, Blue is marginally less cost effective because spares costs are higher. Once the shooting war starts, however, Blue launches the same numbers of weapons, but each weapon is cheaper, so Blue's combat costs are less than Red's; hence Red loses out in the cost-effectiveness comparison. The difference between the two is so marked because so many weapons were launched: had the kill probability ($P_k$) been much higher, then fewer weapons would have been fired, and the differences in cost effectiveness would have been reduced. As it is, with Blue and Red each launching some 39 weapons, each force received only two direct hits; but then, hitting a dispersed, moving target is never going to be easy.

Radar transmitter power may be changed for Blue ship, leaving Red ship as an interactive reference model; see Figure 10.22. The x axis shows the results of a series of 20 simulations: in each, the transmitter power was increased over the previous run. It might be expected that greater transmitter power would lead to longer detection ranges and earlier firing opportunities, thereby giving a significant advantage.

As the graph shows, that was not the case. Upon investigation, it transpired that increasing Blue radar transmitted power gave Red an early advantage, because Red's ESM was able to detect Blue radar from further away. Since radar works to an inverse fourth power law, the advantage fell to ESM, working to an advantageous inverse square law.

What Figure 10.22 appears to show is that the optimum radar transmitter power for Blue was the same as that for Red. That would be a false conclusion, however, since it is based on varying only one parameter. If Blue's radar power were varied in synchronism with Blue's radar receiver sensitivity and Blue's ESM sensitivity, then there would be a combination of these three parameters that gives an optimum casualty exchange ratio, and Blue's radar power would not be the same as Red's.

Following this line of argument suggests that an optimum capability might be identified if all the parameters that impact the overall Blue capability could be varied at once; this notion will be examined later.

While the nonlinear dynamic simulation can be used to explore many different aspects, one of the more unusual is to trade off training and technology. The behavior model of the GRM contains elements (ethics, morals, stereotypes, beliefs, and so on) that are important to behavior; yet, not enough is known about their effects to allow sensible representation in such a simulation. Since both Red and Blue forces are identical, it is reasonable to assume that these behavioral influences are the same on both sides. Red and Blue forces will have had the same training, come from the same culture, have the same instincts, and so on. As we have seen, both forces have a limited set of potential behaviors, typified by the rules of engagement, and constrained by training.

The net result of this is that many of the behavioral effectors on each side will either cancel each other out or be irrelevant. Some, however, will not. Doctrine, as we have seen, determines behavioral "norms"—what is expected to happen under given circumstances. Training influences the speed of response and correctness of interpretation.

By using the simulation as a test bed, the effects of training can be represented by changing the time it takes within the model to formulate a decision

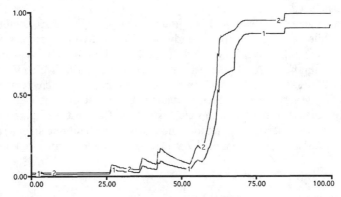

**Figure 10.23** *The effects of training on situation assessment: Blue force, line 1; Red force, line 2.*

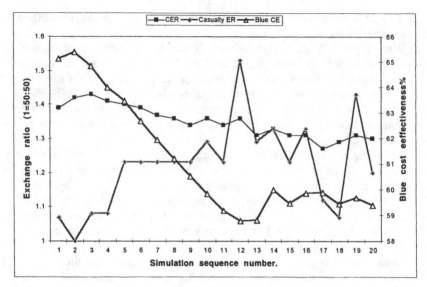

**Figure 10.24** *Technology versus training trade-off; see text.*

and to recognize a threat or opportunity. It is also possible within the simulation to represent "wrong" decisions, that is, where doctrine and rules of engagement indicate one choice, but the commander makes another. So, holding Red's training constant and varying Blue's may reveal the effect of different levels of training.

As an example, consider the impact of training on situation assessment, Figure 10.23. Blue, with poorer training, is both slower to size up situations and establishes an inferior situation picture, or RASP.

A second example might look at the tradeoff between spending money on training or spending it on developing and fielding smarter weapons. If the weapon is smart, the argument goes, then it is easier to use and continual training costs may be reduced.

This idea can be explored by simultaneously varying three factors for Blue force, while keeping Red force steady as an interactive reference: weapon cost; weapon probability of hit, $P_h$; and training level. Figure 10.24 was drawn up on the following basis. A 20-year period was assumed, during which weapons costs rose by 230%, $P_h$ rose from 0.3 to 0.7, and training costs were reduced by one-third. The graph presents three measures: Blue cost effectiveness, which fell sharply because of the rising missile costs; Blue-Red cost exchange ratio (CER), which also fell but only marginally; and Blue-Red casualty exchange ratio which, although variable, did rise markedly as the more accurate weapon took its toll.

*Optimizing the Design Capability*

The test bed of interacting Red and Blue MLFs can be used to improve the design of Blue, that is, to make it more effective, using genetic algorithmic techniques. Instead of varying the performance and cost of one item in the Blue MLF, it is possible to vary many simultaneously and randomly.

The method is cumulative selection; it mimics how nature is believed to work by the genetic selection of offspring that are most suited to the environment. Suppose, as a very simple example, that one gene "codes for" radar transmitter power, while another codes for weapon probability of hit ($P_h$), and yet a third codes for the number of weapons held ready to launch. These three items have little in common except that they all contribute to mission effectiveness. Greater radar power enables a greater detection range but, as we have seen, may render the MLF more detectable. Greater $P_h$ will result in more hits on target, but can $P_h$ be increased, and at what cost? The number of weapons held ready to launch improves the prospect of winning in an all-out exchange of weapons: to run out of weapons would be to lose, but at a cost.

If the value of each of these parameters is changed, then Blue effectiveness will be changed, in several ways. More missiles would cost more money, which would affect cost calculations in any cost-effectiveness or cost exchange measures. Similarly, increasing the $P_h$ might make the weapon more expensive, but on the other hand should reduce the number of weapons launched to achieve supremacy; it would affect measures based on casualty exchange ratios. The radar range affects the maximum range at which weapons can be launched, and so may affect the numbers that can be launched. At the same time, increasing radar power to enhance range, because it makes the MLF more detectable to an enemy, also

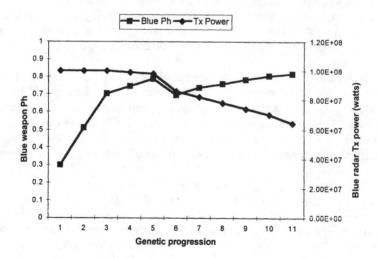

**Figure 10.25** *Optimization: weapon and radar.*

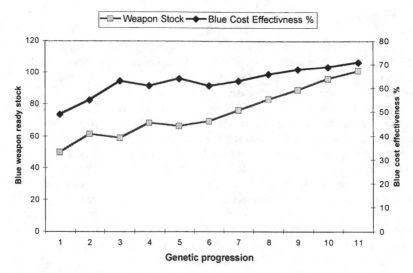

**Figure 10.26** *Optimization: weapon-ready stock and Blue cost effectiveness.*

increases the risk of costly damage from Red weapons. There is evidently a high degree of effect and countereffect to unravel.

One solution is to evolve the ideal combination. Here, the value for which each gene codes is varied randomly on a run-to-run basis, and a record is kept of the resulting values for, say, cost-effectiveness difference (that is, the difference in cost effectiveness between Blue and Red MLFs at the end of a simulated engagement $CE_{diff}$). The combination of gene settings corresponding to the greatest $CE_{diff}$ is also recorded, and then used as the new gene settings for a new set of runs. This process is repeated continually for several generations, until the value of $CE_{diff}$ fails to rise any further. At this point, the gene settings correspond to the values of the various parameters that would create the greatest $CE_{diff}$. This is design optimization.

Some results from the very simple example are given in Figure 10.25: these were derived from 10 generations, starting off with the same settings in Blue as in Red. Holding Red constant as an interactive, dynamic reference, the maximum $CE_{diff}$ from 25 simulated engagements was selected, and resulted in the second plot on each of the two lines. Using the corresponding new gene settings, another 25 engagements were simulated, and again the maximum $CE_{diff}$ was selected for the third plot; and so on.

Not surprisingly, the results in the graph show the Blue $P_h$ rising from 0.3 to 0.8. Less obviously, radar transmitted power falls from 100MW to 64MW. Parallel results shown in Figure 10.26 show Blue weapon-ready stock rising from 50 to 100, in spite of the weapon being expensive (in the model) and in spite of its $P_h$ rising, too. The graph shows a dramatic rise in Blue cost effectiveness from

49% to 71%: what is not shown is the value of $CE_{diff}$, which rises from zero (with Blue and Red equal at the start) to over 38% after 10 generations.

Since Red force, being unchanged, represents the original design, here is a way of significantly improving the designed capability of–in this case–a military force. To undertake the task properly would require significantly more than three "genes" to be employed. In this type of problem, 20 or 30 genes might be more usual, and the number of generations would be in excess of 30. Moreover, optimizing the design using "self" as the dynamic interactive reference goes only so far. The optimized design should be pitted against a variety of typical and some nontypical foes, using different tactics, sensors, and weapons, in many different situations. General purpose modeling tools such as STELLA may prove cumbersome for dealing with such problems, and special programs may prove more appropriate.

During the activity of design optimization, some factors were not included, on the basis that they were likely to cancel each other out as being the same in Red and Blue. Ethics was one such factor. This may be expedient, even reasonable, but it is unsatisfactory. It is sensible to reintroduce such factors to both parties, and to vary Blue parameter values while holding Red's steady. It may be that the effects are minimal, in which case it is reasonable to assume that leaving the parameter out in the first place was a good choice. On the other hand, the effects may be significant, suggesting that research into the subject parameter is called for. The outcome of such research should enable future models to represent the parameter and its effects explicitly.

*Caution*

The results presented in all of the examples above are speculative. The process is one of exploration and learning, rather than precise engineering. However, using

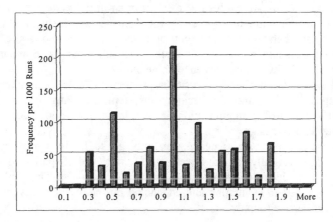

**Figure 10.27** *Casualty exchange rate variability.*

figures allows us to explore the potential for counterintuitive effects that emerge from the model.

Results may also be marginal, bearing in mind that the model has been set to particular strings of random numbers. When the seeds restraining the full random nature of the simulation are removed, the model produces quite different results on a run-to-run basis, and many of the effects such as saving money on training and by buying COTS equipment may become overwhelmed.

Figure 10.27 shows the effect. With the model operating in full random mode – which affects mostly weapons performance and weapons effects – and with Blue and Red nominally equal (CER = 1 in the graph), the histogram shows the spread of results for the casualty exchange ratio over 1,000 runs. The median and the modal value are both 1, but the spread is wide, with only 21.5% of results occurring at the nominal value.

## Summary

Using instantiated interacting GRMs in the manner described is a powerful way to develop, evaluate, and optimize system designs in dynamic operation. The method is scientific, holistic, synthetic, and–particularly–organismic. Where counter-intuitive behavior emerges, sources and conditions can be traced, allowing designs to be modified. Because of the way the model is constructed, it is possible to relate the effects of such things as equipment failure rates, spares holdings, morale, training, experience, and decision making processes to overall capability.

The optimization process is particularly powerful in another respect. The conventional way to design a system such as the MLF is to identify a variety of existing or planned subsystems and equipment, bring them together, and interconnect them. The emergent properties, capabilities, and behaviors of the whole system are largely defined by the initial choice of parts.

With the optimization process, the key performance characteristics of the parts and subsystems are evolved as a set from an initial system design. Each is designed to operate and interact to create the requisite overall emergent properties, capabilities, and behaviors. Some contribute less than expected; some more. All contribute their ideal respective amounts to ensure success. Some parts will require less capability and performance, some more.

The output from the optimization process, then, is a matched set of specifications for each and every major subsystem, including sensors, weapons, command and control, communications, armor, maintenance, logistics, and so on. This is total system design.

This method of systems design can only go so far in a military context, because it does not fully explore the effect of terrain, and of spatial disposition. It is useful as a precursor to such deeper exploration, however. The method is also particularly suited to evaluating and enhancing business competitiveness, where no such considerations apply.

## ADVANCED DESIGN METHODS

Some problems necessitate the use of spatial models and spatial dynamics. An approach to addressing such issues invokes the use of intelligent cellular automata (ICA) to represent entities or systems, moving across "landscapes [7]". The automata are called intelligent because they are able to interact with the landscape (which may be a multilayer, multidimensional computer landscape) and change their behavior in consequence.

As a simple example, an ICA representing a soldier need not walk into water, but would find a way around water to reach an objective. The soldier/automaton would also seek to associate with fellow soldiers in the process, and the group of soldiers would "see" an enemy when the local terrain presented a "line-of-sight" view. Having seen an enemy, the group would act according to rules of engagement: take cover, move to engage, engage, evade, withdraw, report, and so on. The performance of any and all of these activities would be affected by terrain, surface conditions, weather, and visibility. So, the use of ICAs enables the creation of spatial models with spatial dynamics and realistic behavior.

An approach to the design of complex systems has been developed (by A. Christensson, L. Cobb, D. K. Hitchins and A. E. R. Woodcock) that brings together ideas of optimization and the use of ICAs operating across landscapes to create a powerful tool for the analysis, design, and optimization of large-scale, international issues. In brief, the approach establishes a problem space; conceives a potential solution; explores the solution using nonlinear dynamic systems modeling; identifies key "genes;" creates a specific ICA/landscape model using information gleaned from the nonlinear dynamic systems model and features of the specific landscape–geography, demographics, the political situation, resources, threats, damage, and so on; establishes a nominal solution using genes to code for different elements of the solution; optimizes the solution; and solves the problem scientifically, according to Ackoff.

### Famine Relief

As an example of the type of complex issue that can be addressed using this approach, consider the issue of famine relief to a country racked by civil war, which has just experienced a severe natural disaster—a hurricane:

> *The country's infrastructure is shattered. Remote villages are cut off, without food or fresh water. There are few stocks of food. Transport is available, but roads are out and bridges are down. Added to that, insurgents are wandering the countryside, looking for resources and terrorizing the inhabitants.*

In such a parlous situation, what is the best way to go about bring relief to the country, which is facing famine?

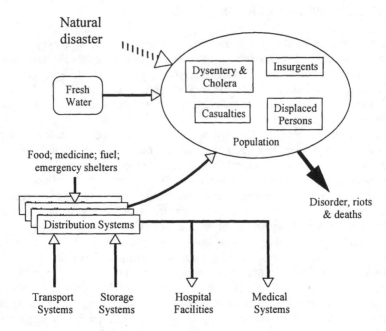

**Figure 10.28** *Disaster relief.*

The situation is represented graphically in Figure 10.28. Outbreaks of dysentery and cholera are inevitable, unless clean, fresh, potable water is made available. Storage, transport, and distribution systems must be installed or restored, and will need protection and security. Key bridges will need to be rebuilt, or temporary structures erected. Medical facilities will be needed to deal with isolated communities, where the sick and injured cannot reach a doctor or a hospital, and so on.

Genes are identified that code for characteristics of the disaster relief "force:" bridge-building facilities, numbers of trucks, numbers of mobile hospitals; numbers of doctors, and so on. A nonlinear dynamic systems model is constructed as a learning laboratory to explore aspects of the problem; Figure 10.29 shows a high-level map of the model.

Figure 10.30 shows detail of the transport module shown in Figure 10.29. Like the other modules, it is relatively simple. Vehicles are dispatched, at the bottom of the model, and return later either still serviceable or in need of repair. The time they take for their journey to delivery relief resources, and the likelihood of needing repairs, are influenced by the onset of the hurricane and by its severity. A choice exists, on the one hand, between supplying the indigenous transport organization with spares so that it can repair and maintain its own transport, and on the other hand supplying Humanitarian Relief Organization (HRO) trucks. Spares, if they are available, should cost less than trucks, and providing spares

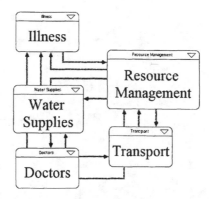

**Figure 10.29** *High-level disaster relief model.*

helps the stricken nation to help itself. Both HRO trucks and vehicle spares would be candidates for genes.

Like transport, the other modules are developed and tested separately before integration to make a full model. As it stands, it is useful; users of the simulation can see how various factors might interact. For instance, repairing bridges to aid transport may also trigger a mass exodus from badly affected areas that will exacerbate transport and medical relief problems. However, the main limitation is the absence of specific, geographically defined, political influences, damage and threats in the model.

The next stage in the development of a solution design is the construction of a simulation which contains a map of the area, and the superposition on the map of the population, road and rail links, washed out bridges, political and geographic boundaries, and so on, together with ICAs; principally representing actors in the relief exercise.

In the case of transport, genes might code for the number of HRO trucks, for instance; an increase in the gene value would correspond to an increase in the number of trucks, and, of course, in the consequent cost. The result of the increase might be that more food reaches some outlying district and, as a direct result, lives are saved. All of which would depend on the roads being passable, with bridges in place, food available to load on the trucks, local insurgents under control, no fresh floods, and so on.

Each aspect of relief operations is fraught with complications, ifs, buts, and maybes. Once the model is constructed and tested, the optimization algorithm can be employed. Trying many different combinations of gene values eventually results in just one or two specific combinations showing the best result, where "best" may mean most lives saved per cost, for instance, or perhaps the most lives saved within a critical period. Overall, the result will identify the best combination of resources, how they should be located, how they will reach the disaster victims, and how the most lives can be saved.

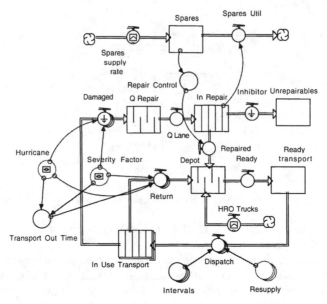

**Figure 10.30** *Transport module: detail.*

The optimum gene values correspond to the optimum design of the relief process, at least under the conditions pertaining in the country that are represented in the model. It is also possible, with care, to vary the conditions to represent what might happen in the near-term future; winter is imminent, another hurricane is due, and so on. Using this enhanced approach it is possible to suggest how the relief process might need to evolve with the changing situation in the country.

The model may be used in several ways:

- To train disaster relief managers:
    - The simulation may be used, without optimization, encouraging managers to propose how they would go about organizing relief.
    - Their inputs are entered into the simulation, which is then permitted to run for a simulated time of, say 1 week.
    - The managers are then presented with the new situation, showing the result of their actions. They introduce new requirements for relief, and the process repeats.
    - If they are good managers, they will save many lives and waste little money.
    - If, for instance, they pay less than sufficient attention to providing potable water, then epidemics of cholera may be expected, with which they will then have to cope.
- To design disaster relief programs:

o By using the optimization routines and varying the situation, a disaster relief team may identify the best strategy for cost-effective relief, or for saving the most lives in the least time, regardless of cost.

o The program could also be used in real time, in the field, feeding in emerging situation data, and running the model on a daily/weekly basis to see what the best tactic is likely to be in the near term.

It would be a mistake to believe that any simulation is able to accurately predict future events; the best that can be hoped for is an indication of likely trends. The kind of simulation model outlined above is useful as a training aid, perhaps even as an adviser, but in the final analysis the planner and manager must come to their own judgments. This kind of model can be improved by continual updating in the field, so that predictions of what should be the best relief plan can be continually adjusted in the light of outcome. Since each set of starting conditions is unique, such experience-based predictions can still never be perfect, but they may improve with time.

## Afghanistan

The method described above can obviously be brought to bear on the issues facing Afghanistan. The problems facing that country are enormous, and the United Nations has a duty of care, which it is addressing as best it can within its resources.

Any simulation approach to such a problem faces immense difficulties:

- The nation, self sufficient in food less than 50 years ago, has been devastated by famine.

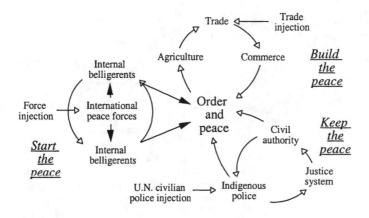

**Figure 10.31** *Peace operations: CLM.*

- There is a severe shortage of water: farmers are drilling deeper wells to gain farm water; as a result, shallower village wells are drying up.
- Heroin is made from the staple poppy crop.
- The infrastructure has been destroyed: there are few passable roads and virtually no rail links.
- The land is ethnically diverse and is divided into some 32 provinces, with warlords flouting the authority of the fledgling central government.
- The threat of resurgent warfare hangs in the air.
- Crime is rampant.
- The police force has been decimated, and courts are held infrequently, if at all.
- Domestic housing has been razed by warfare.
- Women have been subjugated.
- Women doctors, nurses, and teachers have been isolated and prevented from practicing.

Not only is it difficult to see how to cope with the situation, but it is also vital that any solution be effective. The lives of a nation are at stake and, moreover, the peace of a region may depend on a good solution.

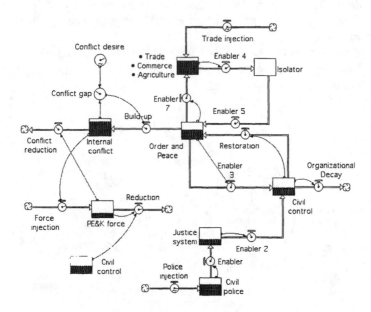

**Figure 10.32** *Initial peace operations model.*

The U.N. has developed an approach to addressing such major issues, called "peace operations." Peace operations are comprised, in principle, of three phases:

- Peace enforcement–the imposition of peace by force giving sufficient time for a negotiated settlement between warring parties;
- Peacekeeping–the maintenance of peace, with a reduced military presence acting largely in a policing role, while the organs of state are reconstituted and preparations are made, hopefully, for democratic elections;
- Peace building–the process of building societal and economic capital so that the once war-torn country can return to the family of nations as a full and contributing member.

A conceptual model of U.N. peace operations, as they are expected to work, was developed: first as a CLM Figure 10.31, and then as a corresponding STELLA model; see Figure 10.32. The model is shown with its animation effects. A half-full reservoir of order and peace is seen at right of center. At left, the injection of the peace enforcement and peacekeeping (PE&K) force reduces internal conflict. When civil control is restored, it will reduce the demand for PE&K Forces.

At center bottom is the restoration of the rule of law, with an injection of civil police enabling the justice system to restart, thereby raising the potential for restoring civil control.

The top right shows trade, commerce, and agriculture being kick started with

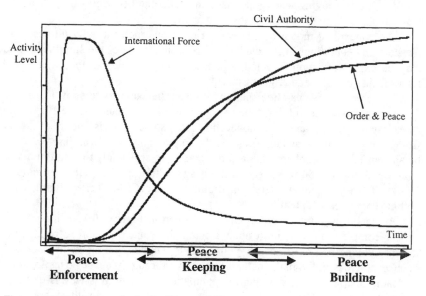

**Figure 10.33** *Results from simulation at Figure 10.32.*

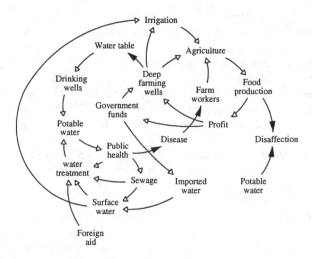

**Figure 10.34** *Water and sanitation in Afghanistan.*

trade injections that, once started, can be stopped, as the factors should be self-sustaining and grow once there is a peaceful and ordered environment.

Figure 10.33 shows the result of running the grossly over-simplified model at Figure 10.32, and the results are not unreasonable. The international force grows initially, only to tail off in size as civil authority is restored, along with peace and order. Politicians much prefer the three phases to be non-overlapping but, as the graph suggests, judging when one phase ends and another begins may difficult in practice: at best, it will be a qualitative judgment.

The conceptual model of UN peace operations took virtually no account of the state and conditions in Afghanistan. The basic outline of the simulation model may be expanded, supplemented, and instantiated to address the extent of the issues facing the UN–and Afghanistan.

The process of expanding the model takes the same route as in RSM; issue symptoms are sought and developed into a causal loop model. Figure 10.34 shows an example, one of many. The model shows that some water is imported: it comes from a lake in Turkmenistan, to the north, and supplies are dwindling–water will be an ever-present problem. The lack of potable water is likely to result in disease, which could affect farm workers, hitting agriculture as the country attempts to recover from years of warfare and drought. If the CLM seems complex, it is nothing compared with reality.

A second example of initial CLMs is given in Figure 10.35. This may also appear complex, but it is built upon a chain showing, from left to right, births, infants, youths and adolescents, young adult females, young adult males, and older people. As individuals mature, they move through the categories from birth to death. One reason for drawing the population model in this way is that infant mortalities are very high in Afghanistan. Experts believe that there are several

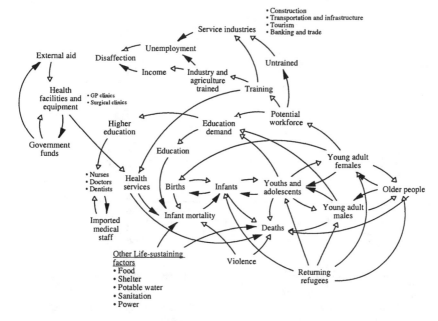

**Figure 10.35** *Population, infant mortality and education.*

ways to reduce the mortality rate, shown in the CLM, including, particularly, the education of young mothers. Young adult females are picked out in the CLM because, having been subjugated, experts believe that there is a need to help them re-emerge, to be educated and trained, not only in childcare, but also as doctors, teachers, and more generally as contributing members of society.

CLMs were developed for conflict, civil unrest, economic factors, restoration of the rule of law, energy and power, industries, manufacturing, agriculture, media, narcotics, quality of life, and others. In each case, the CLMs were based on information specific to Afghanistan, or to regions within Afghanistan. Care was taken not to import Western cultural notions accidentally. For instance, the economy of Afghanistan is fundamentally different from that of, say, the U.S. or a European country. One apparent exception, the education of women, is a cultural issue within the Islamic world, with different nation states adopting different stances.

The CLMs were brought together in a large STELLA model, representing the different situations in all 32/33 provinces in Afghanistan. The model was constructed as separate modules; each being tested and proved independently before integration. The level of representation in each module might best be described as organic; for example, in the rule of law, when organs of civil control are restored–police forces, lawyers and judges, courts, prisons, and so on–progressive restoration of civil control is presumed to follow. All organs must

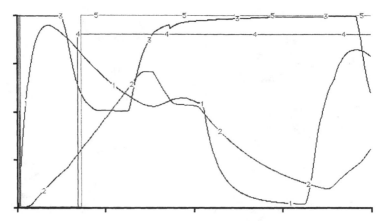

**Figure 10.36** *Afghanistan social dynamics. Line 1: civil conflict. Line 2: international peace enforcement. Line 3: civil control. Line 4: availability of potable water. Line 5: availability of food.*

be restored for full civil control. Conversely, the loss of any organ, due perhaps to local conflict, would threaten civil control in that area.

During the process, a number of indexes were created, including a quality of life index, showing how the people of different regions within Afghanistan might be expected to feel about their respective situations. To exhibit a high index value, there would be for the population at large: food, potable water, shelter, power, employment, heath services, infrastructure, service industries, and low ethnic tension. Deficiencies in any one of these would lower the quality of life, increase disaffection, and increase the prospects for conflict and warfare.

Figure 10.36 shows a typical simulation scenario unfolding. At the start there were shortages in food and water that triggered violence. Civil control broke down, and international peace forces were brought in. The unrest was quelled sufficiently to restore civil control, which caused violence to diminish, with the Peace Forces still present but in decreasing numbers. Continued water shortages eventually caused a fresh outburst of violence and local warfare, and the cycle started to repeat.

The model could be used for a variety of purposes, but it was set up particularly as a learning laboratory, so that the effects of aid and inward investment could be explored. There are so many demands for aid and investment that choosing how to allocate funds is a major problem. Investing differentially in any one area may affect other areas favorably and unfavorably. The objective of the overall investment aid plan is to restore Afghanistan to self-sufficient independence. The learning laboratory is designed to help planners to formulate effective plans.

The next stage in the development process is to transfer what has been learned from the STELLA model into a map-based ICA model, which is better able to represent individual regions, infrastructure, power and energy distribution, agriculture, and so on. Within this environment, the role of optimization is less obvious. Perhaps it may be employed to discover optimal investment plans.

Overall, the enterprise is ambitious, but within the scope of the concepts, methods and tools described above. This is systems thinking, concept development, systems design, and systems engineering rolled into one.

## SUMMARY

The chapter has introduced and given examples of scientifically based approaches to the development of open systems concepts and designs. The methods have employed tools and concepts introduced in previous chapters; in this chapter, they have been used in combination, creating what might be called methodologies. Care has been taken throughout to avoid reduction, and to maintain the principles of synthesis, holism, and organismic construction. Risk analysis, often applied after the design, if not after the event, is seen in this chapter as an intrinsic part of the conception and design of a system; not as an add-on.

Designing a system is feasible only in the context of other systems with which it is interacting. This is because the actions and reactions of other systems affect and enable the system being designed. It is no more possible to design a system in the absence of its interactive context, than to design an engine without knowing what fuel, or energy source, it is to use, or what load it is to drive.

That these approaches are radically different from the more common reductionist methods used, for example, in the design of software, should give food for thought. Approaches based on functional decomposition, often applied to systems engineering, are reductionist by definition and, although convenient and seemingly obvious, do not and cannot afford a sound result. Nor are they mathematically sound; recombining functionally decomposed parts should, but does not, result in the original whole.

By avoiding reduction, system concept and design are able to look outwards rather than inwards. One evident result is the opportunity and ability to address systems on virtually any scale; hence the international relief examples. However, the ideas and methods are scale independent, and are as applicable to the design of a new radar system as they are to a UN intervention.

## ASSIGNMENTS

1. Develop a business strategy for the scenario on page 270 by adopting Ackoff's approach of dissolving the problem.
2. Apply the GRM to a rowing eight, using Figure 8.32 and Figure 8.33 as a guide, and develop a dynamic model of the annual Oxford and Cambridge University boat race, held on the River Thames in London,

England each spring. What factors would you expect to contribute to one boat beating the other?

3. Research the human immune system, and conceive the high-level design of an autoimmune system for a personal computer. Identify the strengths and weaknesses of your design, and suggest why personal computers do not have such devices.

4. Propose what modules you would create in developing a full model to represent peace operations in Afghanistan. Develop an $N^2$ chart with the modules you propose on the leading diagonal, and identify in the other spaces on the chart, in words, what each module contributes to the others.

5. Supposing a full simulation model of peace operations were available, describe how it might be used, by whom, and to what advantage.

## REFERENCES

[1] Hitchins, D. K., "MOSAIC Concepts for the Future Deployment of Air Power in European NATO." In *Advances in Command, Control and Communication Systems* pp. 50–83, C.J Harris and I. White (eds.) IEE Computing Series 11, London, England: Peter Peregrinus, 1987.

[2] Hitchins, D. K., *Putting Systems to Work*, Chichester, England: Wiley, 1992, pp. 214-239.

[3]Sun Tzu, Wun, *The Art of War*, London, England: Hodder & Stoughton, 1981.

[4] Wee, Chow Hou, Lee Khai Sheang, and Bambang Walujo Hidajat, *Sun Tzu: War and Management*, Singapore: Addison Wesley, 1991.

[5] Ackoff, R. L., *Creating the Corporate Future*, New York, NY: Wiley, 1981.

[6] Herrigel, Eugen, "Zen in the Art of Archery." In *The Little Zen Companion*, p.25, Schiller, David (ed.), New York, NY: Workman, 1994.

[7] Dockery, J. T. and A. E. R. Woodcock, "Modeling combat adaptability with cellular automata," *Proc. 1990 Command and Control Research Symposium*, sponsored by the Basic Research Group, Joint Directors of Laboratories, McLean, VA, 1990.

# Chapter 11

## Classification of Systems Engineering

*The energies of our system will decay, the glory of the sun will be dimmed, and the earth, tideless and inert, will no longer tolerate the race which has for a moment disturbed its solitude. Man will go down into the pit, and all his thoughts will perish.*

The Foundations of Belief, *A. J. Balfour, 1848-1930*

### DEFINING SYSTEMS ENGINEERING

Architecture is defined as the art and science of creating buildings. Systems engineering may be similarly defined as the art and science of creating systems.[1] The process of creating a system starts at the point where a problem or issue is identified and a solution is postulated in the solution space.

There may be several, or many, potential solutions to a problem. This book emphasizes the scientific solution; yet, within the scientific solution, there may be art, too. Conceiving solutions to problems is creative, requiring perception, imagination, understanding, and judgment. One solution may be more aesthetically appealing than another, although both may be equally justifiable and effective in other respects. Often, too, in the practice of system design, choices may be made in the absence of solid information about the future in which the designed system will exist and function. There is something of an art in making such decisions; evidently, some people are better at it than others.

Systems engineering extends from the solution concept, through the life of a system, to its demise and replacement. For every system there is a life cycle; see Figure 11.1. The figure shows the seven ages [1] of "system," by analogy with Shakespeare's seven ages[2] of man. There is a separate "age" for transition to use; those systems new in operation take time to achieve their full potential, just as recent college graduates take time to become useful employees in a company or

---

[1] There are many more detailed definitions, but almost universally they try to define how systems engineering works; not what it is.

[2] These are: infant, schoolboy, lover, soldier, justice, pantaloon, and second childhood—"sans teeth, sans eyes, sans taste, sans everything."

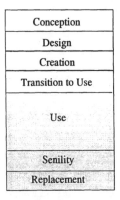

**Figure 11.1** *The seven ages of "system."*

business. There is an expectation of technological systems that they will operate fully, efficiently, and effectively from the moment of delivery. The more complex the system, the less likely is this expectation to be realized. The figure shows that the age of use, or operation, may be longer than any of the other ages. This is a generalization, of course: some systems, such as missiles, or space shots to see comets, are one-shot systems, with only transitory periods of use. In general, however, it is expected that the effort of conception, design, and creation will be rewarded by a long and useful life for a system.

**Extending System Useful Life**

One way of extending the useful life of a system is by continually upgrading it; see Figure 11.2. The figure shows the life cycle as before, but in modified phases: conception, design, creation, proving, and into operation. During operation, the system is expected to perform "satisfactorily." Should the environment, including other systems with which the SOI interacts, change, however, the system may no

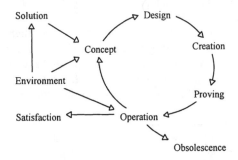

**Figure 11.2** *Extending useful life by continual upgrade.*

longer be providing the ideal solution, and may need to be upgraded. This may be achieved either by taking the system out of service for a period, during which changes are made to restore/enhance its effectiveness, or by a process of continual enhancement, through modifications or procedural changes.

The upgrade procedure may be undertaken, not by the original creators, but by the operators themselves. The extension of useful life by continual upgrade is common in aerospace, where systems may be particularly complex and expensive. Indeed, the eventual system, like the proverbial Viking's axe,[3] may end up significantly different than it was at the outset. Eventually, the system may not be economical to upgrade, and may be termed obsolescent. Even as an obsolescent system, it may provide useful service[4]. In some other arenas, the aging system is simply replaced.

## Characterizing Systems Engineering

Certain characteristics of systems engineering differentiate it from other practices. First and foremost, it concerns itself with systems. In this it is quite distinct from engineering, which concerns itself with technology of various kinds. Systems, as defined earlier, may, but need not be, technological, and may contain technology, but need not.

Systems that are created through systems engineering should, therefore, be holistic in concept, synthetic in design and organismic in structure; if they are to be scientifically sound, they should also be optimal, and provably so. The organ that creates a system–the systems engineering process (or, rather, the person or team that executes that process)–is also a system in its own right, and should therefore satisfy the same criteria.

These characteristics of system engineering mark it out as different from "systematic engineering," in which processes are decomposed into unrelated sub-processes and activities, to be carried out systematically as part of a plan. While the conduct of systems engineering practices may be systematic, in the sense of thorough and painstaking, they should not be founded in functional decomposition, if they are to be judged system scientific.

Throughout, the processes of systems engineering concern themselves with emergent properties, capabilities, and behaviors of the whole, and of the parts and subsystems that, together, constitute the whole. This outward looking approach is the touchstone of synthesis, making it possible to manage complexity, create nonlinear dynamic systems, and achieve goals.

Systems engineering, then, is not really engineering in the conventional sense that relates only to machines and other manufactured things. It is, however,

---

[3] The term "Viking's axe" stems from the no-doubt apocryphal tale of the museum creator who, when asked if a Viking's axe on display was the original item, replied: "Well, it has had six new heads and seven new handles, but, yes, it is the original axe."
[4] The Douglas Dakota DC7 aircraft is a fine case in point.

engineering in the sense of the planning and bringing about of something with ingenuity.

Even the planning aspect may be approached in more than one way. Planning may involve the creation of detailed analysis and plans before creation occurs. On the other hand, planning may involve acceptance that "getting things right first time" may not be practicable, but that continuous improvement will eventually achieve the desired goal. Continuous improvement may, then, form part of the overall plan. This is kaizen, the philosophy of continuous improvement, much favored in some Japanese circles.

Looking back at the NASA Gemini and Apollo projects, both approaches can be perceived in retrospect. Each project, each manned launch, was meticulously planned beforehand, and executed in detail. However, results from each mission were used as lessons to continually improve later missions, which is not inconsistent with kaizen.

## THE FIVE LAYER SYSTEM STRUCTURE

There are many different ways to go about creating systems, with some being more appropriate than others according to situation. It is useful to classify the different approaches to systems engineering. One way to classify systems is to create a classification structure, with different compartments/categories, and to classify particular systems by determining into which category they fit. See Figure 11.3.

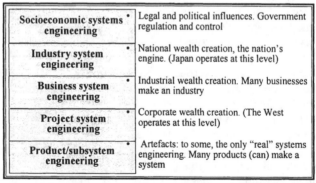

| | |
|---|---|
| **Socioeconomic systems engineering** | Legal and political influences. Government regulation and control |
| **Industry system engineering** | National wealth creation, the nation's engine. (Japan operates at this level) |
| **Business system engineering** | Industrial wealth creation. Many businesses make an industry |
| **Project system engineering** | Corporate wealth creation. (The West operates at this level) |
| **Product/subsystem engineering** | Artefacts: to some, the only "real" systems engineering. Many products (can) make a system |

**Figure 11.3** *The five-layer structure of systems engineering.*

The five-layer model [2] is intended primarily to identify different levels of systems engineering, as shown in the figure, but can also be used to classify systems. For example, the design and manufacture of a new racing car would sit most comfortably at the second level, the project system level. Why? Partly because the new car will be made up of many subsystems (power unit, transmission, suspension, chassis, and so on), so it does not fit at level 1, but mostly because the whole design and manufacture exercise will be a project, that is, it will have a start, a duration, and a clear end-point after which it will be stopped. The exercise does not constitute a business system, as it will not be a business. Racing several cars with teams of drivers, mechanics, and support facilities to win races and prizes may constitute a business, and creating that competitive business might be classified as business systems engineering, or enterprise systems engineering.

Following the car theme, Japanese car manufacturers such as Toyota bring together many businesses to create agile, lean, volume manufacturing systems, spread out originally across the nation, later across the globe; constituting global industry systems. These systems include much more than vehicle artifacts. There are marketing systems, accounting systems, communication systems, transportation systems, and many more; all coordinating and integrating their many separate businesses into industrial leviathans.

Socioeconomic systems include the regulatory systems within which industries and businesses operate, make profits, are taxed and prosecuted for transgressions. In this context, governmental and legal systems are at level 5, the top level, but the regulation "trickles down" and is experienced at all levels.

Socioeconomic systems incorporate much more than regulation and taxation systems. They also tend to influence culture, social behavior (as in the insidious spread of "political correctness"), social dynamics, and, through the selective investment of treasury funds, the growth and demise of businesses, industries, and societies.

This last example shows that the nesting evident in the five-layer structure is not entirely complete. Although many projects contribute to a business, for instance, there is more to a business than the projects.

It is also true to say that most levels incorporate socioeconomic systems. A business will generally include groups of people who form interacting social systems to make money. Human resource (HR) management is usually comprised of a group of people who work together and who form a social system, with its own system behavioral characteristics. Production may similarly form one or more social systems, with different shifts forming discrete social groups. Each shift-as-a-social-system will interact with HR-as-a-social-system on personnel-related matters.[5]

---

[5] Readers with experience in factories may recognize the traditional tension that exists in many factories between engineers and human resource management, or personnel. Where this exists, it is likely to be due in part to the two groups forming two quite different social systems, with different system behavioral characteristics, objectives, and values.

## LEVEL 1: ARTIFACT SYSTEMS ENGINEERING

There are many ways to go about systems engineering at level 1. The outcome is, or should be, a tangible product, the system of interest, that not only meets its intended purpose, but also does so within constraints imposed upon it. In the context of systems engineering, constraints imply rather more than just meeting functional goals. As an open system, the product (SOI) will complement and interact with other systems within a containing system. The purpose and value of constraints upon the SOI will be determined in terms of its degree of contribution to its containing system's objectives, in concert with its sibling systems.

For example, suppose the SOI is an analog amplifier module in a rack-mounted receiver system. The containing receiver system will have been designed and partitioned into modules, such that the emergent properties, capabilities, and behaviors of each and every module are determined. Not only will these properties constrain the solution space for the amplifier module, they will also define its relationships with other modules, rack mounts, backplanes, and so on.

The difference between engineering and systems engineering at this level is that, with systems engineering, the SOI is conceived, perceived, designed, developed and created as an interactive part of a greater whole, rather than as (e.g.) an amplifier created to work to a performance specification.

Readers who are in the business of creating such artifacts may think that systems engineering is rather obvious in this context–why would anyone do anything[6] different? It is not, however, the way in which many people work.

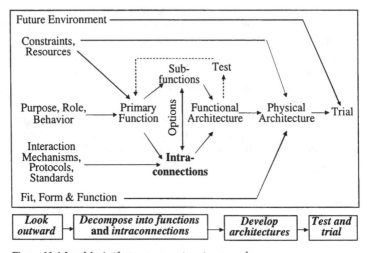

**Figure 11.4** *Level 1: Artifact systems engineering example.*

---

[6] If this is the way in which they normally operate, then they are probably systems engineers in mental attitude, if not in title!

It is not uncommon for something like an amplifier, a circuit, a mechanism, and so on, to be created using a breadboard, or brass-board, layout. The artifact is conceived, designed, developed, and tested as though it were alone in the world. Later, the resulting breadboard may be converted into something that can be produced; perhaps as a rack-mounted card. Postproduction problems may emerge when it is discovered that the new card does not work reliably, perhaps because the components, having been rearranged to suit production, are interfering, both with themselves and with other components on other cards. Difficulties may arise in use because there is insufficient test access, or because the board uses unique, hard-to-acquire, components, and so on.

The systems engineering concept is broadened in Figure 11.4, where a more general approach is outlined. The general process is shown at the lower part of the figure: look outwards first, to establish the containing system and sibling systems, and the constraints and demands they impose; elaborate the purpose and resources into functions (that perform processes) and intraconnections; develop architectures; test and trial.

There is much, much more to artifact systems engineering, but it tends to be technology specific, and beyond the scope of this work. Essentially, once the subject matter becomes technology specific, it becomes engineering and ceases to be systems engineering.


## LEVEL 2: PROJECT SYSTEMS ENGINEERING


### The Classic Systems Engineering Procedure

A conceptual approach to level 2 systems engineering is shown in Figure 11.5; this conceptual model has been in use, successfully, since the 1950s and deserves the epithet "classic."

The classic approach starts with the problem, and researches to understand "the need"–an important distinction from "the want;" what is wanted may not necessarily solve the problem. Solution design options are generated, together with criteria by which they may be judged, so that a "good" solution can be found; the preferred system design, or design(s), if several design options are to be pursued. This is the systems engineering problem-solving paradigm in operation.

The design is then partitioned into manageable parts: sometimes-functional subsystems; sometimes-physical partitions; and, sometimes functions contained one in one within physical partitions. Partitioning necessitates the creation of interfaces between the partitions, so that the process becomes one of elaboration rather than decomposition. This interface infrastructure may be viewed as another subsystem.

Parts are developed, or otherwise acquired, before being progressively tested and brought together within a separate simulated test environment representing the operational environment that the solution will operate within. Once the complete system has been proven within its simulated environment, it is commissioned (that is, put into operation), and subsequently supported and upgraded in operation.

Conceptually, the process should result in a holistic, synthetic, organismic system solution. Moreover, the result should be optimal; the use of the SEPP is designed to identify the optimal solution from among a range of solutions. So the process is also scientific, by Ackoff's criterion.

There are several potential pitfalls:

- The chosen range of potential options may not span the optimal solution. In this event, the SEPP would find the "best of bunch," but it may fall short of optimal; for example, "best value for money."
- The partitioning process can be difficult. In a purist systems environment, the partitions might be functional subsystems, but this is not always the most practical way to proceed. Software may be required in several partitions. Software is not a system, but a set of instructions for hardware; because of its special nature, it may be deemed necessary to develop all of the system software together, cutting across the discrete subsystem boundaries. Off-the-shelf parts may be temptingly available that do not fit the ideal partitioning schema, but that would reduce development delay and provide a pragmatic solution quickly.
- The development process may incur specification "drift." Drift may occur when a part is being developed in relative isolation; the resulting

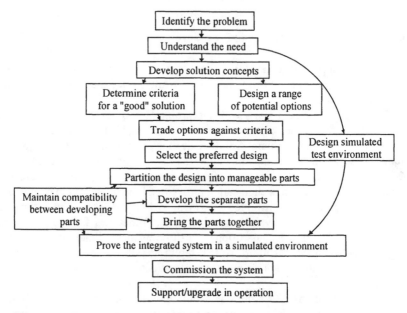

**Figure 11.5** *Classic level 2 systems engineering conceptual process model.*

item may be not quite what was called for. On its own, this may seem unimportant, but it may cause difficulties when the part is married to the other developed parts. Notice in the figure at left the importance of maintaining compatibility between the developing parts.

- The simulated test environment acts as a policeman, toward the end of the creative process, signaling whether the integrated whole is a sound solution to the original problem. There are at least two issues:
    o Years may have passed since the original problem was identified and a solution proposed. During that time, the environment and the interacting systems that the simulation represents may have changed significantly. Should the simulation represent the original environment, the current environment, or some future environment where that is known or predictable?
    o The simulation could be either static or dynamic. For instance, suppose the system were a new, airborne track-while-scan radar, capable of simultaneously tracking, say, 20 targets. A static simulation would provide radar returns representing, perhaps, 100 moving targets, which the radar should be able to detect and display, together with up to 20 of the targets being selected and tracked according to some rules. This would be a static simulation. Suppose now that the radar was placed in a motion simulator representing the situation when it was airborne, including reduced temperature and pressure, vibration and buffet, and so on. As the radar is taken through simulated climbs, dives and turns, it should still be able to detect, display, and track. This second arrangement would be a dynamic simulation. It is more thorough, but also more sophisticated and expensive.

## Process-Oriented Systems Engineering

There are several, archetypal process models for project-level systems engineering [3]. The following models are conceptual only; they do not show how a process is activated, controlled or resourced.

### Waterfall

A classic waterfall model of systems engineering is shown in Figure 11.6. The model presumes that the process starts with Understanding the Requirement, and finishes with Commission and Deliver, so it represents only part of the overall systems engineering life cycle process. Note that it follows the same general route as the conceptual model of Figure 11.5.

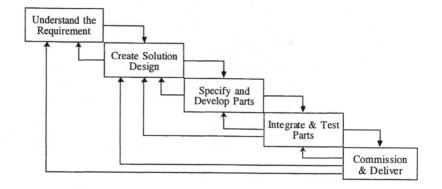

**Figure 11.6** *Waterfall model, showing phases, feed forward, and feedback, or rework.*

In the waterfall model, each process stage is completed before going on to the next. The lines and arrows indicate feedback between successive activities and from "downstream." The term "waterfall" arose from a view of the activity sequence as being like a multistage waterfall with work accumulating in each pool before spilling over into the next, downstream pool.

Partly because of the practice of completing each step before proceeding further, and partly because of the feedback, the waterfall model is considered cautious and thorough by some practitioners, and time consuming by others.

Spiral/helical

The spiral, or helical, model of the systems engineering process is shown in Figure 11.7.

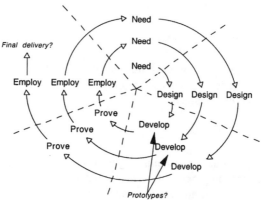

**Figure 11.7** *Spiral, or helical, model.*

This process model is invoked when the best solution to the problem is unclear at the outset. The idea is to create a prototype of the solution; one that can, perhaps, be tested and used in a simulated or, preferably, in a real situation. Deficiencies in the way the prototype performs, or behaves, are observed, and used to devise an improved prototype, which is similarly trialed. Each revolution of the spiral invokes a repetition of the basic waterfall model. The cycle is repeated until a prototype is developed that satisfies the need; this is then used as a template for the design and creation of the full solution.

Simultaneous/concurrent

Simultaneous, or concurrent systems engineering may also use a disguised waterfall model. A criticism of the waterfall model is that it takes a long time, since each and every process "thread" must be kept in step. So, if one part takes a short time to design, it is not considered appropriate to go forward to the next stage until the designs for all the parts are finished and reviewed as a complete design set. There are many good and sound arguments in favor of this practice, [7] but it is not quick.

Advocates of simultaneous/concurrent systems engineering seek to reduce the time it takes to market by pursuing two main themes: telescoping sequential activities, and team design.

Telescoping activities can occur in several ways. Where one activity requires an input from an upstream activity, it may be practicable to start work using interim results from the upstream activity. This procedure contains elements of risk, but it can be effective if done with care. If the various process threads are mutually dependent at many points in the process, however, there is a risk of avalanche. Late changes to the output from an upstream activity may cascade through later activities, requiring them to be amended or updated. This could extend, rather than reduce the overall process time. In addition, it is a piecemeal, rather than holistic process.

Team design, the other arm of simultaneous/concurrent systems engineering, employs contributions from development, manufacture, integration, commissioning, installation, operation, and maintenance. These contributions are designed to anticipate and prevent feedback and delay caused by omission in earlier phases of design and development. (Team design in this manner has been the de facto approach to waterfall-based systems engineering since the 1950s, and is the basis of multidisciplinary systems engineering.)

Another approach to concurrent engineering is that used by NASA in the early 1960s. If the parts of a system are designed such that each functional part is also a discrete physical part, as in the case of Apollo, then each part can be designed and developed in parallel with, but entirely separate from, the others. This offers organizational and management benefits. If one of the parts is not

---

[7] These include keeping control of the overall process, and maintaining a balance between the parts.

developing according to plan, then more effort can be brought to bear on that part without any other part being affected. If all parts are made of comparable magnitude, complexity, and difficulty–which is not always practicable–then the overall time to create may move toward a minimum, without prejudice to quality.

For this approach to work, there must be an effective early design phase in which the whole system is partitioned into discrete functional-physical entities, together with their mutual interfaces and embedded software–if any. In principle, partitions can be developed in different ways, using different processes and technologies, provided that their respective emergent properties, capabilities, and behaviors remain within specification. They can also employ dedicated, self-contained teams, which create their own culture, discipline, enthusiasm, and morale; experience over several thousands of years[8] shows teams to be relevant in producing good-quality systems quickly.

In practice, the development of software in particular has hampered and militated against this approach in recent years. There is a view that software development should be centrally controlled, to ensure that the highest quality and standards are maintained across the board. This view and practice, reasonable when viewed from the standpoint of the software creator, militates against the creation of self-contained teams for each partition.

## Chaos

A quite different approach removes virtually all control from the creation process and allows each and every thread to proceed at its own pace. Given sufficient effort, there should be no gaps between activities and, provided each thread is distinct and has zero or very few lateral branches, the whole process should come to a close in short order.

Chaos [4] may sound like a recipe for–well–chaos, but it can be most effective under particular circumstances, and with certain provisos. The members of the systems engineering team must be experienced, capable, and probably multidisciplinary. The creation plan must be thorough and sound. Progress against the plan must be recorded and evident to all. Without any control, there must be incentives to encourage activity.

The following is not an apocryphal story. A project team, trying to meet the deadline for the completion of a complex design project, has some six weeks of work to complete, but just three weeks to do it. The team leader, aware that "adding people to a late project makes it even later," approaches senior management with a novel suggestion. He proposes to reward every team member with a substantial cash award if the project comes in on time. All members get equal shares. One minute over time and no one receives any bonus. As there is a penalty attached to being late, management agrees.

---

[8] For example, the Great Pyramid of Khufu, Egypt?

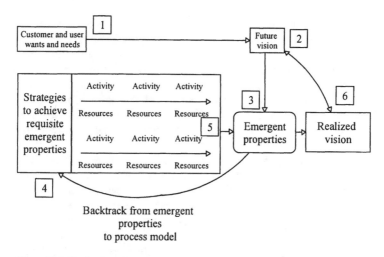

**Figure 11.8** *Goal-oriented systems engineering process concept.*

The team leader calls the team together and announces the deal: team members are enthusiastic, but point out that management is part of the problem; not part of the solution. The team leader accepts the point, and becomes one of the team. Each and every team member sets to work communicating with others working on related activities, and starting on the next "available" activity as soon as his current activity is completed. There is no imposed control, but each of the team members knows the plan, knows what needs to be done, and knows how to do it. The activity pattern is chaotic, but the best possible application of manpower to tasks is achieved and the project comes in on time–just.

Chaotic systems engineering is not for every situation. However, there is a lesson in the tale. The introduction of stages and phases, and the imposition of control and coordination such that managers may satisfy themselves that they are in control and can demonstrate progress to customers and accountants; all contribute to delay and cost.

### Goal-Oriented Systems Engineering

A six-step conceptual process for goal-oriented systems engineering is shown in Figure 11.8. The notion apparently reverses the approach, by first agreeing with the customer and user about what they both want and need, and by using this understanding to create a future vision of the nominated system in its future environment; interacting with other, future systems. The future vision is described in various ways–models, pictures, simulations, CONOPS–such that architect[9] of

---

[9] I use the term "architect," because this may be the way ancient architects, such as Hemon, architect of the Great Pyramid of Khufu, went about creating pyramids and tombs for their pharaohs.

the future system has a full understanding of what the customer and user want and expect. In principle, there are no optional solutions, only one future vision.

This approach of describing the future vision necessarily emphasizes emergent properties, capabilities, and behaviors; after all, these are what will be visible in the mind's eye looking, as it were, from a distance. How big, what shape, how oriented, what color, what is inside, what will it do, how will it work? The last question, how will it work (CONOPS), inescapably leads to essential functions, processes, and features. Step 3 accumulates the emergent properties, capabilities, and behaviors of the future vision.

Step 4 identifies strategies for achieving the emergent properties, capabilities, and behaviors of the future vision. To be practical, these strategies will draw upon available manpower, skills, technologies, and so on, or will identify that the future vision cannot be entirely achieved unless such facilities are acquired.

The various strategies are then elaborated into activity streams, step 5, and provisionally resourced, that is, men, materials, machines and money required to undertake and complete the activity in a given time are identified. The various activity streams are harmonized, both to smooth peaks and troughs in demand for resources, and to avoid one activity stream interfering with, or needlessly undoing, the work of another.

If all goes according to plan, the end-point of the various activity streams should be the requisite emergent properties, capabilities, and behaviors, and hence the achievement of the future vision, step 6.

Unlike earlier process concepts, Figure 11.8 can be reasonably entitled a rudimentary process model, since it represents the resources necessary to activate and support each element in a process; seen here as streams of activities.

Goal-oriented systems engineering may seem rather obvious, and it has particular advantages in that it contains no element of decomposition. One problem that has arisen, and which militates against the employment of such a straightforward process, is the arising of a "distance" between the architect of the system and the customer and user. This has come about because of the introduction of competition, and the need to be seen to be fair. Instead of creating a future vision, customers and users have taken to expressing what they want in words. These so-called "statements of requirement" seek to describe all aspects of the system solution, so that organizations can compete in bidding for and creating the system solution.

Unfortunately, the written word is a poor vehicle for describing future visions of complex systems. It is also poor at conveying the context of future systems. The idea of customers describing the future vision of a system also indicates that they have already identified the problem, conceived a variety of solution concepts, and selected a preferred solution. In other words, they have already executed a significant, and highly creative part of the systems engineering process. In describing the outcome of their work in words, the so-called requirement specification, they inevitably create a major divide in the overall systems engineering process. People who may be unfamiliar with the solution context and unaware of the true nature of the problem, will undertake the remainder of the systems engineering process, using as their source data a description of the future vision that is, at best, vague and sketchy, and, at worst, enormous in volume and detail, as well as inconsistent and incomprehensible.

For these reasons, some systems engineering organizations, upon receipt of a customer's requirement specification, go right back to the beginning of the systems engineering process. They identify the problem space, create solution concepts, compare, contrast, and trade off between the concepts, and then create their own preferred solution concept and conceptual design; hopefully in concert with the customer and, particularly, the user. In effect, they revert to goal orientation, and undertake the full process themselves.

**Unprecedented Systems Engineering**

Unprecedented systems engineering arises where there has been no system before. This is a rare situation. For the most part, any new system supersedes a previous one, even where a technological system replaces or augments a manual one.

Designing unprecedented systems is exciting and creative, partly because of the lack of precedent and constraint, partly because of the challenge, and partly because of the need to conceive everything–which is why much of NASA's work is so fascinating.

## LEVEL 3: BUSINESS SYSTEMS ENGINEERING

Figure 11.9 shows an outline concept for a business systems engineering model. The model shows two–there could be many–projects in operation; each drawing on the organization's resources of manpower, materials, machines, and money [5] to implement the projects. The outcome of each project is some product or service

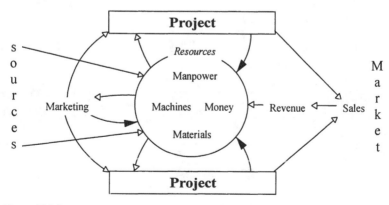

**Figure 11.9** *Business systems engineering conceptual model.*

that sells for money in a market. The revenue from sales goes into the central coffer, replacing what was spent on the projects, and hopefully leaving some over as profit. At left, marketing is seen as instigating new projects, while external sources provide fresh resources, at a cost. So, although marketing and sales are often treated as though they were one, their roles and purposes are quite different.

There is a value added chain concept, in which value is added to the materials that enter at the left such that the product leaving at the right has greater value. As part of this concept, the business as shown provides a marketplace for the sources coming in from the left. The model also shows that projects may compete internally for resources.

Although the model refers to projects, signifying a clear end to a set of tasks, the model could equally apply to continuous-process businesses such as chemicals, pharmaceuticals, smelting, packing, assembly, and so on.

Figure 11.10 shows a more specific model of business systems engineering; this time clearly concerned with continuous-flow assembly, and the production of products for a market, at right. Parts are supplied at the lower left, pass through supply assurance, are assembled in production/assembly, and are sold; this is the value added chain.

A supply channel is identified that conceptually has width and depth. Depth refers to the volume of a particular product's sales, while width refers to the variety of different products being manufactured and sold in parallel. Revenue is

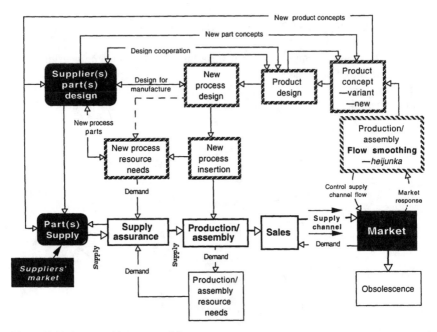

**Figure 11.10** *Conceptual business model.*

related to the depth multiplied by the breadth.

To maintain and enhance the volume passing through the supply channel, it will be necessary to replace products for which the demand is falling, with new products. At the top right of the figure can be seen the process of conceiving and designing replacement products, which are inserted into the flow at bottom center.

In this kind of organization there is a concept, *heijunka* [6], of production smoothing. It can be shown that the existence of resources trapped in the supply chain is minimized by steady flow rates; if the rate either rises or falls, there may be a rise in trapped "work in progress," with a consequent threat to the unit production cost (UPC). In the figure, flow smoothing can be achieved by increased marketing to counter the fall in demand for products, or, alternatively, by reduced marketing to reduce increasing demand. When marketing effort can no longer sustain the sales rate, it is time for a new variant or replacement product, and the old product becomes obsolescent.

There is a high degree of cooperation between the business and it suppliers; they are involved in new product concept development and design for manufacture, including the provision of facilities needed for the assembly of parts. The organization of the business is such that it is self-sustaining. The development and insertion of new products to replace old is not so much an event as a continuous process. The whole is evidently synthetic and organismic; since it automatically drives itself toward the optimum volume in the supply channel, it may be classed as system scientific, too.

**Evolutionary Systems Engineering**

A quite different view of, and approach to, business systems engineering is sometimes referred to as evolutionary acquisition; see Figure 11.11. In the figure, a complex and expensive system is delivered to a customer in three phases. Phase 1 provides basic capabilities, and the users can become familiar with the system and its method of operation without being overwhelmed by the full potential capability. During their period of familiarization, they may find that there are defects, deficiencies, or ways of working that need attention. The supplier is advised of this, and incorporates changes into the phase 2 delivery, which is designed to add significant additional capability. The process is repeated and the customer eventually takes delivery of phase 3.

There are potential advantages to this arrangement. First, the customer is able to spread payments for the system, which may be too expensive to purchase at one time. Second, the creating organization is able to spread its workload, and essentially to take its time over later phases. On the other hand, the whole design must be undertaken to a reasonable depth right at the beginning, so that the necessary "hooks" are incorporated, on which to hang later phases. Moreover, it is important for each delivery to add something of real advantage and significance to the customer; otherwise he may elect not to go any further. It is also possible that, over time, the number and extent of deficiencies uncovered, or even "invented,"

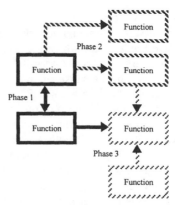

**Figure 11.11** *Evolutionary acquisition.*

by the customer, may rise out of control as he or she becomes more familiar with the new system; creating a rising challenge to the supplier.

Typical of the kinds of systems delivered in this manner might be police and military command information systems (CISs), air traffic management systems, and indeed many near-real-time information, decision, action systems.

## LEVEL 4: INDUSTRY SYSTEMS ENGINEERING

Many large-scale organizations undertake industry systems engineering, in fact, if not in name. Vehicle manufacture, global telephone systems, national transport systems, national health systems, and even schooling may be viewed as industry-level systems engineering. Schooling may be classified in this way, because it generally consists of a number of separate institutions, or businesses, that work together to provide education for those from the age of five or six to age 60 or beyond. The product is a society of people educated to many different levels according to age and ability. A national railway system may similarly comprise many separate businesses working together to convey passengers and freight.

### Mass Production

Mass production has been in vogue since the start of the 20th century; Henry Ford popularized it, particularly, for motor vehicles. Mass production is classified as industry level systems engineering because it works on such a large scale that a number of businesses must cooperate to provide and assemble the many different parts in a typical product; see Figure 11.12. It is the manner of their cooperation and coordination that is of particular relevance to systems engineering.

First-tier suppliers supply parts or subassemblies to a lead company; one that sells directly to the open market. The lead company assembles the parts; perhaps

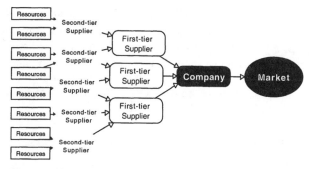

**Figure 11.12** *Mass production concept.*

on some chassis, or in some cabinet. Second tier suppliers provide first tier suppliers with the parts, or sub-subassemblies, to make the subassemblies. There may be several tiers, and more than one supplier may provide individual assemblies to the same specification.

The objective of mass production was, and is, to provide low-cost goods, using economies of scale to reduce overheads, and to amortize the cost of automation machinery across a large number of identical products.[10] The lead company usually designs the complete end product, and partitions the design into a number of discrete parts; each carefully specified. It then invites potential suppliers to make these parts, using the competitive situation to drive down the price. Potential suppliers do likewise down the chain to their potential suppliers.

Once a chain of contracts is agreed, manufacture and assembly can begin. The process builds from the left in the figure. For each of the suppliers in the chain, there is a need to manufacture in volume, to minimize overheads, and to exploit economies of scale. Because of the competition at the start, it is reasonable to expect that each supplier is making only a limited profit per part supplied. Once the mass production chain is in full operation, it would also be reasonable to expect that suppliers might try to extract a higher price for their products, or that they might attempt to cut production costs, or both.

Vehicle mass production, as conceived by Henry Ford, had one other feature of particular note. The various tasks that were needed to assemble, say, a motor vehicle, could be broken down into simple steps. Many of these steps could be automated. Some steps, however, could not; these were the province of the mass production worker, who found himself or herself undertaking repetitive simple chores that were to become a byword for boredom. The result was labor difficulties, with strikes for more pay, which inevitably resulted in rising production costs, and the introduction of more automation.

Mass-produced product prices are generally based on cost build-up: the costs of parts, labor, machinery, operations, sales, overheads, and so on, are added

---

[10] Hence: "You can have any color of car you like, so long as it is black:" the famous quote attributed to Henry Ford.

together to calculate a cost per product, to which a percentage is added for profit to establish a market price. The motivation of the mass production industry is to make as many of a product as possible, so reducing the cost per item, and making it more competitive in an open market. The hallmark of many mass production industries in the second half of the 20$^{th}$ century was large areas outside factories where manufactured goods were stored because, having been manufactured, they could not be sold immediately, or perhaps at all.

## Lean Volume Supply Systems Engineering

Relatively new on the scene compared with mass production, is lean volume production; (see Figure 11.13), which developed mainly in Japan after World War II. The two approaches to engineering supply systems could hardly be more different.

As with mass production, there is a lead company interacting with the market, supported by tiers of suppliers. The lead company identifies a product, together with a price, that should sell well in the market. Using reverse engineering, a price for various constituent parts is determined. The lead company works with its suppliers to establish how the parts can be made within those respective price targets, allowing each supplier to make a reasonable profit. Should a supplier subsequently find that it can manufacture parts for less, then it is encouraged to do so; reduced costs will be passed on to reduce the cost of the end product, and some of the savings will go to the supplier as increased profit per part. Suppliers, then, are motivated to reduce their price; not raise it.

Unlike mass production, where a subassembly may be made up from many, separately sourced parts, suppliers are encouraged to conceive, design, create, and provide complete subassemblies. This overcomes problems of incompatibility between parts due to different physical characteristics, such as thermal expansion, the coefficient of friction, and so on, which can bedevil mass-produced products where each part may have been separately sourced.

Lean volume supply works on a basis of market pull, as opposed to mass production's production push. Market pull, for private vehicles, means that a new car is not made until it is sold. Upon receipt of the order, the lead company informs all of the support organizations up the chain, of the order and of the need to make the specific parts for that order to time, and in sequence. This implies a sophisticated information system, "wiring" the various elements in the industrial system together so that the whole is synergistic. It takes time for the order to be filled; however, for a new motor vehicle, made to order, the time may be as little as two or three weeks.

Unlike mass production, there is no need for vast storage facilities for unsold goods; any work-in-progress should be associated with a sale. Lean volume supply has other features, too, to reduce the amount of material, parts, and subassemblies in the process, since they all constitute a cost. Parts are transferred between factories at different locations. The farther apart the factories, the longer

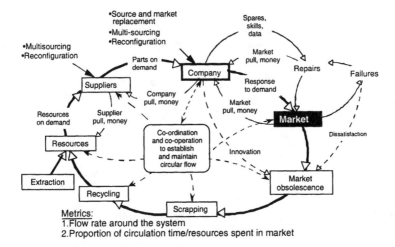

Figure 11.13 *Level 4: Lean volume supply systems engineering.*

it takes to transfer them, which constitutes a cost. This is compounded if the transport, a large vehicle perhaps, waits at a factory for a full load to accumulate before setting out. Lean volume production will employ small vehicles, even motorbikes; transporting individual parts as soon as they become available. This is part of just in time (JIT), designed to provide parts just as they are needed, and to avoid any build-up of inventory waiting to be processed.

In Figure 11.13, the supply chain has been formed into a loop to account for recycling, which is becoming a dominant theme for consumer products. Not everything can be recycled, but there is value to the manufacturer in the process, and it will reduce the mountains of hazardous waste that are the expanding feature of rubbish dumps and landfill sites. The figure also shows the following:

- Associated businesses repairing products that have failed in use;
- The influence of defective products in creating customer dissatisfaction, which will result in the customer buying from a different supplier;
- The influence of innovation, which operates by making products already with the consumer seem outmoded and obsolescent. This is a key feature of such supply systems, since it creates demand and, effectively, spins the wheel, that is, makes material flow around the loop. The faster the spin, the more the revenue;
- The anticlockwise flow of market pull and money, against the clockwise flow of material;
- Some central coordinating influence, which may be an organization, or could be government regulations, or both;
- Putative metrics, by which to judge a lean volume supply system.

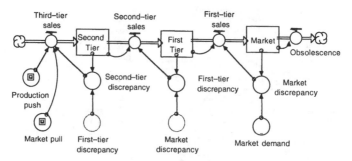

Figure 11.14 *Mass production versus lean volume supply: STELLA model.*

## Comparing Mass Production and Lean Volume Supply

A simple STELLA model may be used to highlight some of the differences between mass production and lean volume supply; see Figure 11.14. The model shows goods passing through two tiers into a market, at right. At bottom right is a converter that presents market demand, which can be varied. Market discrepancy calculates the difference between market demand and the goods available in the market, and uses that difference to draw goods out from the first tier, that is, first tier sales in the model. This process is repeated, using first market discrepancy as a driver, and then first-tier discrepancy. At the left are converters that can be selected either to use the discrepancies to determine third-tier sales (the third tier is assumed, not shown), or to use a fixed value as would be the case for mass production.

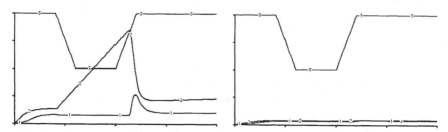

Figure 11.15 *Production push versus market pull; inventory. Line 1: material in tier 1 reservoir. Line 2: material in tier 2 reservoir. Line 3: variable market demand.*

The simulation is evidently not a full systems model, but is sufficient to represent a phenomenon. Results are shown in Figure 11.15. At left is the result of the simulation, selecting production push; characteristic of mass production. Line 3, which is the same in both graphs, shows that market demand dips by some 50% for a time during the simulation run, representing a temporary dip in demand. Line 2 shows a build-up of inventory in tier 2 during the dip, as it continues to produce parts for tier 1. Tier 1 does not accept the produced parts from tier 2, as it has no

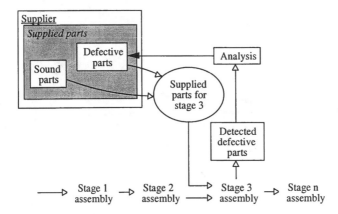

**Figure 11.16** *Kaizen in lean volume supply.*

prospect of selling during the dip. Even after the market demand is restored, the level of inventory in tier 2 does not fall to its previous value, because it is making slightly too much compared with demand. The area under the curve for line 2 is proportional to the cost of inventory for the second-tier supplier.

At right is the result for market pull, characteristic of lean volume supply. The inventory in neither tier is affected by the dramatic change in demand, which simply resulted in proportionally fewer orders being processed.

Although this simulation underscores the advantages of lean volume supply, it is not always so one-sided. Similar research shows that mass production can make greater sales than lean volume supply systems during periods of dynamic, rising demand, and that during such buoyant periods the cost of inventory is sustainable.

Another aspect of contrast between mass production and lean volume production has already been alluded to: quality. There are many approaches to improving quality. One approach, developed by Taiichi Ohno of Toyota is especially noteworthy.

Ohno set up a novel assembly line for a motor vehicle, with each assembly stage provided with an alarm. Should an assembly stage experience difficulty in fitting a part, the alarm was raised, stopping the whole assembly line. This was revolutionary: normal practice in mass production was to tag the fault for post assembly rectification, but to keep the assembly line going at all costs, since the dominant view was that there could be no profit unless products continued to come off the line. Critics of Ohno's approach believed he would never get any vehicles off the end of his line.

By stopping the line, Ohno had at his disposal a workforce of skilled operators who were able to tackle the problem en masse. They worked with the supplier of the offending part to eliminate problems, since the suppliers were viewed (in fact and in contract) as part of the team; actively involved in the success of the assembly process. The process is illustrated in Figure 11.16, which

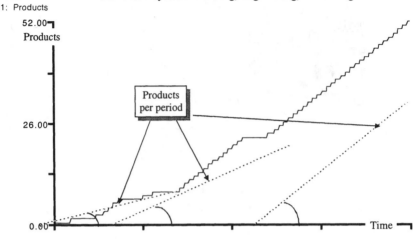

**Figure 11.17** *Kaizen in production.*

shows the process of detecting and eliminating defective parts for stage 3 only, of an n-stage assembly line.

Once the problem was sorted, the line restarted, only to be stopped again by the next defect. Slowly, each successive assembly stage was cleared of defects until finally the first vehicle made it to the end of the line. While this had taken some time, it had also ensured that all parts fitted properly, and that problems with non-fitment had been eliminated at the source. The next vehicle went through without problem, as did the next. More defects arose in time, as the suppliers' product quality varied, but the procedure that Ohno had introduced was able to accommodate such issues, which diminished with time. Post-assembly defect rectification proved unnecessary.

Figure 11.17 shows the result from simulating a three-stage assembly system using the "Ohno method." The stepped line shows output of products from the end of the assembly line, with periods–where the line is flat–indicating where time is taken out to address defect issues. The lines marked "products per period" indicate the slow build-up characteristic of such assembly lines.

There are many other points of comparison and contrast between mass production and lean volume supply; some are shown in Figure 11.18. Today, some U.S. industries exceed the lean volume supply performance of the best Japanese industries, while some Japanese industries have experienced difficulties and have learned from the U.S. As we have seen, the startling success of lean volume supply systems has inspired the U.S. to adopt the process for defense procurement.

Understanding the significant differences between, for example, mass and lean volume production suggests that there are many ways to go about systems engineering, and that prescribing particular ways as "proper" and others as "improper" should be unacceptable. There is a particular onus on those developing standards for systems engineering, in that they do not inhibit new and creative

| Mass production | *Comparison* | Lean volume supply |
|---|---|---|
| Profit | *Objective* | Survival |
| Free | *Competition* | Between circles |
| Free market | *Regulation* | Indiginization |
| Production push | *Assembly* | Market pull |
| Cost plus | *Pricing* | Market minus |
| Adversarial | *Contract* | Synergistic |
| Specialist | *Defense* | Homogeneous |
| Hire and fire | *Labor* | Jobs for life |
| Specialization | *Skills* | Multiskilled |
| Lowest bid wins | *Suppliers* | Vital source—protect |
| Supplier stocks | *Inventory* | Nobody stocks |

**Figure 11.18** *Comparing U.S. mass production with Japanese lean volume supply* [7]. *(Indiginization implies that core capabilities and facilities should be within national boundaries of the home nation, that is, Japan.)*

approaches by introducing Procrustean[11] standards. If we are to learn anything from the development of lean volume supply, it is that "one size does not fit all," and that prescriptive standards may seriously inhibit systems engineering, rather than enhance it.

### Nesting at Levels 2, 3, and 4

Tiers in a supply chain, as shown in Figure 11.19, show three levels of systems engineering, nesting one within the other. Level 2, project systems engineering can be seen within both the lead business and the first-tier supplier. In this situation, project systems engineering conceives, designs, develops, proves, and inserts a new product, or variant, into the left-to-right stream of products going to market. There may be several such projects running at the same time; perhaps at different stages in the process.

Shown between the two businesses is a joint integrated product team (IPT). Since the new product design will be made of parts, or subassemblies, provided by first-tier suppliers, it is sensible to form joint teams to design both the whole and the constituent parts of a new product together.

Level 3, business systems engineering, is evident in the two businesses that form links in the supply chain. The businesses are concerned with creating new and variant goods to supply to their respective markets. Integrating the various businesses, together with their manufacturing capabilities and their market, constitutes level 4 industry systems engineering.

---

[11] From mythology, Procrustes made all visitors fit one size of bed. If too long, he cut off their feet; if too short, he stretched them. Hence, Procrustean standards impose conformity, regardless of need or suitability of the subject.

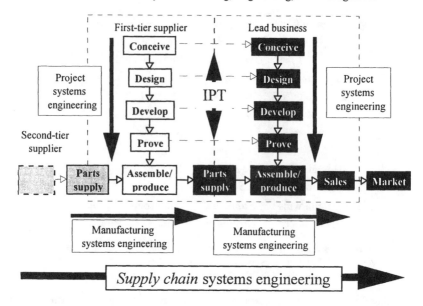

**Figure 11.19** *Level 4: Industry systems engineering with levels 2 and 3 "nesting."*

Each of the three levels provides the environment, and hence the constraints as well as the purpose, for the level below. The whole is evidently synthesized, and, since the various parts support each other, it is organismic, too. Taking the whole supply circle into account, it may be viewed as holistic, as well as being a system of systems.

Using kaizen, or some equivalent process, the whole may evolve toward an optimum state. Such a state might envisage the minimum use of new resources, relying instead on recycling. It might include minimal pollution of the environment, not only from an ethical standpoint, but also from the pragmatic position that the whole system exists within the environment it may be polluting. By using Ackoff's criterion, therefore, the design of the whole may be system scientific, that is, optimized.

## LEVEL 5: SOCIOECONOMIC SYSTEMS ENGINEERING

Level 5, socioeconomic systems engineering, operates at the highest level; nationally and globally. However, it is also evident on much smaller scales, within regions, districts, and societies. Figure 11.20 presents an instance at national level, in the form of an $N^2$ chart [8].

Selected sectors of a socioeconomic system are shown on the leading diagonal. The sectors interact and exchange with each other through the interfaces. So, farming industries provide food to society, and society provides human resources to service industries, raw materials industries, and manufacturing industries.

The whole forms a complementary set, with each sector providing what others need, and receiving from others what it needs; see Figure 6.2 on page 110. Interchange activity expends energy. In different cultures, the manner of interchange may be different.

In the now-defunct U.S.S.R., for instance, interchanges between sectors were planned[12] and prescribed, so that each sector was required to produce predetermined amounts, and process at predetermined rates, to provide to other sectors. History tells us that, despite the most careful planning and control, this arrangement–which was an instance of socioeconomic systems engineering–failed to work effectively.

In free-market, free-enterprise situations, such as apply in capitalist countries, the same figure would apply, but the interchanges are not controlled. Industries and businesses within sectors make and sell products and provide services in an uncontrolled, ad hoc manner to satisfy demand. Companies spring into existence, operate, and die in a scintillating fabric of chaotic change. As a result, the whole may be dynamically stable, although the point of stability may wander erratically over time.

This is a different approach to socioeconomic systems engineering, one in which national governments have a major role to play. Generally, their role is to enable, rather than control or limit, the multitudes of interchanges. Similarly, their role is to encourage new companies, industries, and enterprises to spring up, rather than to artificially extend the life of failing companies. Thus, stability of the whole depends upon the chaotic, uncoordinated, ever-changing nature of sector components and intersector interchanges–see Systems Life Cycle Theory on page 107 and following–and on the avoidance of dominance,[13] which threatens the continual generation of variety.

The $N^2$ chart of Figure 11.20 can be modeled using STELLA, or some similar tool, or a custom-made program. Figure 11.20 could represent one nation. Creating other, similar $N^2$ charts for other nations allows the various charts to be interconnected to show the transfer of goods and services between nations as exports and imports. It is possible, in this manner, to build up a global socioeconomic system representation, and to create a nonlinear dynamic simulation. In effect, this process creates an environment in which our "own nation" can be seen to operate and flourish–or not. Varying the socioeconomic paradigms can then suggest which paradigms offer the best promise for the future. This is a system-scientific approach to socioeconomic systems engineering seeking, as it does, some optimum solution.

---

[12] Hence, of course, the "planned economy."
[13] That is why governments in free-market economies seek to prevent monopolies, while monopolies are the stuff of life in planned economies.

| Raw materials industries | • Energy • Metals • Woods • Plastics • Composites | • Dated skills | • Domestic raw materials | • Fertilizers |
|---|---|---|---|---|
| • Machinery • Knowledge • Power | Manufacturing industries | • Dated skills • Power • Machines | • Domestic products/materials | • Farm machinery • Power |
| • Skilled people • Recyclable raw material | • Skills • Logistics • Machinery | Service industries | • Power • Food • Distribution • Transport • Communication | • Power • Fertilizers • Pesticides • Husbandry |
| • Human resources | • Human resources | • Human resources • Dated skills | Society | • Human resources |
| • Recyclable resources | • Recyclable machinery | • Foodstuffs • Dated skills | • Food | Farming industries |

Figure 11.20 *Level 5: Socioeconomic systems engineering.*

## SCIENCE-BASED SYSTEMS ENGINEERING

The various patterns, paradigms, and classifications of systems engineering in this chapter may be appropriate in the relevant context, but would they create sound, effective, and optimal solutions to their respective problems? To be sure, some of the processes have within them the means of creating optimal solutions – see Figure 11.5, for instance–but there is no hard evidence that the processes would work as proposed. We are, after all, dealing with complex, open, interacting systems, with their propensity for counterintuitive behavior.

It should be important that a systems engineering plan is provable, or at least that one systems engineering plan is demonstrably better than another. At the present time, however, proof is not required, and the effectiveness of one systems engineering method compared with another has to be taken on trust; live experiments, where alternative systems engineering processes compete under laboratory conditions, are impracticable.

In the absence of hard facts, it seems that the choice of systems engineering process may be determined by fashion, or because "we have always done it that way." Conferences and workshops have been held to establish the advantages of systems engineering. Results have been inconclusive: many such activities have conspicuously failed to agree what systems engineering is, and–particularly–what it is *not,* a topic that pundits discuss endlessly.

So, what would it mean to prove that a systems engineering method and plan were sound and optimal, and how could it be done?

The only practicable tool in the circumstances is simulation. If we could show by simulating processes and outcomes that the systems engineering process was itself, say, cost effective, and that the product, also a system, was optimally effective, then we would have the basis for proof. [14]

A simulation, especially if it was in the form of a learning laboratory as presented in earlier chapters, would allow exploration of different situations, optional systems engineering processes, and so on, would allow the researcher to show whether a selected approach was feasible, let alone optimal, and how it might be improved.

What is cost effective at the start of a project, under starting conditions, may no longer be cost-effective later, when environment and situation have changed, as they will almost inevitably do during an extended project. This suggests that scientific systems engineering should employ dynamic simulation throughout a project, in which the contemporary environment and situation are represented and the optimal solution, both in terms of the systems engineering process and the product, is continually recalculated. [15]

The simulation concept is illustrated in Figure 11.21. A sociotechnical system is shown at left, as an instantiation of the generic reference model; this is the creating system. Three environments are shown: a business environment, within which the creating system operates, undertakes work, and creates system solutions within time and budgetary constraints; a resource environment, from which the creating system derives trained manpower, tools, machines, facilities, and so on,

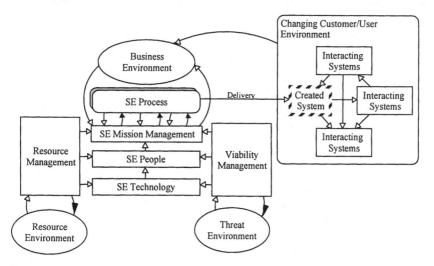

**Figure 11.21** *Science-based systems engineering concept.*

---

[14] Using simulation as proof is becoming accepted, even for operational weapon systems, owing to the cost, risk, and impracticability of conducting clinical trials of weapons (e.g., smart torpedoes).

[15] Using such an approach is similar in concept to dynamic programming, in which the optimum solution is sought at each step along the path to a solution. In like manner, it offers the optimum solution for traversing a dynamically changing landscape.

necessary to operate and sustain the creating system such that it can satisfy business constraints; and a threat environment that threatens system viability.

Above the creating system is the systems engineering process; the mission of the creating system is to execute the systems engineering process, activity by activity, objective by objective, until the product–a created system–is delivered to a customer/user organization at right.

There are optional systems engineering processes. Employing different systems engineering processes will cause the project duration to extend or be reduce, cost more or less money, and take more or less risk. The time taken to complete any of the optional systems engineering processes will be affected by the availability of resources, and by the robustness of the creating system in rebutting threats.

One way to compare different processes would be to create a simulation in the style that was introduced in Operational Design Approach on page 280, with two identical creating systems competing, both initially with the same systems engineering process (e.g., waterfall). See Figure 11.22. Using identical systems engineering processes, the competing creating systems would perform identically.

With Blue and Red creating systems and their systems engineering processes matched, the model may be used to see the impact of changing any one factor on the outcome; for example, choice of tool or method, training level, degree of requirement fault eradication, resource reductions, and so on. Where only one factor was changed, any resulting difference in outcome would be related to that factor alone, allowing users of the nonlinear dynamic simulation to understand the consequences of planning actions. This method also allows a creating system and its systems engineering process to be evolved and optimized using cumulative selection; see Design Capability Ratcheting on page 279.

Holding Red creating system and its process (say, waterfall) constant, a competing process (e.g., concurrent) would be inserted into the other simulation,

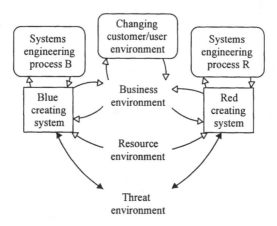

**Figure 11.22** *Competing systems engineering processes.*

and the results compared. This parallel approach has particular advantages, as it allows the two creating systems to compete for resources (skilled manpower, tools, facilities, and so on) from the same pool, but at different times and in different degrees.

Overall, then, such a simulation provides its users with the opportunity to compare different systems engineering processes; to see how long different design schemes might take to create, and at what risk; to determine how much resource the whole process would need, what were the probabilities of overrun or under run, costs, and so on. Moreover, using the cumulative selection approach introduced earlier, it would be possible to identify "genes" to code for critical parameters such as resources, tools, skills, and so on, and to evolve an optimal solution.

In the right form, such a simulation would provide good service throughout projects, enabling users to predict the overall impact of late changes, of inadequate resources, of defects and deficiencies, and of rework. Users could also, if necessary, change horses in midstream, by switching to an alternative systems engineering process secure in the knowledge that the change would be beneficial rather than a speculative gamble.

Curiously, although systems engineers may simulate the operation of a created system in its future environment, it is not standard practice to simulate the operation of a systems engineering process in its future environment, and it is certainly not standard to represent the project team executing the systems engineering process. Since, as we have seen from the five-layer structure, the project-level systems engineering process is at the heart of business and industry systems engineering, too (see Figure 11.19 on page 334), creating an effective representation of systems engineering at level 2 would provide the enabling capability for similar simulations at levels 3 and 4, too.

Science-based systems engineering, then, is not in common usage. The approach is straightforward, however, which raises the question: why not?

## SUMMARY

Systems engineering is defined, simply, as the art and science of creating systems. A unique, five-layer structure is introduced to classify systems engineering as: artifact, project, business, industry, or socioeconomic. Examples at each level showed that the levels "nest" one within another. Project systems engineering is discernable within business systems engineering, and business systems engineering can be seen within industry systems engineering. At industry level, the phenomenon of lean volume supply systems engineering was highlighted by contrast with mass production. The socioeconomic level was viewed at the national and, briefly, at the global level.

The concept of science-based systems engineering was introduced as a means of proving that a proposed systems engineering process would achieve the goals proposed. The means of achieving a reasonable proof is seen as nonlinear dynamic simulation, not only of the particular process (e.g., waterfall, concurrent, and so on), but also of the creating system that is executing the process. Not only would such a simulation be used to choose the ideal systems engineering process, but it would also serve in place of a PERT-type network, as a real-time project management tool that is capable of interrogation, and of determining the optimum path to project closure from any point in the live process.

## ASSIGNMENTS

1. At what level in the five-level systems engineering system would you classify a supermarket chain? Develop a systems engineering model for a supermarket chain, including its international/global sources of food supply.
2. In a lean volume supply chain, or circle, show how varying the production rate increases work-in-progress (WIP) inventory.
3. For a systems engineering standard to provide useful guidance to the creators of systems, yet not inhibit creativity and innovation, what subjects should the standard address, and what should it not address, and why.
4. Scientific systems engineering, as defined in this chapter, is not in vogue. Suggest why this might be.
5. You are a director of an innovative, independent systems company, seeking to increase its turnover and profit. You are tasked by the board with establishing a case for scientific systems engineering and a case against, and with making a recommendation to the board about the "way ahead," for the corporation. You are also tasked to present your findings, in terms the other members of the board can understand (!), in no more than five pages.

## REFERENCES

[1] Hitchins, D. K., "Managing Systems Creation," *IEE Proceedings, Vol.133, Pt. A, No. 6*, 1986, pp. 343 – 354.

[2] Hitchins, D. K., "World Class Systems Engineering," *Engineering Management Journal*, Institution of Electrical Engineers, 1994, pp.81 – 88.

[3] DeGrace, Peter, and Leslie Hulet Stahl, *Wicked Problems, Righteous Solutions*, Upper Saddle River, NJ: Yourdon Press, 1990.

[4] Raccoon, L. B. S., "The Chaos Model and the Chaos Life Cycle," *Software Engineering Notes*, Vol. 20, No. 1, 1995.

[5] Jenkins, G. M., "The Systems Approach," *Systems Behaviour*, Milton Keynes, England: Open University Press, 1972.

[6] Womack, J. P., Jones, D.T., and D. Roos, *The Machine that Changed the World*, New York, NY: Rawson Associates, 1990.

[7] Ibid.

[8] Lano, R. J., *Operational Concept Formulation*, TRW Defense and Space Systems Group Publication, 1980.

# Chapter 12

## From Systems Thinking to Systems in Operation

*Let me have men about me who are fat;*
*Sleek-headed men and such as sleep o'nights;*
*Yond' Cassius has a lean and hungry look;*
*He thinks too much: such men are dangerous*
*Julius Caesar*
William Shakespeare, 1564-1616

### SYSTEMS THINKING: ABOUT SYSTEMS ENGINEERING

#### Beyond Technology

Curiously, systems thinking does not seem to be applied to systems engineering, at least not on a regular basis. Systems thinking certainly addresses the product of systems engineering, the system that is to be delivered to some customer, but rarely is it directed toward the creating system itself.

To be sure, there are different paradigms for systems engineering, as we have seen in the foregoing chapter, but these seem to be the subject of assertion rather than exploration. For example, consider the statement: "Our organization employs concurrent engineering; it saves time and money." This might be described as the "machismo" approach to systems engineering and to project management since, in the wrong circumstances, concurrent engineering (or any other arbitrary prescription) may cost more time and money; not less.

To presume that one systems engineering paradigm is always better than another regardless of circumstance is unwise. To presume that any of the classic paradigms is the ideal solution in a particular circumstance may be unwise, too. Paradigms are archetypal, that is, they serve as an ideal world model. In practice, variations from the ideal may be advantageous, the more so as conditions and situations may be changing continually throughout the creation process.

The five-layer structure showed that systems engineering could be identified at different levels. Since any system to be created at any level can be described

Advanced Systems Thinking, Engineering, and Management

using the same GRM as a template, then there should be a process model for creating systems that applies at each and every level. A generic systems engineering process model, one that explains just what systems engineering is, and how systems engineering is conducted, would operate at any and all of the levels.

Some of the systems to be created may be without technology, or may be such that technology plays a minor, insignificant role. At level 2, for instance, creating a new medical practice with premises, doctors, nurses, treatment facilities, records, and so on, is a task worthy of systems engineering, yet technology is not really evident.

In a second example, this time at level 5, the government may have concerns about the processing of asylum seekers trying to enter the country during troubled times. Many of these unfortunates are genuine cases, but some may not be, and it is even possible that terrorists may attempt to infiltrate as bogus asylum seekers. Creating a new system for processing asylum seekers might involve: containment facilities, with accommodation, food, clothing, medical, counseling, and so on; investigators, with facilities for tracing origins and bona fides; education facilities; entertainment facilities, especially for those being "contained" over extended periods; religious facilities; facilities for dispersing successful applicants to suitable locations, finding them jobs and accommodation; and facilities for dealing with bogus applicants, too. The whole has to be fair, humane, fast, and effective. Creating such a system would be a task worthy of systems engineering, yet once again modern technology is not really in evidence, and this kind of system has existed for many years without any technology.

It follows that systems engineering is, essentially, not about technology. It is about creating solutions to problems; the solutions may invoke technology, but equally they may not. Systems engineering, then, stands apart from technology, and the discipline of systems engineering is separate from the technological and engineering disciplines. This suggests that systems engineering is not multi-disciplinary, at least not in the sense of bringing electrical, electronic, information, mechanical, aeronautical, civil, and other engineering disciplines together and creating (hoping for?) some engineering amalgam. However, there is a common theme to each of those disciplines, generally masked by their different "languages," which may be systems engineering, and which–again–is not technology.

Systems engineering has been defined earlier as the art and science of creating systems. Perhaps it should be described, too, as the art and science of creating purposeful order.

### Always Two Systems: Creating and Created

Systems engineering is always about two distinct systems–creating and created–see Figure 12.1. The figure shows a containing system at left–in this case a business or enterprise–that is operating a number of projects, each at different

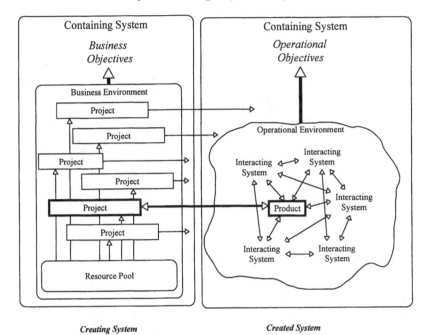

**Figure 12.1** *Two systems: creating and created. Each project creates one or more products (or services) for an operational environment, only one of which is shown at right.*

stages of progress. The projects draw upon, and hence compete for, resources within the business organization, such as trained manpower, tools, information systems, communications, security, and administrative facilities. The business has business objectives, concerned with survival and profitability, time and budget, and so on. The value of an individual project to the business is in the degree to which that project contributes, along with other projects, to the business objectives.

One of the projects is shown as creating a product which, at some future time, will be inserted into an operational environment. There, it will interact with other systems to solve a problem. This is, in effect, the solution space. The set of interacting systems will exist within its own containing system, which will have operational objectives. The value of the new product will be in how much it contributes, together with the other interacting systems, to its containing system's operational objectives. The figure illustrates the quite different interests, values and objectives of the two containing systems. The project must take cognizance of the product in its operational environment if it is to create an acceptable product and to encourage further business.

Figure 12.2 shows the impact of the future product on the current process, as the project seeks to create a product that will be ideally suited to the containing system; and the customer. In the figure, the "system to be created," the product, is

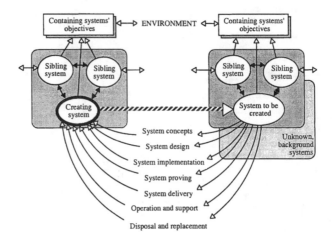

**Figure 12.2** *Impact of product on process.*

seen interacting in some future environment with other systems in a containing system; in this case, part of the solution space may be unknown.

This is frequently the case for some defense projects where, for reasons of security perhaps, the project team is denied access to some sensitive information. Nonetheless, information from the solution space guides many factors that go to make up the product system's life cycle, as shown in the figure. Some of these factors may influence the choice of technology, but most will influence the systems engineering plan, complementary operational systems (maintenance, training, and so on) and project management timeframes.

Each of the two systems, creating and created, has its life cycle. These can be seen as orthogonal; see Figure 12.3. In the figure, the creating system must be

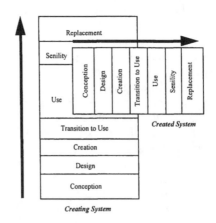

**Figure 12.3** *Intersecting life cycles.*

created first, before it, in its turn, manifests the created, or product, system. Of course, the creating system may create more than one product system; it may create product system after product system, and each product system may differ from the preceding and succeeding systems.

It is not uncommon practice to optimize the product system. The creating system may be optimized, too, although this is less common. The two systems may be optimized in different ways, to achieve different criteria. The product system might be optimized so as to minimize the life cycle cost, or cost exchange ratios, for instance, since these are of importance in the operational environment. The creating system might be optimized for maximum cost effectiveness, on the other hand, since this characteristic would contribute to the goals of the creating business organization.

Where each product system is either intrinsically different from its predecessor, or is going to interact with a different set of operational systems, or both, the features that are fed back from the solution space–see Figure 12.2–will be unique. The processes undertaken by the creating system will be tailored to these unique features both in content and duration, so the processes will also be unique. It follows that the creating system requires reoptimization for each product system it creates–unless, as in some level 4 systems, it makes batches of identical systems for a mass consumer market.

As Figure 12.3 suggests, the creating system may still operate when it has entered its age of senility. Like aging people, a senile creating system would repeat itself, creating the same kind of system according to the same plans, using the same paradigm largely regardless of the solution space; this is the Procrustean approach to systems engineering. To avoid this pitfall;[1] as well as for the purpose of reoptimization, the creating system may need to continually redesign itself, using new ideas, new methods, new techniques and tools, and new people.

## IDENTIFYING THE STREAM

### The Source

If technology is not at the heart of systems engineering, then what is? When a so-called process model is presented, what is it processing? If we were talking about a chemical plant, then chemicals would be being processed. For systems engineering, it is largely information that is being processed; not just any information, but information about the emergent properties, capabilities, and behaviors of solution concepts and systems. This goes to explain why systems

---

[1] The phenomenon of continual repetition is sometimes referred to as "groundhog day," after the film of the same name, in which a township was caught in a time loop.

engineering is not essentially about technology; it is possible to describe what a system does, and even how it does it, without mentioning the technology.

Figure 10.11 and Figure 10.12 showed subsystems in the human body and in a home respectively. It is entirely possible to describe the nature and operation of any of the subsystems without reference to their technology.

> The system circulates blood around the body to provide oxygen essential to the workings of organs and tissues, and to collect waste products from their activities. There is an organ within the cardio-vascular system, the heart, which pumps blood through the lungs and through the arterial network, to continually re-oxygenate the blood, and facilitate the circulation.

There is no technology: the heart, lungs, arteries, and so on, could have been mechanical, electro-mechanical, or biological–it is irrelevant at the level of description. Similarly, the subsystems of the home need not mention technology when describing systems for food preparation, waste disposal, power supply, insulation, and so on. Instead, they may describe the properties of these systems, what they do, and how they work.

Using emergent properties, capabilities, and behaviors to describe systems is using information, but of a very particular kind. The description of the cardio-vascular system importantly captures the dynamics of the subsystem operation; were the system to become static, not only would the whole body cease to function, but at a more immediate level, the cardiovascular system would stop. Its organ, the heart, is a complex muscle that also needs re-oxygenated blood to function, so it supplies itself with oxygen via the coronary artery in a neat, bootstrap[2] operation.

Dynamics are vitally important in describing the emergent properties, capabilities, and behaviors of open systems. The top speed of a motorcycle is affected by fuel; if there is no fuel at all, the speed will obviously be zero, but if the fuel is of low or high, calorific content, performance will be affected accordingly. Inflows and outflows can significantly affect system emergent properties, capabilities, and behaviors. So much so, that it is helpful to coin the term "*dynamic* emergent properties, capabilities, and behaviors," or DEPCABs (dep-cabs).

DEPCABs first appear for the whole system during solution concept development. The solution concept, fully described, will include the DEPCABs of the proposed solution, together with its CONOPS, and a similar description of other, interacting systems and their environment(s)–the solution space. A description of a solution concept is of little value without the relevant context, including solution system inflow and outflow rates and patterns. Such descriptions may be in words and pictures, but these cannot convey the dynamics of any solution space effectively, so the description should also include dynamic models

---

[2] Some electrical generators provide current to their own field coils in a similar bootstrap technique; there is little new under the sun.

and simulations – of the complete solution space. This is the source of the stream that flows through systems engineering.

## Partitioning the Stream

One objective of systems engineering–as we have seen, there may be business as well as customer/operational objectives – is to create a product or process that exhibits the DEPCABs identified during the conception process. For anything but the most trivial solution, the putative product or process will be partitioned into "manageable chunks;" that is, parts that can be created relatively easily.

The whole solution system, described by its DEPCABs and CONOPS, is then elaborated to reveal subsystems; each with its own DEPCABs and CONOPS. This can be achieved in several ways. One approach is to identify so-called mission functions first, using the CONOPS as a guide. Figure 10.1 on page 248 presented an interceptor's CONOPS in the form of a causal loop model, for instance, and it is possible to infer a set of mission functions:

- Search for new target, implying visual search, search by radar, or perhaps some other form of search;
- Detect target;
- Locate target, implying knowledge of its relative position in three dimensions;
- Identify target, that is, categorize it as friend, foe, or neutral;
- Create airborne recognized air picture, a cooperative function between several interceptors, airborne command posts, et al.;
- Allocate target to interceptor, a co-operative function to avoid target overlap;
- Engage target, approaching the target in such a way as to facilitate attack;
- Neutralize target, that is, disable, deter, force down, divert, and so on;
- Search for new target–back to the start, assuming the interceptor to have sufficient fuel and weapons.

In addition to mission functions, there are resource management functions. Two have just been mentioned: fuel and weapons. There will be many others: redundant systems, oils and lubricants, engine starters, recording media, and so on. There are also viability functions; those necessary to establish and maintain the system's capability even without a mission. In the example of an interceptor, the vehicle may patrol for considerable lengths of time, and on many occasions, without making any intercepts. It requires functional systems to maintain itself, to maintain the environment for the crew (if any), to evade and defend itself against threats, and to interconnect its many systems so that they operate seamlessly

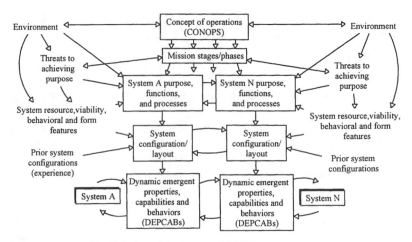

**Figure 12.4** *From CONOPS, through functions, to DEPCABs.*

together. It may also engage in practice interceptions (PIs), to evolve and improve its capabilities, and these improved procedures and practices may be imprinted into functional systems.

Working from the CONOPS, then, and using the GRM to point to the need for resource and viability functions, it is possible to generate a set of functions that the system must possess to satisfy its purpose. These functions can be grouped into subsystems; each with its own DEPCABs and CONOPS; see Figure 12.4 The process of partitioning can be as much an art as a science, and there are many influences to be considered.

In the case of the interceptor, for instance, it was for many years the practice to buy-in subsystems from various manufacturers. So, there might be a navigation system from one manufacturer, a radar system from a second, an automatic flight control system from a third, a weapon aiming system from a fourth, and a communications system from a fifth. These would be interfaced to create an integrated avionics suite. However, each of these separate subsystems has its own power supplies, its own processors and memories, and its own chassis, boxes, and cabling. There is also replication of, for instance, gyroscopes and accelerometers in several systems.

An alternative elaboration and partitioning might conceive a shared, single processing system and memory, one triple-redundant spatial reference system, and a series of small modules rather than large boxes to create an avionics suite with the same overall DEPCABS. Well, it would be almost the same: the alternative version would be smaller, lighter, more reliable, easier to maintain, and would deprive the original manufacturers of the separate subsystems of much of their income. It may also require them to reveal some of their ways of designing and making complex systems to competitors, which would be resisted. Partitioning can prove to be a sensitive task.

In any event, this smaller, more tightly coupled alternative need not be the best solution. In an ideal world, elaboration might result in a number of subsystems; each of which could exist and operate as a discrete physical entity – as could the major subsystems of the Apollo missions. There may be some repetition between the various subsystems, but these enable each subsystem to operate independently; easing development, testing and operation, and reducing the risk of single-point, catastrophic failure.

This example points to a more general rule. There are two archetypal ways to group and instantiate subsystems. One archetypal method is to create discrete self-contained mission functions, with their own, dedicated resource management and viability management features. These discrete groupings are realized as physical entities that, by virtue of their intrinsic facilities, are able to operate with little or no support–although they would be of little use operating on their own.

The alternative archetypal grouping method is to create an integrated set of resource management and viability management features, and to support mission functions from this set.

A well-known instance of the archetypal choices at work can be seen in computing, where an overall computing task may be undertaken by allocating a number of discrete computers to different functions, with each computer sharing the results of its labors with others through some network. Alternatively, a single, large computer with a large random access memory, high-speed symmetric multi-processor, and a multiple hard disk configuration, may undertake the overall task. Each approach has pros and cons.

## Confluence

In any event, the integration of the parts with their separate DEPCABs must result in a whole system with the requisite overall system solution DEPCABs. Because of interflows between the parts, however, they need not sum linearly, or algebraically, so that nonlinear dynamic simulation may be necessary to prove that the parts will sum to the whole.

Design goes further than subsystem DEPCABs; it goes on to optimize the overall system's DEPCABs by adjusting those of the subsystems, and also delineates the physical layout of the subsystems.

Information about the various subsystems becomes the basis for detailed design and development. The whole process may be shifted down a hierarchy level, so that any one subsystem can be viewed as being "at the top;" interacting with other subsystems in their respective environments. The subsystem of interest may be elaborated into sub-subsystems as before.

For a large and complex system, this process may be pursued through several hierarchy levels; a point will be reached, however, where the system, subsystem, sub-subsystem, and so on, has to be described and instantiated in more specific terms. The DEPCABs of a navigation system for a journey across the Sahara desert may be exhibited by a combination of an electronic map display, vehicle-mounted global positioning system (GPS) receiver system with solar batteries, and human navigator. Alternatively, a Bedouin on a camel, following the stars, the sun and moon, and ancient routes between waterholes, may exhibit the same DEPCABs–and find vital water as well. Again, the answer need not be technology, or if it is technology, it need not be a particular technology.[3]

Once subsystem DEPCABs are established, the subsystems may be bought in, developed from scratch, or created by modifying existing parts. In any event, from this point on, the matching set of subsystem DEPCABs is being tangibly realized. The original DEPCABs of the whole system solution in the system space are the guiding light, as Figure 12.1 shows.

So, the systems engineering process concerns itself from start to finish with information about emergent properties, capabilities, and behaviors, even if that information may occasionally concern technology or people, or more often a combination of the two. Figure 12.5 is a generic systems engineering behavior model, presenting the foregoing stream of information about emergent properties, capabilities, and behaviors diagrammatically. Since the model is generic, it should apply at any of the systems engineering levels, one to five, and for any system: social, sociotechnical, or technology only.

The behavior model follows the usual conventions, with the stream of activities flowing vertically down the center. At the left of each activity is the resource necessary to activate it. At the right of each activity is the product of the activity. As can be seen, resources and products are data and information.

At top left is the problem space, while at top right is the solution space, into which the instantiated solution system will be inserted. Not shown in the figure is the possibility of moving further down the hierarchy of systems before instantiation. Instead of instantiating a candidate subsystem, it may be elaborated into sub-subsystems using its own CONOPS as a guide; this would be a second-level elaboration. For a large, complex system, this process of successive elaboration may be repeated to lower and lower levels before instantiation.

A mission to Mars might necessitate a large, complex system. The overall system would be partitioned into a number of parts; several of them vehicles in their own right. Each of these vehicles would be a system and may be partitioned into major subsystems: propulsion, attitude control, life support, flight control, and so on. Each of these major subsystems may be partitioned into sub-subsystems. Life support might be partitioned into: air temperature and humidity, atmospheric content regulation, atmospheric pressure management, crew exercise facilities,

---

[3] In Victorian England, children were given toy theaters for Christmas, costing a few pounds or given away by shopkeepers to favored customers. The theaters were operated by levers and cams, so that characters would come out of the wings, to gesticulate, rotate, and so on, and then retire. Children would speak characters' parts, thereby entertaining parents and themselves. An undergraduate competition recently invited students to create similar theaters, which they did using microprocessors, small electric motors, drive chains, and split gear trains, at a cost of several thousand dollars. This technology was all the undergraduates knew. "If the only tool you have is a hammer, soon everything starts to look like a nail."

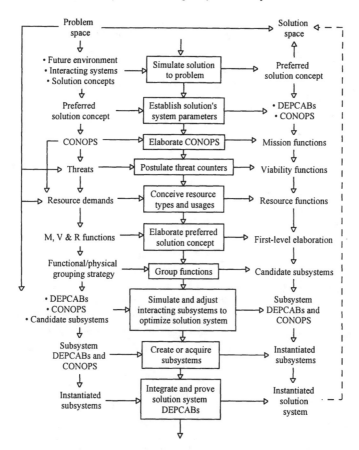

**Figure 12.5** *Generic systems engineering behavior diagram. DEPCABs: dynamic emergent properties, capabilities, and behaviors. CONOPS: concept of operations. M, V & R: mission, viability and resource.*

food preparation, sanitation, artificial gravity,[4] and so on. Each of these systems may in their turn be elaborated yet further, and each may have its own system design team. Eventually, however, instantiation must occur, but within the constraints that have been cascaded down from the original problem space via each level of elaboration.

Note from the figure that optimization occurs in several places. It is overt in the eighth step, but is also present at the start when one solution concept is preferred to others. Optimization goals are furthered during the subsystem instantiation process, when subsystems with appropriate capabilities and

---

[4] For long duration missions, vehicle rotation may provide an alternative to gravity, thereby preventing the onset of osteoporosis, muscle wastage, and other conditions associated with prolonged living in zero gravity.

corresponding costs are acquired or created. So, if overall cost effectiveness is an optimization goal, then each subsystem should have the appropriate DEPCABs, including cost, to achieve that goal. Note, too, that the act of partitioning and grouping creates an infrastructure of interfaces, interconnections, and interrelationships. This infrastructure is viewed as another subsystem in the figure, with its own DEPCABs.

## THE END-TO-END LIFE CYCLE PROCESS

### Outline Level 2/3: Conceptual Process

The life cycle of a system may be divided into two parts: the creative phase, and the operational, or in-use phase (see next chapter). Figure 12.6 shows a conceptual process model for the creative phase, as it might be conceived at systems engineering levels two or three. Note that the concept presents as a data state model [1]; that is, the processes are in the arrows, and the boxes contain data and information; this is consistent with the stream being filled with information.

The conceptual model is presented in three rows: starting from the problem situation, the top row creates one or more system solution options; the middle row designs the creating system to meet the business goals of the creating

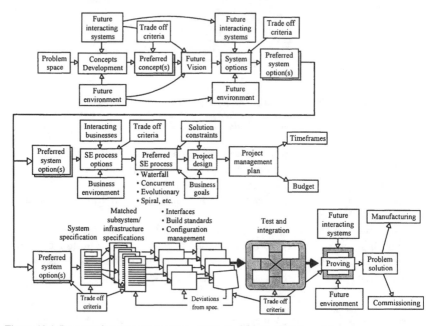

**Figure 12.6** *Conceptual systems engineering process model at levels 2, 3, and 4.*

organization; and the bottom row creates the product system to solve the problem.

The top row identifies future interacting systems and the future environment; these are part of the problem and solution spaces, but have been separately highlighted because they reappear on the bottom row as an essential adjunct to proving that the product system is a robust, optimum solution.

The middle row shows, in similar vein, the interacting businesses and business environment, since these influence the design of the systems engineering process. The design of the systems engineering process is then incorporated into the project management plan, along with the resources necessary to enable the process, to form a budget for the systems engineering project.

The bottom row shows the preferred system option (there may be more than one) being specified, formalizing the whole system DEPCABs and CONOPS. Subsystems are elaborated, as shown above, and similarly specified. A functional/physical grouping strategy will have been employed to create appropriate subsystems, which are then developed in parallel strands as shown in the figure. The choice of subsystems will dictate how much coupling, if any, connects the strands. As the subsystems develop, their actual, as opposed to specified, DEPCABs will emerge; where these deviate from specification there may be rework, intersubsystem rebalancing, or even redesign. The subsystems are then progressively tested and integrated to create a potential system solution that is proven in a dynamic test environment.

Note the use of trade off criteria throughout the three rows. Appropriate criteria are used to seek the best option at decision points throughout the project.

From the above, it follows that the term "systems engineering," which is often used in connection with row 3, really applies to all three rows. The relationship between project management and systems engineering is also explained: systems engineering designs the project; project management runs the project; and systems engineering conceives, designs, specifies and proves the solution system under the direction and leadership of the project manager.

Figure 12.7 shows a conceptual process model for a typical project. In this case, the processes are in the boxes, while the arrows indicate the direction of information flow. The figure runs from bottom to top, starting with identifying the goal and the initial goal strategy; that is, how it is proposed to achieve the goal, which will contain elements of business and customer objectives. A typical goal might be to create a high-integrity, through-life cost-effective system solution within budget and time constraints. The goal strategy might involve the development of prototypes, buying in subsystems from an existing source, maintaining a waterfall approach to ensure solution integrity, and so on. The goal strategy is subject to continual review as the project proceeds.

The central stream of processes moves up to the point of designing and specifying the system, and specifying the subsystems. These are then viewed as being developed in the central box; the activities in the box are not systems engineering, except for subsystem design which is systems engineering at Level 1.

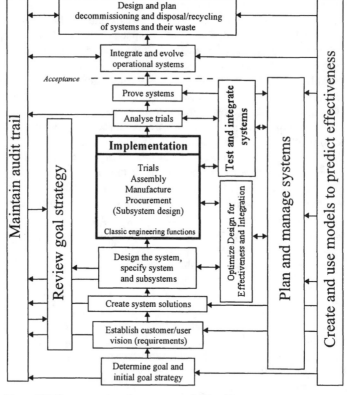

**Figure 12.7** *Systems engineering process model: level 2.*

The products from the central box are then integrated, trailed, and proven, before being accepted by the customer and integrated into operational use with other operational and support systems. Finally, at the top, the operational system comes to an end, is decommissioned, and is disposed of, or recycled.

Boxes to the left and right of the main stream are continuing activities, managing the flow, providing resources to energize the flow, and auditing to anticipate future difficulties.

**The Systems Engineering Shadow-board**

Systems engineering may employ a wide range of tools, techniques and methods; see Table 12.1. The table is referred to as a shadow-board after the practice of good car mechanics and aircraft fitters who paint mock shadows of their tools on the board where they are stored when not in use. This enables them to see at a glance if any tools are missing and unaccounted for at the end of the day, and to mount a search for lost items.

The tools are arranged in columns from left to right, following a typical project cycle from operations analysis (concept development), to installation and commissioning. Since different practitioners use different terms, the principal tasks executed during the cycle are indicated in the second row. Subsequent rows show a range of typical tool and model categories.

## System-level Elaboration and Specification

As shown in Figure 12.6, there are two systems to be designed and specified: the creating and the created. The same principles apply in both cases: specification of the whole systems presents DEPCABs and CONOPS; specification identifies the DEPCABs and CONOPS of subsystems and connecting infrastructure, but goes no deeper.

**Table 12.1**
The System Engineer's Shadow-board [2]

| Operations analysis | Requirements analysis | System design and specification | Project engineering | Integration and test | Installation and commissioning |
|---|---|---|---|---|---|
| Solution feasibility and performance | Requirements consistency and completeness | Options, interfaces, trade offs and specifications | Configuration, compatibility, interchange | Test environment | Customer acceptance |
| Scenario models | System models (physical, mathematical, simulation...) Relationship models (e.g., $N^2$) | | | Dynamic simulation Hardware in the loop | |
| System boundary models | Requirements tools and specification tools | | | Threat simulation | Acceptance models |
| | Human engineering models and tools | | | | |
| | Logistics models | | | Subsystem simulation | |
| Risk models | | RAM/ FMECA* | | | |
| Functional elaboration | Networks and architectures | | Configuration management tools | | |
| | System prototyping | | Interface control tools | | |
| | Functional/physical mapping | | Data management tools | | |
| Dynamic, nonlinear simulation of creating system executing creation process | | | | | |
| Cost, planning, and scheduling tools and models | | | | | |

* RAM: Reliability, availability, and maintainability. FMECA: Failure modes, effects, and criticality analysis.

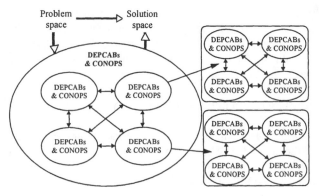

**Figure 12.8** *Specifying systems and subsystems.*

The structure is explained in Figure 12.8. At top left is the problem space, feeding both the specification of the whole system solution–at left–and the solution space at top center. The specification of the whole system is elaborated to only one downward level of hierarchy, that is, to subsystem level and no further.

Further elaboration of two only of the subsystems to sub-subsystem level, has been shown at the right; these would be separate specifications. In each case, their respective problem and solution spaces would be found in their containing, whole system solution specification.

To produce this specification "waterfall" with integrity, it is necessary to prove that the subsystems, when realized according to their specifications and integrated according to their (specified) infrastructure, will create the whole system solution, and that the solution will operate dynamically in its intended environment as specified. Smart tools are available that can help designers to establish that proof. Essentially, the tools enable a dynamic, nonlinear simulation of the system solution in its solution space, using specifications for the system, and its parts as input. Should any part of the specification be missing or incorrect, then the simulation will show errors, or may not run at all.

**Integration, Test, and Proving**

The process of proving that a potential solution will actually work correctly when delivered is of vital importance in many fields of endeavor. When it is discovered during operation, or even test flying, that an aircraft cannot recover from a stall, it is too late. Finding that some pharmaceutical process has excessive variance only when patients die is clearly too late. Arriving on the surface of Mars only to find out that the environment erodes the fabric of astronaut space suits may be too late, also.

Proving is an essential "end stop" in the creative systems engineering process. The proving process acts as a police officer, ensuring the absolute integrity of the delivered product. Proving may occur as a formal demonstration of the system solution operating dynamically in a representative environment, including threats and opportunities. It may also occur progressively during test and integration of the various parts of the system, as they are brought together after development.

A sound strategy for integration and test will see it being conducted within the creating organization, where problems can be addressed using available skills and facilities. It could be an expensive mistake to assume that the various parts of the system will simply fit together and work correctly, first time, in situ.[5]

It is also sound practice to undertake the procedure progressively, testing one part and then a second, before bringing them together and testing their ability to work together. Further parts are added individually, and testing occurs at each addition. Without such a cautious procedure it may be difficult-to-impossible to unravel the cause of a defect or maloperation. All of this may appear obvious, until you realize that testing parts of the complete solution for correct operation may necessitate representing the missing parts, those yet to be completed, as well as the dynamic solution environment. Additionally, specifications for correct operation of only "part-builds" of the system are needed.

If all goes according to plan, the test of the final addition will be the test of the system solution. This may prove to the creating organization that they have done the job, but they may still have to demonstrate compliance to the customer.

### Focusing Developing Emergent Properties

An essential feature of the system creation process is the continual refocusing of the developing emergent properties of the solution or product system. Emergent properties, capabilities, and behaviors may "wander" because:

- Developing parts are not specified adequately, or correctly;
- Developing parts are not developing according to specification;
- The need has evolved, with the customer, perhaps, introducing late changes; or
- The environment, or the interacting systems in the solution space, have changed since the solution system was originally conceived.

The last bullet, often overlooked, arises because the emergent properties, capabilities, and behaviors of a system are partly due to the inflows it receives. If the environment or interacting systems have changed, such that inflows to the solution system have altered, then the DEPCABs of the solution system may have altered, too. For instance, suppose an interceptor aircraft were designed to participate in a network-centric air defense barrier, and, specifically to exploit

---

[5] This was the mistake that resulted in the Hubble space telescope mirror problem that was discovered only when in orbit.

information provided through the network, then loss of, or even changes to, that information[6] could affect the interceptor's performance.

Whatever the reason, wandering subsystem emergent properties, capabilities, and behaviors may present a problem–or not, as the case may be. It is entirely possible for the characteristics of a subsystem to differ from those specified, without the whole system being affected. Just as a man may function with a mechanical heart, any system may function with an altered subsystem, provided its emergent properties, capabilities, and behaviors do not materially change those of the whole system. However, there will be a threshold, beyond which subsystem variation will render the whole system invalid. In this situation, either the offending subsystem will need correction, or the whole set of interacting subsystems will need rebalancing. As before, nonlinear dynamic simulation would be a valuable, even essential, tool in this eventuality.

## SYSTEMS ENGINEERING GOALS AND OBJECTIVES

Figure 12.6 on page 354 showed that designing the process was a major task for systems engineering. The place to start design is with the goal or goals, since we are, as always in systems engineering, addressing two systems, and it is important to do this early; before commitment.

Consider first, the customer and user objectives.[7] In many situations, the customer is separate from, and supposedly representing, a user organization. Customers may take it upon themselves to interpret what they believe the user wants or needs, and to isolate the user from a particular project. Since the customer is he or she who accepts the product system and pays the money, it may seem prudent to satisfy the customer and forget the user. That might be good politics or economics, but it would be poor systems engineering to provide a system that did not meet users' needs. Furthermore, user dissatisfaction would undoubtedly reflect adversely on future business with that customer, so in the long run, it is poor economics, too.

User objectives may vary, too. Experience shows that among a user population, a proportion (perhaps 10 to 15%) will have an imaginative approach to the needs they express for some new system. The remainder may well present as future objectives; those of the systems they are currently engaged with operationally. In defense, for instance, new technology may permit quite new ways of operating, but not every operator can see the point or how to exploit the new opportunity. It is all too easy to become locked-in to the current way of

---

[6] One of the risks in such enterprises is of enemy deception, putting bogus information on the network, or overloading it with spurious data.
[7] I use the terms "objective" and "goal" loosely. In general, achieving several objectives may result in achieving a goal, or the final objective in a sequence of objectives may be termed the goal.

operating.[8] (One way to overcome this problem is to include some of the more imaginative users on the systems design team.)

There is also the question of feasibility. If the user wants to be able to "cloak" his platform, that is, make it invisible, should this constitute a reasonable objective? To be sure, it may not be impossible, but such capabilities are presently at the interface between science fiction and advanced research–although, of course, the humble flatfish can render itself invisible on the seabed by pattern matching its background. All in all, it is prudent to look warily at the customer objectives, and to establish some means of ascertaining the user's objectives directly, and not through some surrogate.

In some situations, the customer does not express objectives. In lean volume supply systems, the supplier may create innovative new products and test the market by producing them as speculative ventures. The customer buys if the product appeals. This is the way to create markets, rather than simply satisfy market demand.

The Sony Walkman, for instance, would never have been produced had it depended upon statements of customer or user objectives. Instead, the approach was to produce something new, and try it out. The rest, as they say, is history.

This approach is something of a way of life among volume suppliers of consumer products. Instead of market surveys seeking to find out what the public would buy, the market is showered with innovative new products, some of which take off while others founder. It is a dynamic, roller-coaster approach to marketing and merchandising. From the nature of products that survive, it is possible to infer customers' objectives, but since these are subject to whim and fashion, objectives may have changed before the analysis is complete. A similar approach to defense procurement of COTS equipment from competing, lean volume suppliers has been outlined in earlier chapters.

Figure 12.2 on page 346 showed how customer objectives could impact on the project design, and hence on business objectives. As ever, business objectives must be to survive and make a profit, which generally means executing the business within a sensible timeframe: overruns cost money. Timeframes influence the choice of process archetype; as we have seen, waterfall systems engineering may be expected to take longer than concurrent systems engineering under some circumstances. However, the customer's delivery plan may influence timeframes, too; evolutionary acquisition would have a profound effect.

Overall timeframes are related to the nature of systems engineering, too. We have also seen earlier that the amount of time spent eradicating errors in specifications inversely affects the overall project duration: the more time spent

---

[8] A classic example concerns the AV-8B, the vertical, short take off, and landing (V-STOL) aircraft. Originally, as the Harrier, only U.K. air forces used it. When it was procured for the U.S. Marines, U.S. pilots discover that the jets that gave it its vertical take off capability could also be angled downwards during turns to make them much tighter. This so-called vectored in-flight flying (VIFF) was not, seemingly, discovered by the U.K. pilots because they had been taught, and had become used to, one particular way of flying.

eradicating errors up front, the less the overall project duration and cost. In designing the project, such factors should be taken into serious account, as failure to do so may prove disastrous. Despite this, it is far from unknown for corporate financial executives to reduce the overall cost of a bid[9] simply by cutting, say, 10% off the duration of each and every project activity, and hence, supposedly, off the bid price.

Given a robust set of both customer/user and business goals, it is straightforward to work back from the goals toward a start point, identifying milestones by which key stages should be reached. Intervals between milestones show the time available to achieve the milestone objectives. The processes, activities, and resources necessary to reach the milestones in the estimated time can then be postulated and tested.

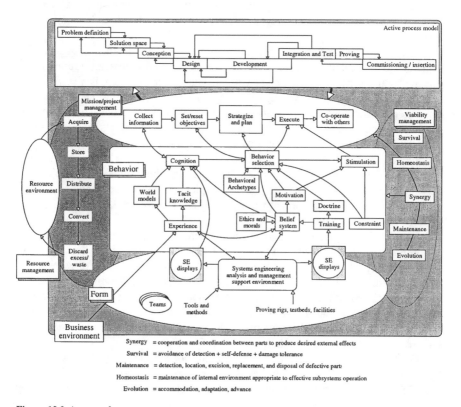

**Figure 12.9** *A system for systems engineering.*

---

[9] This is separate from the practice of quoting a low price for an expensive, prestigious project, knowing that the customer would never proceed were he to be aware of the full cost. Such practices are not systems engineering, although customers could become much smarter were they to use systems engineering simulation models to estimate the true costs in advance.

Much may depend on the resources available, particularly in terms of skills, experience, and current practice levels, together with tools. If robust plans and estimates are needed, and they usually are, a scientific approach will be needed to keep track of, and balance, the competing influences.

## DEVELOPING STRUCTURAL SUPPORT

### Process Resourcing

Figure 11.21 on page 337 and Figure 11.22 on page 338 showed how the problem may be approached. It is a classic problem in system thinking, and can be addressed as have others in earlier chapters.

Figure 12.9 shows the creating system based on the GRM. There would be two or more of such models; each with a different process model in the top layers, so that the process options (waterfall, concurrent, helical, and so on) may be compared. The form model shows the teams, tools, and facilities. The behavior model shows the people undertaking the various missions, where a mission might be to execute one of the process model activities. The figure shows that behavior depends upon the skill, training, and experience of the practitioners, and upon the doctrine, or systems engineering philosophy, pursued by the creating organization.

The mission management layer shows the usual activities, working on elements of the process model at the top. At left is the resource management module that is continually extracting resources (trained, experienced practitioners, tools, methods, documentation, data, and so on) from a resource environment. The parallel model, not shown, will access the same resource environment, thereby representing more than one project running in an organization at the same time.

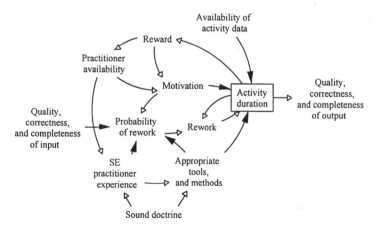

**Figure 12.10** *Factors affecting systems engineering activity duration: CLM.*

The viability management layer is shown at right. In this situation, these functions boil down to local management activities: ensuring that things work, that people connect, that progress is coordinated, and that morale is maintained.

**Understanding the Process as a System**

The process model in the top layer of Figure 12.9 is active, meaning that each element connects to, and interacts with, other elements. Some of the factors affecting activity duration are shown in Figure 12.10, from which it can be seen that there is more to be considered than a simple time estimate.

Although tools, particularly information tools, may be much in evidence during systems engineering, they do not "do the work." Instead, they generally record the work that is done by the practitioner. Systems engineering is a cerebral activity, so motivation is a factor in the length of time an activity takes. As the figure shows, this is a two-edged sword: motivation may make the work go faster, but it may also increase the probability of rework, either because speed leads to mistakes, or because the motivated practitioner is more conscientious. The probability of rework is increased, of course, if information entering the process from upstream contains incipient errors.

There are also reasons for process activity to falter, or stop. If data is needed,

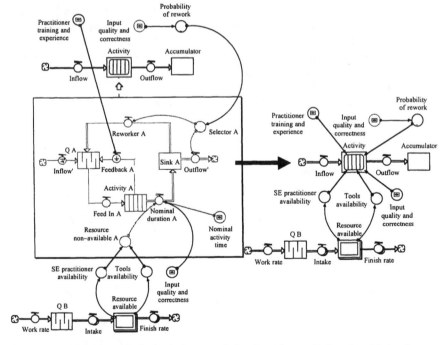

**Figure 12.11** *STELLA model of individual process. Left: submodel open. Right: submodel closed.*

but is not available, or if tools are occupied and queues form, then the process may grind to a halt–in fact, if not visibly. (The motivated worker always looks busy!)

The CLM of Figure 12.10 may be represented in a STELLA model (see Figure 12.11). The figure shows the model twice, expanded at the left and closed at the right, where the central box is a submodel. At left, the expanded submodel shows how the CLM is realized. There are two bases for rework: first, there is rework due to the lack of practitioner training and experience; second, there is rework due to errors and poor quality (e.g., incompleteness, inadequacy). The expanded submodel also shows control of nominal activity time: the time taken for one activity to pass through the conveyer under ideal conditions. The conveyer may be held up by either a lack of practitioners or a lack of available equipment, tools, and so on. These possibilities are represented in the simple queuing model at the bottom, which is an array model; addressing the availability of all resources.

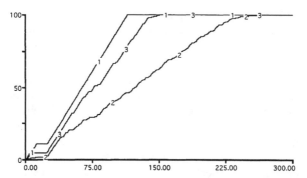

**Figure 12.12** *Process times showing variations due to practitioner training and experience, input quality and correctness, and practitioner and tool availability.*

The STELLA model may seem complicated at first, but it is simpler when closed up at the right in the figure. The various controls are designed to connect to the appropriate parts of the full model in Figure 12.9, thereby enabling the top row of that model to connect to the lower rows. The STELLA model may also be used to explore the separate and combined effects of the different controls as part of a learning laboratory.

Figure 12.12 shows the results from a learning laboratory. One hundred tasks

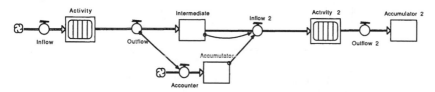

**Figure 12.13** *Coupling activities.*

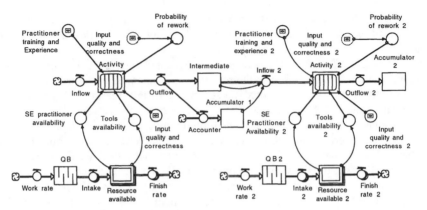

**Figure 12.14** *Coupling sequential resourced activities.*

are loaded instantaneously into the inflow, with the aim of seeing how long it takes for all activities to be undertaken, and the results posted in the accumulator. Line 1 of the graph shows the results, with the last of the 100 items being completed in 113 time units.

The step in the rising line is caused by temporary nonavailability of a practitioner or a tool. Line 2 takes some 258 time units, as a result of poor training, experience, and practice, and because of poor information input quality; line 3 shows an intermediate result where only one of these is deficient.

Once the activity model is established, full, interactive process models can be built by coupling the sub-models together to form networks–see Figure 12.13 for a simple example. Each submodel is connected to its resources, and receives information from the previous submodel, still containing, perhaps, errors or omissions that have yet to be resolved. So, the upstream quality affects downstream duration and cost, and the probability of rework, at the macro as well

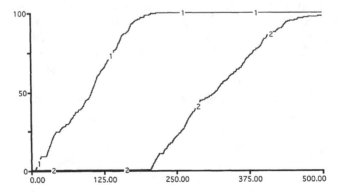

**Figure 12.15** *Coupled, sequential activity. Line 1:Accumulator 1. Line 2: Accumulator 2.*

as the micro level. The so-called "hooks" for connecting the activities into the resource and behavior elements of the full model of Figure 12.9 are shown again in Figure 12.14.

Figure 12.15 shows one set of results for a run of Figure 12.13 and Figure 12.14. In this instance, the pair was set up to work in waterfall fashion, that is, the second process did not start until the first had completed. It would be straightforward to have the second activity start before the first had completed, to represent some degree of concurrent processing.

Rather than constructing a full creating system model such as that of Figure 12.11 for each process, it may be tempting to take a shortcut–see Figure 12.16. Two activities are shown, Activity A and Activity N, the latter being some way downstream.

Suppose that part of Activity A's task is to eradicate specification errors, and that the more time spent undertaking Activity A, the more errors are corrected. Then Activity N will have fewer errors to deal with, and should take less time. The model shows at the top how Activity A's error search time generates the proportion of errors detected, and how this in turn reduces the duration of Activity N.

Results from running the simple simulation are shown in Figure 12.17. At the top, a plot of the two activities' durations shows, as expected, that as Activity A (x axis) extends and finds more snags, Activity B (y axis) shortens. At the bottom, plotting Activity A (x axis) duration against the sum of Activities A and B (y axis) shows there to be an optimum duration for Activity A to minimize the sum. In this particular instance, it appears that allowing about 5.75 time units for Activity A (x axis) will result in a minimum overall duration for the two activities taken together of about 8.8 time units (y axis).

### Calibrating the Creating System

However, there is rather more to it. Looking again at Figure 12.16 shows that the proportion of errors detected has been represented by a graphical function. The reason is that the relationship between activity time and the proportion of errors found is complex and uncertain, as we have already seen in the CLM of Figure 12.10. To use a model such as that of Figure 12.16 requires us to be able to

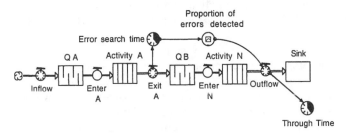

**Figure 12.16** *Simple process interaction model.*

"calibrate" the creating team, so that we can establish the relationships between resources, skills, and activity durations. We might do this either by reviewing past projects, or by running a more sophisticated simulation, such as Figure 12.9, and transferring the data from that model into simpler models such as that of Figure 12.16.

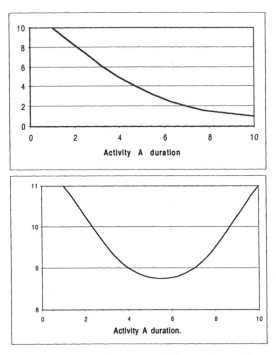

**Figure 12.17** *Activity interactions. Upper chart: activity N duration. Lower chart: sum of activity A and activity N durations, showing a minimum condition.*

Calibration is widely used in the prediction of timeframes and costs. The U.S. DoD uses a cost/risk identification and management system (CRIMS) that, amongst other things, employs earned value and technical performance measures as part of project risk management. Essentially, this procedure employs historic information to predict the duration and cost of current and future activities, processes, and projects, and hence the degree of cost and risk inherent in a project plan.

Some systems engineering organizations similarly calibrate their own software development performance as part of cost estimating. For systems engineering per se, calibration may prove difficult at the project level, owing to the interactions between activities and processes shown above. The interconnectedness of the various activities and processes indicates that they form a single system, such that altering one part affects the others. Because of this interactivity, and as we saw above, there may be potential to create an optimum process; one taking perhaps the shortest time for the least cost, or one that is most cost effective–provided there is flexibility in the design of the process. This optimum solution would best be found using a nonlinear dynamic simulation, in the style of Figure 12.9 on page 362.

### Tailoring the Systems Engineering Process

So, instead of choosing to follow a waterfall process, a concurrent approach, or any other archetype, the opportunity exists to design a systems engineering process that is uniquely tailored to a particular project, with its timeframes, delivery schedules, costs, resource limitations, operational environment, and so on. Moreover, as conditions change throughout a project–and they do, continually–it would be possible to reoptimize the process on the fly, taking account of the contemporary state of progress, resources, changed circumstances and the new factors.

Over the course of a project, the nonlinear dynamic simulation would guide and monitor the project, and would create an effective audit trail with each re-optimization run. Here, then, is a new, and potentially effective way to conceive, design, implement, and manage a systems engineering project or program.

What would be needed for this type of optimization? We have already seen a not dissimilar approach taken in the design of the mobile land force, see Optimizing the Design Capability on page 293, where the parameters of the force were evolved in STELLA using cumulative selection. That process would undoubtedly work for systems engineering projects, too, using a model of the type shown in Figure 12.9. A specific program would be advisable, however, since STELLA and other, similar tools, are not designed for cumulative selection.

The result of such modeling is not only the best process within given constraints at the time of modeling, but also the resources required, and the impact on timeframes and budget of having less, or more.

## ACCOMMODATING LEGACY AND CHANGE

One of the problems besetting the would-be system creator is the continuing existence of the system that the creator would like to enhance, or replace; these pre-existing systems are referred to as "legacy."

A scientific systems engineering approach to legacy systems would be to ignore them in the first instance. In conceiving a system solution to a problem, the usual procedure would be to create an overarching system solution within a system space, and to form a CONOPS, which can be worked through, step by step, to show that the system solution would work. One system solution might associate with several CONOPSs. Different putative solutions would be generated; each with one or more CONOPS. The competing solutions would be modeled, compared, and optimized; see Figure 12.5 on page 353.

CONOPS is then used to identify mission subsystems, and the GRM is similarly used to generate essential functional and support systems. At this point it may be possible to marry legacy systems; those that the customer or client wishes to retain, into the developing solution concept. If a legacy system can serve as either a mission subsystem or as a functional/support system, then the issue is resolved to advantage, since the legacy system already exists, requires no development and is already trusted by the user and customer–or else it would not be retained as a legacy.

A perfect match, such as that of the previous paragraph, would be fortuitous. Often, however, legacy systems are less than a perfect match. They may provide some of the facilities; provide more than is needed; provide something different, but be capable of modification; and so on. Where a legacy system provides more than is required in the new situation, choices will be needed about the additional capabilities: should they be retained but not used, or deleted? Sometimes it is more expensive and risky to delete, and better to leave the legacy system acting as a whole, self-contained subsystem; only part of which is used in the new situation.

**Police Information System**

Figure 12.18 shows how legacy systems may be turned to advantage. Above the dashed line in the diagram is a legacy system used originally for police operations and crime records. Its technology and operating system are unsuited to the latest

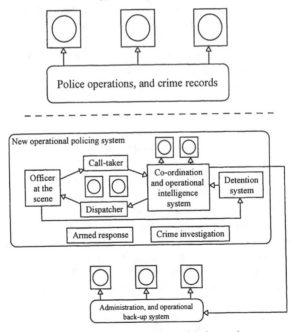

**Figure 12.18.** *Legacy in a police command and control system.*

policing methods, which employ video, digital photography, rotating 3-D graphics of buildings and services, and other graphic information. In the lower figure, a new system is introduced that is able to handle these new features and facilities. The legacy system has been converted for use as a police administrative facility, where it is used for correspondence, filing, personnel, building security, intranet, web site mastering, and other matters.

The new operational policing system was conceived as having a fallback facility to guard against loss of capability due to failure, fire, sabotage or information warfare (viruses, worms, and so on) If such events occur, the legacy system provides that fallback facility, retaining much of its original hardware and software. Data from the new operational system is copied to a new memory system in the legacy system via an optical link, so that the legacy system is kept up to date (a few seconds behind, actually), even when being used in its administrative role. The legacy system is in a different location, has a different operating system, and has separate power supplies, so is unlikely to be affected by whatever befalls the new, primary system.

Care is needed in setting up such systems. The database schema, data dictionary, and some listing and labeling procedures in the new system need to be compatible with those of the legacy system, so that the latter may recognize and handle the data it receives. Operators of the new system need practice, too, in reverting to the legacy system in times of difficulty. None of these factors is insurmountable. The net result is a benefit to the user and customer.

**The New Interceptor**

A national government decides that it needs a new air defense interceptor. To economize, it proposes to create a new variant of an existing reconnaissance and tactical strike (RATS) aircraft. Recognizing that RATS sensors are unsuitable, government also prescribes the use of particular air-to-air radar equipment, currently in late development under government contract. To emphasize economy, government also prescribes that the new interceptor, the air interceptor (AI) variant, must have 85% commonality of equipment with the RATS variant.

The RATS aircraft, or at least an unspecified 85% of it, is made up of imposed legacy systems. Unlike the police command and control system above, this situation is rather more difficult to turn to advantage. The reason becomes evident when working through the CONOPS for any likely solution. Fundamental characteristics of AI fighters are speed and agility. These in turn depend largely upon power-to-weight ratio[10]; a typical AI fighter would be able to accelerate in vertical flight because its full engine thrust exceeds its combat weight. This is not the case for RATS aircraft, where the emphasis is on flight duration, relatively

---

[10] There is more to it, of course. Agility is related to wing loading, for example, so that wing shape and wing area come into the equation. And we should not forget the AV-8B's ability to swivel its thrust nozzles in the turn.

low specific fuel consumption, steady flight at low level for weapon delivery, and the ability to follow, and hide behind terrain, also at low level.

As a direct result, the new AI aircraft cannot operate like a conventional interceptor. It dare not engage in close combat with competitive fighters because it would be outmaneuvered and destroyed. A different CONOPS is needed to solve, resolve, or dissolve this legacy-dominated problem space.

If the new variant cannot "mix it" with other fighters, it evidently must operate at arms' length; keeping intruders at bay. There are limits to this philosophy, of course, since the air-to-air missiles with which the AI variant will be fitted have a limited maximum relative range (MRR).[11] However, it may be possible to conceive a solution in which the aircraft operates more as a missile carrier than a combat aircraft; able to fire off missiles in several directions against several targets simultaneously. This might get around some limitations by transferring agile performance from the aircraft to the missile.

The feasibility of a new CONOPS based on such ideas would require a different approach to interception. The crew would be provided with multitarget allocation through a tactical data network. Using a semi-active homing missile, the new radar would need to track, lock-on to, and illuminate several allocated targets simultaneously. (Alternatively, and significantly better, would be to fit a long-range, fire-and-forget missile, if such a device existed.) Once the interceptor had fired all its weapons, it would be vulnerable and would need to retire from the combat arena, to be replaced by others. To give it speed in the combat zone, it might also operate from height, diving into the zone and using kinetic energy as a (temporary) substitute for raw engine power. Fire-down capability (snap-down, to use the jargon expression) would be advantageous, too, allowing the interceptor to stay higher than the intruder.

As can be seen, the process of creating a workable CONOPS generates ideas that may or may not be acceptable. The AI interceptor would operate within a network centric combat environment. The new radar would illuminate several targets in parallel, implying capabilities for both the radar and the missile that may require modifications. The best that can be devised is a compromise, limited by the imposition of legacy systems on the potential design and CONOPS of the solution system.

This is systems engineering with one hand tied behind its back, which is far from unusual. However, within the constraints, it is still possible to find a solution system, with its CONOPS, and to find a local optimum condition that works around intractable legacy systems. In this example, the result would be novel, and nonlinear dynamic modeling would be helpful; not for just one AI variant, but for a mixed group of fighters defending against a capable and mixed set of intruders—a so-called many-on-many scenario. With ingenuity, it might just turn out, given the right tactics, that the new AI aircraft complements other, more

---

[11] Maximum relative range is the distance in front of an aircraft that its missile travels before the aircraft starts to overhaul the decelerating missile.

conventional fighters to give an improved overall force capability. Should we, perhaps, have been viewing the problem[12] from the containing system viewpoint all along, that is, from the air defense force as a whole, rather than from the individual aircraft?

## SUMMARY

By looking across the five levels of systems engineering, the issue is seen to be fundamentally not about technology, but about information relating to emergent properties, capabilities, and behaviors. This information flows in a hopefully unbroken and uninterrupted stream from the process of conception through the development of future vision into design, specification, creation, proving, and beyond. A generic behavior diagram is presented for the systems engineering of any system.

At systems engineering levels 2 and 3, the creative process involves two distinct systems: the creating system and the created system. Each has its objectives and its own life cycle. An end-to-end life cycle process is presented, showing the creative process from conception to delivery. Three threads are visible: the development of a conceptual solution (the future vision) to a problem; the design of a creating system to create the solution, which is tantamount to project design; and the process of creating the tangible solution. Systems engineering is seen in all three threads. In particular, systems engineering designs the project, while project management manages the process of creation.

Designing the project is a crucial stage, since it affects both the business and the deliverable product. The project is seen as a series of processes that are undertaken by the resources (men, machines, materials, and money), of the creating system. A model of the creating system acting upon the process system is presented as a basis for nonlinear dynamic systems simulation. Such a simulation could be used, not only during the design of the project, but continually throughout the project to reschedule and reoptimize as conditions altered. This would mean that projects need not employ particular archetypal systems engineering paradigms, but could tailor the process to the subject and the situation.

Lastly come legacy systems–those systems that may be said to have two or more lifetimes. Legacy is undoubtedly a restriction on creative systems engineering, but in suitable cases that restriction may be turned to advantage. Brief examples are given of legacy in police and aerospace systems.

---

[12] Or perhaps, that was the job of the government customer, who stated both the requirement for, and the constraints upon, the new aircraft. Systems engineering requires integrity. There is no point in putting the onus on customers; after all, they go to systems engineers because they believe systems engineers know how to solve their problem.

**ASSIGNMENTS**

1 . Compare and contrast the alternative archetypal partitioning/grouping strategies on page 351 with respect to the following:

- Ease of conception, design and development;
- Ease of operation;
- Security, supposing the task to be sensitive;
- Vulnerability;
- Resilience and recovery from outage;
- Integrity;
- Overall cost.

2. If systems engineering is not about technology, then how can it exist at level 1, the systems engineering of artifacts? Explain and justify–or not.

3. You are asked to create a new system by employing existing parts wherever practicable, to save time and money. You identify a module of the software program Microsoft® Word (used to create this book) as a likely candidate, but the module has wider capability than is needed for your application. Do you: a) use the whole module unchanged and make no use of the unwanted capability, b) attempt to modify the module so that it only does what you require, or c) expand your solution to take advantage of the bonus capability? (Assume, for the exercise, that b) is feasible.) How would you ensure that the Word module worked correctly outside of its normal environment; that is, without interconnections to other Word modules?

4. List the items and topics you would include in a description of a future vision for a new Mars lander system, and justify the reasons for their inclusion.

---

**REFERENCES**

[1] Lano, R. J. A., *Techniques for Software and Systems Design*, Upper Saddle River, NJ: TRW Series on Software Technology, Vol. 3, North Holland Publishing, 1979.

[2] Hitchins, D. K., *Putting Systems to Work*, Chichester, England: Wiley, 1992.

# Chapter 13

## Operational Systems Engineering

*O world! world! thus is the poor agent despised. O traitors and bawds, how earnestly are you set a-work, and how ill requited! Why should our endeavour be so loved, and the performance so loathed?*
Troilus and Cressida, *William Shakespeare*, 1564-1611.

### OPERATIONAL SYSTEMS ENGINEERING: LEVELS 2 – 4

#### Operational Systems Engineering at Level 2

Delivered systems that are in operation may be the subject of systems engineering. There are many systems engineers in such industries as air transport, railways, and military operations. The concern for operational systems engineers at level 2 is not only the initial creation of the operational system per se, but also the creation of complementary systems to support and enhance the operational system. Operational systems engineering is disregarded by some as unimportant, irrelevant, and not really systems engineering. They could not be more wrong.

In Figure 13.1, the operational system is seen at the focus of systems to train its operators and crews, and of systems to maintain, service, modify, and upgrade the operational technology through life. Without these complementary systems, the operational system will cease to operate. Establishing this network of supporting and enabling systems is the role of operational systems engineering. Managing, maintaining, and improving this network are essential to the longevity of the operational system, and to its effectiveness.

Maintaining the status quo is not systems engineering; instead, systems engineering is changing systems such that their emergent properties, capabilities, and behaviors alter to achieve some purpose. So, maintenance is restoring the performance of some part or system by repairing faults and defects, but maintenance is not systems engineering. However, introducing Automatic Test Equipment (ATE) changes the emergent properties, capabilities, and behaviors of the maintenance system, and, therefore, the conception, design, creation, and introduction of ATE is systems engineering. Similarly, conceiving, designing, and

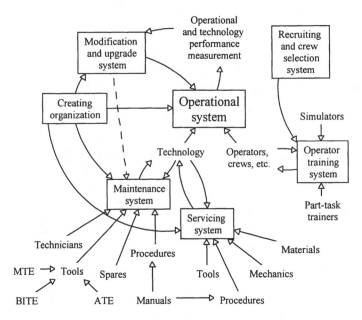

**Figure 13.1** *Operational systems engineering. MTE: manual test equipment. BITE: built-in test equipment. ATE: automatic test equipment.*

creating new servicing systems or crew training systems, would be reasonably classified as systems engineering.

Some aspects of operational systems engineering are unusual, and not to be found in other arenas. Military operations, in particular, present unusual problems that require innovative approaches.

## In-Service Support Systems

During the operational, or in-use phase of the life cycle, operational systems may require support to maintain their integrity and performance. Typically, such support will include maintenance facilities and training facilities. This is true of human activity systems as well as sociotechnical systems, although we may have to be generous in interpreting the term "maintenance."

Consider setting up a major new library service, for example. The staff members will need to be trained, and may need further training throughout their careers as they take on more demanding or technical tasks. If the new facility is large enough, this may warrant a dedicated training system. Similarly, staff members may leave, perhaps to get another job, or because they are unsuited to library work, or because of misbehavior. Replacing staff members that leave for whatever reason may be thought of as maintenance, since it maintains the overall

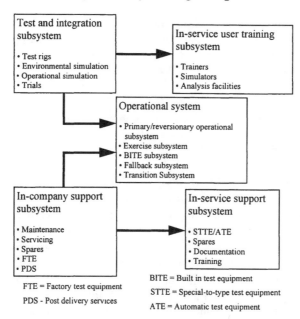

**Figure 13.2** *Project subsystems [1].*

capability of the workforce. Specialist groups, personnel, or human resources, usually conduct such functions.

In the general case, should the organization that is creating the primary operational system also create the in-service support systems? See Figure 13.2, which shows some of project subsystems involved in a command and control project. The in-company support subsystem, as the name implies, supports the developing technology parts of the product system; during a protracted development, these parts may fail and require repair, or need maintenance and servicing. A body of knowledge builds up in the company that is directly relevant to the design of the in-service support system. For example, factory test equipment (FTE) may serve as a design template for in-service test equipment–indeed, if the FTE is appropriately designed, it could become the in-service test equipment.

Similarly the test and integration subsystem may contribute to the in-service user training subsystem. Overall, then, a robust case can be made for treating all of the systems in the figure as part of the project, with the three in-service systems all being provided as one integrated set.

Operational Servicing Systems

Consider a squadron of aircraft on immediate alert standby, that is, poised and ready to go. That state of high alert may be called suddenly and without warning, and may require that the squadron provide, say, 10 out of its complement of 13 fighters as operational-ready. Why not all 13? Because some of the aircraft will inevitably have defects, or may have been disassembled for "deep" servicing. So, how can we service and maintain the aircraft so that 10 out of 13 can be guaranteed ready at short notice, say, four hours?

There are different ways to undertake servicing, as follows:

- *Scheduled servicing*–This is servicing in which the servicing plan for each aircraft is split up, usually on a calendar basis, into major servicing and minor servicing. Major servicing sees the aircraft stripped down to pieces (hence "deep"), and cleaned, oiled, machined, and so on, before reassembly. It might take two weeks or more to recover an aircraft to operations from major servicing, and a day or two from minor servicing.

- *Progressive servicing*–This is servicing where the servicing plan is split into more frequent, but smaller packages of work. These may necessitate some disassembly of parts of the aircraft, but may be designed so as to facilitate fast recovery to operational readiness.

- *Opportunity servicing*–Where an aircraft is known to "generate" faults and defects on a frequent, if not regular, basis, the servicing plan may be split up so as to take advantage of any disassembly or access, to undertake servicing in the area of the repair. This is turning necessity to advantage, and overall may enable most aircraft to be at, or near, operational readiness. However, it relies on frequent failure–not an ideal situation.

- *Opportunity progressive servicing*–As the title implies, this combines aspects of opportunity and progressive servicing. If opportunities present themselves, servicing will be undertaken that is due, or nearly due, in the area relevant to the defect, and will be "ticked off" the progressive schedule. Should no opportunities present themselves in the shape of convenient defects, then progressive servicing will be necessary to ensure the integrity of the aircraft.

The conception, design, and implementation of such servicing systems is a challenge for systems engineering, and the choice of servicing system affects the operational readiness of, in this case, the squadron. Similar considerations would apply to a fleet of airliners that face seasonal peaks in demand, and for railway systems–track and train–facing seasonal peaks at holiday periods.

**Operational Maintenance Systems**

The conception, design, and creation of optimal operational maintenance systems are similarly a challenge to systems engineering. There are many options, and the optimum solution depends on many different factors. Figure 13.3 illustrates some of the issues, by presenting four optional maintenance schemes:

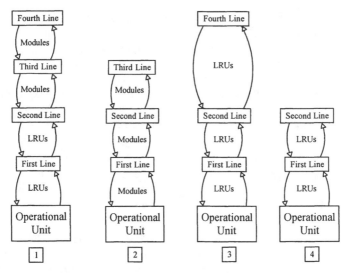

**Figure 13.3** *Operational maintenance schemes.*

- Scheme 1 might be used in aircraft avionics maintenance. The various avionic, electronic, and electrical systems are often manufactured and installed in boxes, referred to as line replaceable units (LRUs), so-called because they can be replaced at the (flight) line; that is, at the aircraft.[1] The priority is to restore the aircraft to operational readiness, so a new LRU is inserted in place of the suspect one. The suspect LRU is sent to nearby workshops, second line, where it is tested, and the fault is diagnosed and repaired by replacement of one of the modules from which the LRU is constructed. The LRU, now serviceable, is made ready for fitting to the next aircraft with a relevant defect. The offending module is returned to third line, a specialist organization, which can repair down to component level, using special test and repair facilities. Once repaired, the module is ready for return to second line. If found not repairable, the module may be returned to fourth line: the manufacturer.

- Scheme 2 might be used where the avionics or other system consisted, not of LRUs, but of line replaceable modules (LRMs). One of the issues with Scheme 1 is the cost; replacing complete LRUs, in which only one module is faulty, is expensive since a number of LRUs must be available to fill up the pipeline of spare LRUs that forms while suspect LRUs are being returned to second line, and diagnosed, repaired, tested, and made ready. Using Scheme 2 reduces this overhead, since the pipeline is filled with modules instead of the more expensive, complete LRU.

---

[1] The term LRU has spread in use and is common currency for boxes that may be replaced in tanks, ships, trains, and so on.

- Scheme 3 is similar to scheme 1, but omits third line. This scheme might be used where the LRU is highly reliable, with a mean time between failures of thousands of hours. In such circumstances, the cost of setting up second and third line repair facilities, provisioning spare modules, and training maintenance staff may be uneconomic. It is cheaper to throw the offending LRU away, and to fit a new one. To avoid making mistakes, however, the suspect LRU might be tested at second line to confirm that it is faulty, before returning to fourth line: the supplier.

- Scheme 4, similar to scheme 3, might be used for COTS equipments that are to be treated as consumables. Second line merely confirms that the item is defective, and throws it away. Even this step may be unnecessary if the item can be diagnosed effectively at the platform, thereby virtually eliminating the maintenance chain for the LRU or LRM.

In practice, a typical aircraft might have 100 to 200 different LRUs. Each will be different in terms of reliability, maintenance needs, and the cost of test facilities and spares, so each might potentially warrant a different maintenance scheme. Conceiving and creating a comprehensive maintenance system and organization for the whole fleet of aircraft is a challenge to systems engineering.[2]

**Operational Systems Engineering at Level 3**

Operational systems engineering may support, and be an important part of, an operational business; see Figure 13.4. The figure shows two operational systems providing a service in mutual competition, to meet a market demand. The size of the demand is driven by socioeconomic factors: for airlines, these would include "feel-good" factors, encouraging tourism, buoyant business performance, encouraging business deals, exports and imports, and so on.

The figure shows that revenue derives from market share, and drives the provision of the complete support system needed to keep the operational system working, and providing whatever service might be involved: airlines, underground railways, road haulage, building services, and so on.

A key issue in such systems is continuing profitability. Fundamentally, there are three reasons for low operating profits:

- High costs of support facilities, crews, and so on, and repayment on capital loans;
- Low revenues that, in turn, may be caused by:
  - o   Low market demand; or

---

[2] One seemingly attractive solution was thought, particularly in the 1960s and 1970s, to be automatic test equipment. The trials and tribulations of introducing ATE would constitute a separate book, however.

381

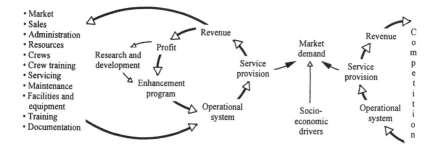

**Figure 13.4** *Operational business systems engineering.*

    o   Low share of market demand due, perhaps to low capacity, pricing policy, or reputation;

• Excessive spending on research and development.

These factors are mutually dependent. A low selling price may attract trade, but fail to cover costs and make a profit. Low spending on support facilities may result in unreliable operational systems, leading to increased running costs, reduced revenue, and poor reputation.

At some point, the business will perceive that it is not doing as well as expected vis-à-vis the competition, and an enhancement program will be triggered. For an airline, this might focus on an updated version of the aircraft already in use, perhaps a "stretched" version able to carry more passengers, or a longer range version able to travel to more destinations and so attract more customers. Alternatively, the organization might decide to re-equip with a different make of aircraft, with consequent major upheaval to the entire support system.

Such decisions are complex and multifaceted, not only in terms of how the support systems will be affected, the disruption, and so on, but also in terms of timing, costs, market image, and so on. The issues are of such complexity that nonlinear dynamic simulation at level 3, business systems engineering, will be highly advantageous. With an appropriate simulation showing both our own operational system (blue) and the competition (red), the performance of competing systems can be shown, both in terms of service provision and in financial terms. As we have seen before, it is then possible to enhance the blue system, and to optimize, say, its cost effectiveness, with respect to the competition.

Using such a model, it is also possible to explore the impact of alternative enhancement programs. Transitional costs may be estimated and, since each option or alternative can be optimized in the model, comparative financial performances over time can be assessed.

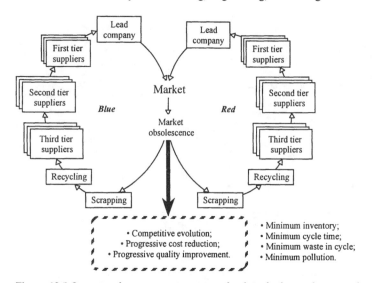

**Figure 13.5** *Operational systems engineering at level 4: the lean volume supply system in competition.*


## Operational Systems Engineering at Level 4

At lower levels, operational systems engineering has addressed the operational phase of a system that started out as a project, and was delivered to some customer. At levels 4 and 5, the system to be engineered is the creating system, and is less "delivered," than "lived in." There is no project, in the sense of a process with a clear end-point. Instead, there is an ongoing process of continual improvement.

The legal system that regulates socioeconomic systems, and that underlies governmental control of social behavior and misbehavior, is not viewed as having a beginning and an end, yet it undergoes continual change as social views of crime evolve, and as incoming administrations seek to legitimize their plans and actions. Similarly, the social system and the nation state are not thought of as having a beginning or an end, but are nonetheless subject to continuous social engineering.[3]

We have already met the lean volume supply circle; see Figure 11.13 on page 329. Figure 13.5 shows two such supply systems in competition within some

---

[3] The term "social engineering" is not in polite use, since it brings to mind eugenics and other unpleasant features of the Nazi regime in pre-World War II Germany. Be that as it may, governments of all shades continually indulge in social engineering; for example, when they change laws on contraception, abortion, cloning, the legal status of homosexual marriage, punishment for specific crimes, housing and agricultural subsidies, and so on. Political correctness is social engineering, too, with some aspects of it specifically intended to restrict freedom of speech. Whether the term is acceptable or not, social engineering is all around us, encouraging us not to notice its insidious presence.

unspecified socioeconomic system (which need not be confined within national boundaries, and therefore may be subject to more than one set of legal rules and regulations). As before, the two systems have been identified as Blue and Red for convenience, and to emphasize the intrinsic element of competition.

Under the discipline of competition, competing supply chains are expected to evolve, become leaner, and produce goods of lower cost and higher quality. As the figure suggests, there is also an ethic in the recycling of consumables. While minimum inventory, minimum cycle time, and minimum waste aim to reduce unit production costs, recycling is aimed at reducing pollution. For large, lean, volume supply systems, this is only sensible, since they would otherwise "foul their own nests," but it is also becoming socially important to control the mountains of waste that people, particularly in the Western world, create every day. As the figure suggests, competition may even arise in the future between circles to recycle each other's waste: at present, so little of most consumables is recycled that the issue has not yet arisen. Circles do, however, compete for each other's customers, through a variety of methods: branding, merchandising, and–particularly –innovation. The essence of lean volume supply success is innovation: the presentation to the consumers of something new and interesting that will convince them that their present product is outmoded and obsolescent.

Competing lean volume systems create another opportunity for the nonlinear dynamic simulation of either or both of the competing systems. Figure 13.6 shows some of the interactions within a lean volume supply system. The figure emphasizes the lead company, but similar activities occur in the supplier organizations; each of which may see its succeeding organization as its respective market. The whole is organismic, with the many parts interacting with each other in a dynamically complex manner.

Upon receipt of an order, the lead company transmits instructions upstream to the supplier organizations. Each organization is instructed about order details; the

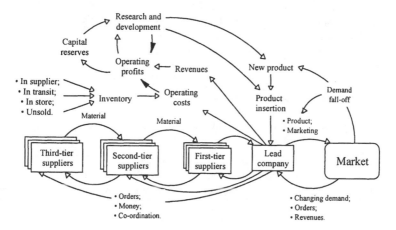

**Figure 13.6.** *Lean volume supply system organics.*

parts that go to make up the order; variations, or special requirements; who is to make what parts; timing; and so on. In cooperation with the supplier, the lead company will already have planned the manufacture and assembly process, including what optional features the order may require, and how to address novel features. The lead company signals the start of the preplanned process.

In principle, each tier undertakes its due processes immediately upon receipt of parts and assemblies from the preceding supplier and, in principle, there are no queues and no delays either in the processes or between processes.

Most lean volume supply systems rely on more than one product. Most have many products or, if only a few products, then many variants. Many different products are being ordered in many different markets at the same time. To accommodate this variety, yet economize on manufacturing and assembly machinery, it may be tempting to manufacture in batches–as in mass production. However, batch manufacture creates queues of orders and inventory, for varieties not in the current batch. The solution is to avoid batching if possible, such that each item[4] being processed may be different from preceding and succeeding items. To eliminate batching yet minimize process time requires that each machine can change its manufacturing or assembly activity in short order. Part of the skill of creating such systems is in the rapid change of the machine's role.

Figure 13.6 shows simplistically, at top center, how operating finances accumulate. A significant point is the ability of lean volume supply systems to build up large capital reserves, and to afford large research and development budgets. These are necessary to continually create new, innovative, and attractive products. Where such systems are employed for defense procurement, the heavy burden of defense R&D expenditure may be transferred from government to supplier. For reasons of scale, it is likely that the lean volume supply system will produce defense products alongside commercial consumable products. Defense is a small customer relative to the enormous demand for consumer durables–white goods and brown goods[5]–satisfied by the largest lean volume suppliers. Suppliers will need the vast commercial throughput to afford defense R&D which, although small in product numbers, may prove expensive in complexity and technological terms. On the other hand, such defense research should feed through into more elegant, sophisticated, and innovative commercial products.

**Underground Railways**

Unlike competing supply systems, it is not always so simple to see transportation systems as being in mutual competition. Take, for example, the typical underground railway, metro, tube, subway, and so on, that most big Western cities employ to move their populations around by the millions each day. These railways are typical continuous operation systems, but they present special problems when

---

[4] This is sometimes referred to, paradoxically, as "a batch size of one."

[5] White goods are washing machines, refrigerators, freezers, etc. Brown goods are TVs, radios, etc.

it comes to upgrading. Consider a complex underground rail network, with several "lines;" that is, several subnetworks; each with its own underground stations, elevators, escalators, stairways, ticketing booths, and so on. The subnetworks are complete railways in their own right, with dedicated track, signaling, and so on, although they will cross other subnetworks, share some lines, and perhaps share some stations, although using different platforms.

Now consider upgrading just one of the subnetworks in a city with a daytime population of several million; the upgrade necessitates major changes to the stations, including elevators and escalators, and the introduction of new, longer carriages necessitating platform reconstruction. The problem facing systems engineering is that of planning and managing the upgrade, with the minimum of disruption to the public.

There is, of course, no right answer; the real world is not like that. But some answers will be better than others, and some will be acceptable, while others may not be. The approach to this kind of issue is to treat it as just that: an issue, to be solved, resolved, or dissolved (see on page 269). If the goal is to minimize disruption, then a number of strategies might be formed, along with criteria which would together indicate a "good" strategy, and the implications of each strategy would be explored and rationalized; possibly using a nonlinear dynamic model of the issue to facilitate effective comparisons.

For instance:

- Should the whole subnetwork be closed so that a large workforce could effect the changes in the minimum time, uninterrupted? What happens to would-be passengers in the meantime?
- Should replacement bus services be provided to accommodate would-be passengers–and hopefully, to retain their good will? With closures, will the increased road traffic above ground permit a replacement bus service to operate effectively? (Note that, although road transport and underground rail transport may not be perceived as competing, the two systems nonetheless interact with each other.)
- Could the whole program be implemented, piece by piece, between midnight and 5:00 a.m., when the rail service is virtually zero, without closing anything down?
- Should half a station (e.g., the up-line or the down-line, plus their respective elevators, escalators, signals, platforms, passages, halls, emergency facilities, and so on) be refurbished at a time, leaving the other half open and operational? That would allow people to travel to interchange stations, and thereby to access the rest of the network.

Criteria for a good strategy include: cost, timeframe, safety of workers and passengers; minimal inconvenience and loss of customers; minimal potential for permanent loss of customers; limits to traffic chaos above ground; and so on.

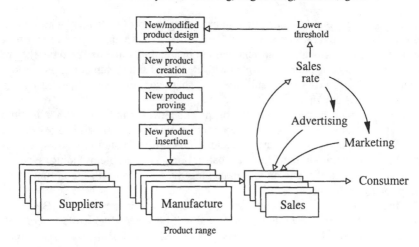

**Figure 13.7** *Heijunka: product replacement.*

## Heijunka

While the classic free and open market may not be subject to engineering, it can certainly be manipulated. We saw earlier that inventory was potentially minimal when lean volume production was steady: rises or falls in the production rate would result in spare inventory. Heijunka, or production smoothing, employed marketing and advertising to oppose falls in demand rate, and reduced such measures as the rate rose again. We also saw that lean volume supply systems depend on innovation to ensure a continuous flow of products, on which their systems are founded economically.

Figure 13.7 shows a simple model of one aspect of heijunka. The figure shows a supply chain with products being sold to consumers. The lead company monitors sales rates, and uses increased or decreased advertising and marketing, as appropriate to maintain a steady consumer demand for the product. The effort to maintain sales level becomes critical as a product becomes less interesting to consumers, perhaps because competitors have brought out attractive alternatives. The lead company will then take a more aggressive marketing position, cutting prices, offering special deals, and so on, while at the same time preparing a new product to succeed the old—now about to become obsolescent. At the crucial moment (in theory), production of the old gives way seamlessly to production of the new; the arrival of the latter in the market place will have been announced in advance to selling organizations and publicists.

Of course, it is not as easy as that. It is difficult to predict just how much advertising and marketing can sustain the sales of an aging product, especially as competitors will attempt to conceal their activities. Demand for the new product cannot be stimulated too early, or too vigorously, or else there will be customer

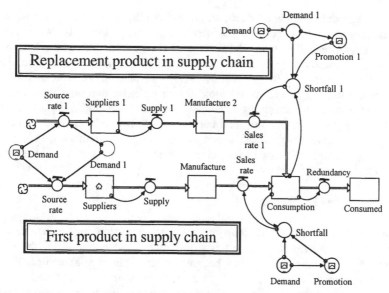

**Figure 13.8** *STELLA model of product replacement.*

disillusionment, and marketing effort will result in people buying a competitor's product.

Figure 13.8 shows a simple STELLA model of the situation, based on Figure 13.7. The model shows an existing product, in the bottom row, about to be replaced by a new product, in the top row. The new product is likely to be a variant of the old product, so that some of the assembly/manufacturing equipment may remain unaffected. As demand for the first product falls away, promotion (advertising, and so on) attempts to compensate, but eventually a shortfall arises. Conversely, promotion increases demand for the replacement product.

If the timing is right, the results may be as shown in Figure 13.9, where line 1 shows the first product sales and line 2 shows the replacement product sales. Note that respective demand is used to set source rate in both lines, as would be the usual case in a lean volume supply system to reduce inventory holdings.

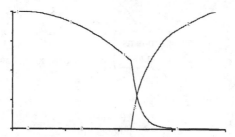

**Figure 13.9** *Heijunka: product switchover. Line 1: old product. Line 2: new product.*

In the model, the new product takes over from the old with only a modest dip in total production numbers, provided that, for a short period, both variants of the product are made side by side. The switchover works–in the model–because of the somewhat arbitrary curves in the graphical representations of demand and promotion. Experience in the real world would enable the lead company to determine what the curves would look like in practice, and how they could be changed by different marketing strategies. Since we are talking about the mass market buying of consumables, real curves would also be influenced by culture, disposable income, fashion, the economy, and so on. Nonetheless, this type of model can be calibrated, used to develop understanding, and even used for short-term anticipation of switchover.

## Market Systems Engineering

Figure 13.10 presents a CLM, showing how continuing flow may be achieved and maintained in the market. At left is manufacture, the output from the supply system. Products are bought and become products in use, at center. Product innovation introduces new products to the market, with new bells and whistles; making the current product in use appear obsolescent and redundant. So, owners of this product are persuaded to buy a new one.

In the 1950s and '60s, it was believed that the way to get a customer to buy a new product was to build obsolescence into the current product, such that it would be of little use or value after a short period. This practice was applied to Japanese cars, and, for a time, they became a byword for shoddy, inferior goods as a direct result. Nor did the technique prove effective. Certainly, the owner of the now obsolescent or faulty product bought a new one, but he or she went to a different supplier, hoping to get a better, more durable product.

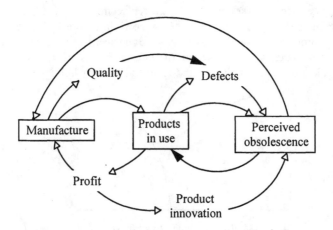

**Figure 13.10** *Market systems engineering CLM.*

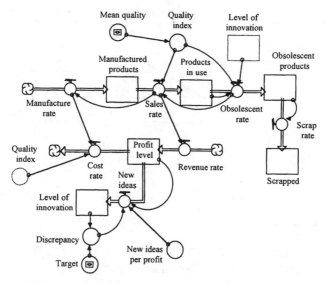

**Figure 13.11** *STELLA model of market systems engineering.*

Improving quality was the next idea: if the quality was really good, and there were no defects or failures, then customers would surely buy. This policy was also successful, but not quite in the way intended. Customers bought, but the goods never wore out, so there were no repeat sales.[6] Profits, after an initial surge, soon fell.

Finally, suppliers hit on a dual strategy: high quality, reliability and durability on the one hand, and innovation on the other. This worked. However, it can be overdone, as anyone will confirm who has bought the latest personal computer or TV, only to find an even newer model in the shops within weeks. Presenting innovative variants too quickly to the market (product "thrashing") turns customers off. There is an optimum interval between new products or variants: if it is shorter, customers will not buy, reducing revenues; if it is longer, revenues fall proportionately.

The market systems engineering CLM can be converted to a STELLA model; see Figure 13.11. More factors have been introduced to make it work sensibly. The top row of reservoirs shows manufacture, products in use, and obsolescent products, and obsolescence is related to quality and to innovation, as before.

Since innovation is funded out of profits (see Figure 13.6) the model shows profit funding innovation, which creates new ideas at a rate determined by new ideas per profit. This parameter suggests that there is a way of turning profit into new ideas, which must be a measure of the R&D group innovative productivity.

---

[6] As a young, single man, I favored nylon shirts: they were smart, easily washed, needed no ironing, and never wore out. As a result of these features, manufacturers stopped making them, and reverted to making shirts that did wear out.

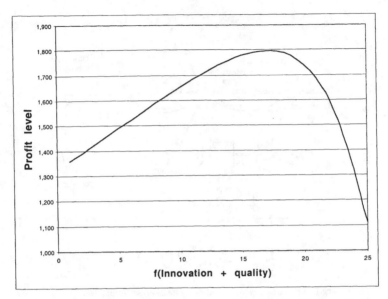

**Figure 13.12** *Effect of increasing quality and innovation on profit (arbitrary units).*

Using the STELLA model, it is possible to explore the effects of increasing quality and of introducing greater levels of innovation, either separately, or together. As the model suggests, they work in opposite senses: increasing quality costs more money in manufacture and increases a product's useful life; increasing the level and rate of innovation also costs money, but at the same time it reduces a product's useful life, rendering it obsolescent earlier.

Figure 13.12 shows the effect of increasing both quality and innovation together. This would be difficult to do in any practical situation, but the model, acting as a learning laboratory, allows us to investigate the possibilities. As the graph shows, increasing innovation and quality together initially increases profit, but a maximum profit is reached, beyond which profit falls away.

This fall-off occurs because quality improvements (represented in the model by increasing reliability) eat into profits, and because the level of innovation is funded directly out of profits. Eventually, a point is reached where quality is as high as it can be, and where further innovation does not attract further sales.

## SOCIOECONOMIC SYSTEMS ENGINEERING: LEVEL 5

Previous chapters have introduced systems thinking at level 5, see Societal Power on page 187; National Energy Strategy on page 190; and Afghanistan on page 301. In each case, it has been necessary to assume certain societal behavioral characteristics.

For instance, in the last example, Afghanistan, it was presumed that, by restoring, retraining, and enhancing the police, the judiciary, and the penal systems, civil control over the population would be restored. That is not an unreasonable assumption, but for civil control to exist there has to be a sense of trust, cooperation, and reciprocity between the various parties concerned. Would that always be the case, or does the behavior of people in social groups vary with culture, ethnicity, political ideology, education, training, sex, and so on?[7] Similarly, assuming that a larger, better-equipped force will overcome a lesser force may be reasonable for the Afghanistan model, but history tells us that it is not always so.

In different societies and different cultures, there may be different degrees of relationship, trust, cooperation, and reciprocity. Without such features, society does not exist; instead there are atomized groups. For systems engineering to be effective at level 5, then, it would not only need to comprehend and account for the ways in which behavior between individuals underlies the behavior of social groups, communities, and societies, but also how such behaviors differ between different cultures and ethnicities. For systems engineering at level 5 also to be scientific it would aim, using Ackoff's dictate, to optimize aspects of society, such as quality of life, freedom from pollution, available energy per capita, wealth generation, and so on, or all of the above.

Socioeconomic systems engineering is the practice of leaders, politicians, lawmakers, and others, both elected and selfappointed. Since there are politicians of sorts at all levels in societies, it is helpful to visualize level 5 at the level above industry level. In other words, systems engineering at level 5 is creating the environment and hence the constraints and goals of industries, and by "trickle-down," of businesses and projects implicitly, too.

Those operating at level 5 would not think of themselves as systems engineers—unfortunately, perhaps, since thoughtful planning and strategizing should be the order of the day. Some evidently see their role as one of control, and while control may have its value, the primary means of regulation in systems is, as we have seen, through dynamic interaction between systems, rather than by cybernetic control.

Realizing political goals is not easy in a democratic society [2]. One way of achieving structural change is through differential funding. Tax money may be used to fund/subsidize some aspect of society that politicians favor, while starving other, unfavorable aspects. In a dynamic society, there will be those who "rise to the bait;" seeing the prospects of earning money, and so achieving the politician's goals. This is not control in any cybernetic sense. Instead, it is a long-range, "hands-off" approach. Since most politicians at government level have neither the time nor the inclination to manage the process of change in any detail, it may

---

[7] Answering this kind of question is the province of controversial disciplines such as sociobiology and evolutionary psychology.

work. On the other hand, simply providing the money may encourage the less scrupulous to misappropriate at least a part of the bounty.

Another approach, again not control in the cybernetic sense, is to set up two organizations: one does the "business," while the other monitors, audits, and criticizes–publicly. This approach may be used with public utilities, public transport systems, and so on.

In totalitarian societies, the state frequently exercises monopoly over essentials such as food supply, power, transport, and so on. In democratic societies, such monopolies are seen as working against the dynamics of the free market, so another form of regulation, exercised by politicians and governments at level 5, is to restrict monopolies. Typically, this works through governmental restrictions on mergers, so that no one organization becomes so large that it can dominate the market. By rule of thumb, it seems, there should be at least four competing organizations in any sector–and the competition has to be seen to be genuine; otherwise, cartels may create virtual monopolies.

**Social Capital**

A contemporary view of societies considers the concept of "social capital [3]." Social capital concerns itself, not just with the people in a society, but also with their mutual interactions, their trust, and cooperation. As with network-centric combat, there is value and power in the network that binds a society together, and that prevents a society simply being a number of unconnected individuals.

The ability of a society to develop and flourish is related to its social capital. When criminal and loutish behavior causes people to barricade themselves in their homes and not go out after dark, thereby atomizing society and reducing interactions, social capital is reduced.

A simple view of social capital might consider the social interactions as existing at one level: on a plane. Such a view might afford high social capital to specialized groups, including terrorists and criminal groups. A broader view would include vertical, as well as horizontal connections, so that such isolated antisocial groups would not be afforded inappropriate status.

Another view would suggest that social capital might be measured, just as we measure corporate capital. This is more problematic. Social capital is not just about the number of interrelationships, or the number of groups, it is more about the realization of actual benefit from relationship.

For instance, suppose a plumber had built up his business over a number of years, and had a long list of regular customers who trusted him and who called upon him in time of emergency. The plumber moves to a new town, and is disappointed to find that he is getting very little business. Whereas there was high social capital in his previous situation, there is little (from his perspective) in the new location. Instead, there is another, established plumber who is trusted and esteemed by local residents.

In a second example, consider a ground-based air defense command and control system, consisting of two rows of radars, such that an incoming, low-flying intruder will pass over one row and then, some short time later, the second. The incoming aircraft is identified using identification friend or foe (IFF), where no response to a challenge indicates that the intruder is an enemy, to be intercepted.

Suppose, however, that the intruder is a friendly aircraft returning from a mission; there is a risk of fratricide, or so-called blue-on-blue casualty. IFF should prevent fratricide, but the system is less than prefect, so there is a chance that an IFF challenge is either not received by the incomer, or that it is received but no suitable response is given.

If the two rows of radars are independent, each incomer runs the gauntlet twice. If the probability of a successful challenge is 0.9, then the probability of two successful, independent challenges[8] is 0.81 (= $0.9^2$, the probability that both challenges were successful). On the other hand, if the first row passes track and identity data to the second row, such that the second row is already expecting a friend, then the cooperative probability of success can be as high as 0.99 (= 1- $0.1^2$, the probability that at least one of the rows challenged successfully). Put in betting terms, the probability of being shot down by one's own defenses has reduced from 1 in 10, to 1 in 100; this is cooperative behavior showing benefit.

Societies with high social capital are cohesive and internally dynamic. Good neighboring will be evident, and there may be a high proportion of people voting at election time. People will keep an eye on neighbors' houses when they are away. Crime levels will be low. In business societies, mutual trust and reciprocal behavior facilitate transactions and deals. At a national level, social capital will also evidence itself in a cohesive, resilient society. And, when politicians fund developments, those undertaking the work will be disinclined to misappropriate the bounty.

Socioeconomic systems engineering at level 5 is at an early stage; to be effective it will need to account for the emergent properties, capabilities, and behaviors of social groups, and the concepts behind social capital appear promising in that respect.

## Historical Systems Engineering at Level 5

Looking at past stable civilizations may reveal the impact of systems engineering at level 5. One such is the New Kingdom of ancient Egypt. Apart from one aspect, the socioeconomic system was engineered very neatly; see Figure 13.13.

The nation revolved around the pharaoh, who had three discrete roles: pharaoh was the all-conquering military leader who subjugated foreigners to bring security and tribute to Egypt; he was the living god Horus, the only one able to

---

[8] Passing through, say, seven independent rows–a more typical situation–reduces the probability of always being recognized as a friend to less than 50%.

intercede with the supreme god Amun-Ra on behalf of the people; and he united upper and lower Egypt, a "political" unifying role.

As Figure 13.13 shows, these three roles had effects that interacted to stabilize the nation. Egypt was fabulously wealthy at this time, largely because of successes in warfare. This wealth allowed many people the time and resources for the pursuit of eternal life, an Egyptian obsession. To gain entry to the afterlife, however, meant that the soul of the deceased (the "ka") had to pass through the Hall of Judgment, a spiritual court where the dead pharaoh, now transmogrified from Horus to Osiris, sat in final judgment. So, if the individual had behaved badly toward the pharaoh in life, it seemed that the chances of reaching the afterlife could be seriously prejudiced. This tended, naturally, to encourage good behavior.

In his role of uniting the two lands, the pharaoh presided over a vast nationwide bureaucracy, inherited from the time of the pyramids, some 1200 years earlier. The pharaoh appointed viziers to travel the land dispensing justice in his name: he appointed mayors; he traveled up and down the Nile with his court, visiting existing temples and endowing new ones; and he was generally highly visible to his people as their spiritual intercessor through the temples, and their temporal ruler through the courts.

The pharaoh was also responsible throughout the nation for Ma'at–a peculiarly Egyptian concept of harmony, justice, and peace, based on the idea that there was a correct way for things on earth, just as there was celestial harmony in

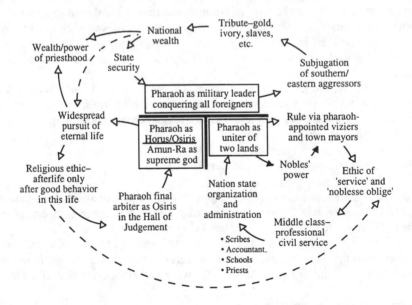

**Figure 13.13** *Level 5 systems engineering: New Kingdom, ancient Egypt, around 1300 to 1200 B.C.*

the stars. This required that those who had wealth should look after the poor, the orphan, and the widow. There was an air of "noblesse oblige," again tied to the idea that entry into the afterlife depended upon pureness of heart. Ma'at meant that the pharaoh and the bureaucracy ruled with a gentle hand.

As the figure shows, the three aspects of the pharaoh's rule locked together to produce a stable nation, except for the ever-increasing wealth and power of the priesthood. Eventually, this would get out of hand, but that is another story.

Ancient Egypt, with its many periods of stability over a 3,000-year span is a rich source of understanding about ways of systems engineering at level 5. It is valuable, not only for understanding the basis of socioeconomic stability, but also for the basis of the collapse of that stability.

## SUMMARY

Operational systems engineering is a somewhat neglected discipline, exercised, however, by a great number of people in transportation, education, the armed services, the emergency services, and many other arenas. Operational systems engineering at Levels 2 and 3 is the provision of active facilities and resources to sustain and enhance the operation of some primary system. Operational support systems include: maintenance, servicing, crew training, recruiting, selection, and training systems; upgrade systems; and many more. These all must be in place and effective for the primary system to achieve operational capability and longevity.

Operational systems engineering at levels 4 and 5 differs, since the systems involved are continuous, and not the outcome of projects. At these levels the systems engineering process is one of continuous improvement.

Heijunka and market systems engineering are briefly explored; particularly as the latter relates to quality and innovation in level 4 lean volume supply systems.

Socioeconomic systems engineering at level 5 is touched upon. Whereas it is possible to undertake systems thinking and to develop learning laboratories at this level, there is less assurance about systems engineering at level 5, which would be the province of government, politicians, and lawmakers. Relevant, sometimes controversial, subjects such as sociobiology, environmental psychology and, particularly, social capital are touched upon, but it is concluded that, although objective systems engineering at this level may be highly preferred to the blind guesswork that rules at present, the potential discipline is in its infancy.

## ASSIGNMENTS

1.  The underground railway dilemma on page 385 poses alternative ways of upgrading and refurbishing the subnetwork. Make a rational choice between the alternatives, and justify your choice. Without building one, describe the kind of nonlinear dynamic model of the issue that you would construct to

assist you in choosing the best strategy. Similarly, without building it, describe the kind of model you would employ to help plan and manage the work within any one station where each side of the station had dedicated elevators, escalators, and platforms, but where there was a common entrance, exit, and ticketing area for passengers at street level.

2.  It was stated on page 386 that inventory was potentially minimal when lean volume production was steady, and that both rises and falls in production rates could increase inventory. Explore, explain, and either justify or disprove this statement.

3.  The economy of the Old Kingdom of ancient Egypt, (circa 2686 to 2181 B.C.), [4] had an agrarian barter economy. Food was plentiful, the population expanded, temples were endowed, and the pyramids were built. Develop a CLM to show how an economy can flourish when there is no money, and where the only "industry" is the building of temples and the carving of religious statues.

4.  You manage a large workshop dedicated to the maintenance and repair of some 150 different, expensive, LRUs from a fleet of short-to-medium range airliners. The current system employs one piece of manual test equipment (MTE), with one full-time technician and spares, to every five LRUs; each MTE is fully utilized. The financial director proposes to replace the lot with one piece of automatic test equipment, to be used for all the LRUs because, he insists, it will save money; after all, one ATE tests and repairs each LRU 15 times faster than an MTE, and only one full-time technician is needed to run an ATE. You find out that one ATE will cost as much as 20 MTEs, one ATE costs as much as four LRUs, and, you doubt whether one ATE will be enough. Assuming reasonable values for any figures you may need, prepare two cases, one for and one against, this proposed change, remembering that a) the financial director's principal interest is money, and b) she is risk averse.

---

## REFERENCES

[1] Hitchins, D. K., "Managing Systems Creation," *IEE Proceedings*, Vol. 133, Pt. A, No. 6, 1986, pp. 343-354.

[2] Putnam, Robert D. *Making Democracy Work: Civic Traditions in Modern Italy*, Princeton, NJ: Princeton University Press, 1993.

[3] Coleman, James S., "Social Capital in the Creation of Human Capital," *American Journal of Sociology Supplement*, Vol. 94, S95-S120, 1988.

[4] Clayton, Peter A., *Chronicle of the Pharaohs*, London, England: Thames and Hudson, 1994.

# PART IV

# SYSTEMS

# MANAGEMENT

# AND

# ORGANIZATION

# Chapter 14

## Managing Systems

*You can't "manage" people. But you can bribe them.*

David Aycock, former CEO, Nucor Steel
(New York Times, *January 14, 1992*)

### MANAGING SYSTEMS CONCEPTS

#### The Need for Management

The need for management is widely assumed to be self-evident. However, management, as a function performed by people specifically employed and trained for the role, is relatively recent in historical terms. In warfare, for instance, the role of staff officer (in effect a military manager) only came into being with Napoleon [1].

A bird's eye view of a human activity system, in Figure 14.1, shows an organization of complementary subsystems with no evident management, at least in the sense of no management organization and nobody "in charge;" the overall system works provided everyone knows what he is doing and is motivated to do it. Does such an organization exist? *Could* such an organization exist?

Psychologists observe a major difference between the sexes when people come together in some undertaking. Among males, there will be a period during which there is more interest in establishing dominance and pecking order than in work. Among female-only groups, initial behavior can be quite different, with no one eager to take charge, and everyone keen to establish relationships. Often female-only groups work successfully together with one or more acting as a co-ordinator, but with no one in charge.

There may be a difference in outcome between such formative, single-sex groups. Among males, he who becomes dominant early on may be aggressive, but need not be overly intelligent. Such dominant males may drive their herd along a less-than-ideal path toward a less-than-optimal goal. Among females, the choice of goal and path may be more discerning, but the drive to make progress may be

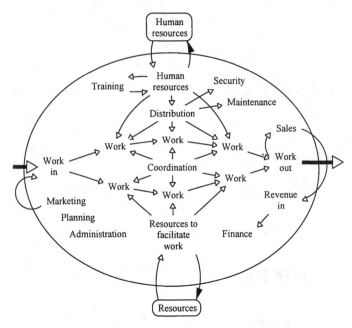

**Figure 14.1** *Work without management?*

less forceful and more consensual. Is it better to reach the wrong goal quickly, or the right goal not so quickly? That rather depends on circumstance. In reference to other generals, Napoleon habitually asked: "Has he luck?" [2] If you substitute luck for brains, males may have it–but only while they stay lucky.

So, while it may be unusual in Western culture, it is possible in principle for organizations to exist without management. In construction, history relates that there was usually an architect who conceived and designed the whole structure, located and acquired resources, and planned and directed the setting out and the building program. He may have had a number of architects working with him who looked after the detail of different parts of the structure. He also had master stonemasons, master carpenters, and other specialist advisers. None, it seems, was a manager in the modern sense of the term. If this seems strange, consider the humble termites that, without plans and management, nonetheless construct, maintain, expand, and defend complex structures. The whole structure "happens" because every termite does its bit according to simple behavioral routines. There are no managers, no plans; just simple, cooperative diligence. Termites know how to build only one kind of structure, however.

Road traffic in congested cities requires management–or so the pundits insist. As a result, we have traffic lights, one-way systems, variable rules of behavior (e.g., "right-on-red" in some U.S. states), and gridlock at peak traffic times. Yet, it is difficult to avoid noticing that traffic lights chop traffic up into uniform lengths, such that one set of vehicles moves along a section of road, then–after a pause–another set of vehicles moves, and so on. The net result is that the busy road carries traffic for only a portion of the time. Furthermore, because the traffic is continually accelerating and decelerating between sets of traffic lights, the distance between vehicles is variable; sometimes great, and sometimes small. Overall, the mean traffic density is relatively low because of this uneven behavior.

Compare that with the traffic chaos evident until recently in, say, Cairo, Egypt.[1] Although there were traffic lights, nobody really took any notice of them. Cars moved into any space, to the right and left of vehicles in front, honking their horns to warn the person in front of their actions. The whole traffic scene really was, as a result, cheerfully chaotic.

However, it was noticeable that the traffic kept moving; there were few stationary gridlocks. There were surprisingly few collisions, too; perhaps because there was no vehicle insurance and owners would have to pay for any damage. And there was none of the high speed racing between sets of lights that is characteristic of traffic control systems. In Cairo, the whole road was used all the time, mean traffic density was higher, and, surprisingly perhaps, the mean speed over a journey of several miles was comparable with, or greater than, that in New York, Paris or London–where the mean speed today is less than 100 years ago in the time of horse-drawn hansom cabs.

I am not suggesting a return to chaos on the roads. However, I am suggesting that the Western obsession with management by control may, like any obsession, lead us into seemingly obvious judgments that may, on closer examination, prove unwise.

## Management by Control

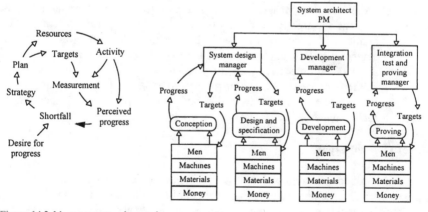

**Figure 14.2** *Measurement and control in management.*

---

[1] Cairo has recently enforced traffic control in much the same form as most Western cities.

A typical management control scheme is illustrated in Figure 14.2. At left is a CLM showing that the underlying management philosophy employs measurement as a control medium. Managers plan, set targets, apply resources, measure progress against the targets, and apply corrections accordingly. At right in the figure, the diagram shows a nominal management structure. At the top of the tree sits the senior manager, with three submanagers reporting to him. The submanagers are responsible to him for different phases of the work activity. Each phase is meticulously planned, and the resources (men, machines, materials, and money in the figure) are applied to achieve progress toward the target.

This approach dominates current business practice. It is, however, based on unsound premises, as evidenced in the many overruns in budget and time that such tightly controlled management structures frequently exhibit. (See Defense Acquisition on page 219 for the adverse impact of control on procurement.) Some of the issues are listed below.

- *Strategy*–Does the strategy account for threats, risks, and the frequent changes in situation and circumstance that are the practical stuff of business and project life, or is it based on some hypothetical ideal world? Is the strategy one that can give way to an alternative strategy later in the project, or does it commit all at an early stage, with no way back?
- *Plan*–Is the plan realistic, and how would you know? Are time allocations to achieve targets and milestones based on historical statistics, drawn from a different situation, with different people, tools and machinery? Does the plan assume that all resources will be 100% available, and that input data will be complete and of perfect quality? Does the plan omit rework? (See Figure 12.10 on page 363.)
- *Measurement*–Measurement is not practicable for: morale, integrity, motivation, creativity, and initiative. Yet these factors will affect the quality and duration of the work undertaken. What effect does working against the clock have on individuals? Do they stretch out the work to fit the planned time, even when they could have finished earlier; or do they skimp on work to keep within a timeframe that is really too short for their individual capabilities, expertise, and level of practice? Do they end up feeling undervalued as just another brick in the wall?

## The "Control Freak"

Control can become compulsive obsessive behavior (COB) for some managers, and as such is a recognized psychological condition, like continual hand washing, or continual locking and unlocking doors. Such people act, psychologists say, out of fear. The so-called "control freak" that we have all met at work is afraid of performing poorly for his or her boss (COB is no respecter of the sexes).

The control freak works at a furious pace, instructing subordinates in great detail about what they are to do. He allows no room for them to use their own initiative because he does not trust them; they are, in his view, weak, incompetent, and untrustworthy. The control freak with several subordinates has to work long hours, since he has to map out the work for each and every one in detail. Nothing must be left to chance, since failure on their part is failure on his part in front of his boss. The control freak lives in a world of fear, distrust, disloyalty, dominance, submission, and feverish activity. He may even convince himself that he alone is keeping the enterprise afloat.

So, the control freak stifles initiative, damages morale, smothers personal development, is unable to delegate, prevents subordinate promotion, and generally damages all those around him. To his boss, however, he may appear as the "steady hand;" one that always gets the job done. Control is an insidious practice.

## Management by Incentive

A different approach is illustrated in Figure 14.3. At left is a CLM showing that incentive may be used to encourage work, activity, and progress. Incentive might be money,[2] but it might equally be praise, the conferment of status, or the simple pleasure of the craftsman; that is, knowing that a job has been well done.

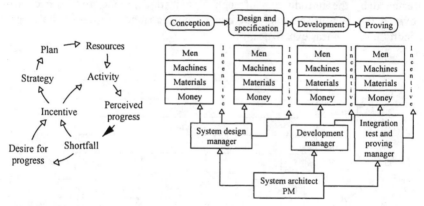

**Figure 14.3** *Management by enabling and incentive.*

At right, the management structure is shown inverted, when compared with that of Figure 14.2. The inversion is intended to show a quite different attitude in management; one in which management's role is to support and enable the people doing the work, not to control them. The rate at which work is done is also shown

---

[2] What appears to be incentive, may be control in disguise; for example, so-called "performance-related pay." It may appear sensible, even obvious, to pay individuals according to how well they have performed, but unfortunately the practice can be divisive; allowing managers to control, favor, and blackmail individuals. A true incentive might be a bonus paid, not to an individual, but to a team. This approach binds team members together in a cooperative effort, and increases social capital within the team.

as influenced by incentives. In this situation, those doing the work are encouraged to use their initiative, rather than to simply do as they are told.

In one successful automobile manufacturing company where this structure and philosophy were employed, workers were given two jobs: doing their job, and doing their job better. At the end of each work session, workers came together to see if they could identify and implement better ways of working. Effectively, they became their own managers, organizing and managing themselves, improving their activities and processes, and consequently making the product better and cheaper.

Integrated product teams (IPTs) have been introduced within such organizations. Such teams in the automobile industry might consist of about five people: an engineer or technician, a contracts person, a finance person, a team leader, and a quality controller. Members are chosen somewhat ad hoc, according to the issue in hand, but the team is empowered to make binding decisions about proposed changes to the vehicle design or construction, without recourse to middle or senior management.

This approach involves delegation of authority to empower the workers who, after all, are the ones who know how to do the job. Middle managers may feel threatened by the introduction of such organizations, since they lose control. Provided they know the job, they can become coaches, however, helping to support those with less experience.

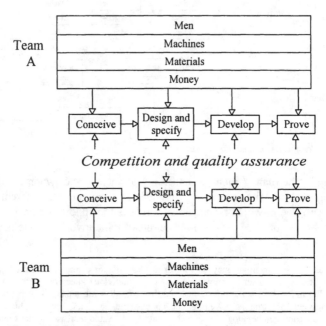

Figure 14.4 *Complementation, competition, and quality assurance.*

**Management by Competition**

We saw in Chapter 1 how management in natural systems was by dynamic interaction, not by regulation or control. With this in mind, a different approach is shown in Figure 14.4. The figure shows two teams, A and B, each addressing the same problem that, in this instance, invokes four sequential activities: conceive, design and specify, develop, and prove. Each team is given appropriate resources.

There are several ways in which these two teams may operate:

1. *Independent parallel competition, with no interaction between the teams.* This type of competitive approach has been adopted in the past for defense procurement, for example, the General Dynamics YF16 and the Northrop YF-17 (later the McDonnell Douglas F18 Hornet). The two YF aircraft were developed independently in parallel, as prototypes for a competitive fly-off. The winner, the YF16, was procured for the U.S. Airforce as the F16, while the loser was modified and than procured as the F18 for the U.S. Navy and the USMC.

2. *As creator and as quality assuror (QA).* One team may undertake the process of solving the problem and creating a solution, while the other team may check the quality and progress of their work. This technique is sometimes used with complex software development. In practice each team may alternate as creator and as QA; this ensures that each is current and has full expertise. It may be preferable to employing full-time QA staff members who, because they are continually checking the work of others, never get the chance to be creative or to work "at the coal face," and so lose integrity and authority in their pronouncements.

3. *Each team may compete independently, phase by phase.* At the end of each phase, the results from the two teams are compared and the better solution is selected, but with worthy inclusions taken from the losing solution. In this way, the two teams both compete and cooperate to produce the best solution to the problem. Additional incentives may be included for the team that wins each phase, where "winning" might include not only having the better solution, but also achieving it in the shorter time and/or for the lesser cost. This "best method" may then become the basis for best practice within the organization, but it can evolve further in competition to improve even further over time.

4. *Each team may work on the same problem, phase by phase, but with the intent of producing alternate solutions.* Some high-integrity systems are designed and created in mirror pairs. Each of the pairs exhibits the same emergent properties, capabilities, and behaviors, yet each is deliberately quite different. So, while one team uses positive logic, the other team uses inverted logic; each team will use different algorithms to calculate the same result; each will employ different operating systems and different hardware; and so on. The two teams are not necessarily in

competition; instead, they cooperate to ensure that they are quite different. The delivered system operates both parts in parallel. If each gives the same output, then there is a very high probability of correctness. The whole is designed to both afford integrity and avoid systematic error. This approach has been used in the creation of high-integrity railway signaling and switching systems.

## ORGANIZED MANAGEMENT SYSTEMS

### Organized Creating Systems

Companies, corporations, enterprises, and organizations have internal divisions. The act of dividing, no matter how well intentioned, raises issues in human affairs. Dividing creates "us" and "them" psychological, as well as material, boundaries that must be overcome. Dividing creates interfaces, interconnections, interactions, and interference.

Yet, left to their own devices, people will form into groups. While six or even seven people may not group, eight or nine people form quickly into two roughly

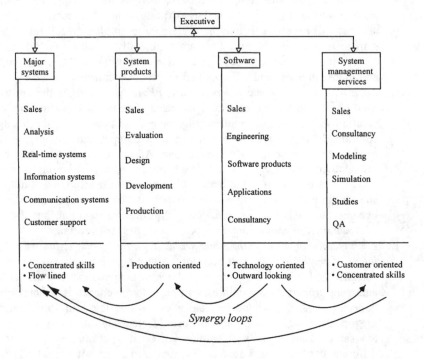

**Figure 14.5** *Synergistic organizational structure. [3]*

equal groups; perhaps another example of "Miller's Seven," [4] that curious limitation of human perception that limits us to processing about seven things in parallel. Of course, we also divide ourselves into groups to undertake work within organizations.

While there are many ways to organize large companies, businesses, and organizations, some may be better than others. Consider the organization represented in Figure 14.5. Depending upon the scale of the organization, it could represent four groups in an enterprise, four divisions in a large company, or perhaps four companies in a conglomerate. In any event, each of the four parts of the organization is different, yet each operates synergistically with the others.

For instance, the software division creates and sells software products to the market, and in this respect is a viable business. But it also creates software for the other three divisions that enables them to pursue their business objectives, too. Similarly, the systems management services division undertakes studies, many of which will result in customers ordering and buying systems, some of which may be produced by the other divisions.

In the nature of things, each division's internal organization will differ. The system products division, for example, will usually focus on the business of production in which it invests much of its resources. Such an internal arrangement would not suit the major systems division, which evidently is able to address many different kinds of system at the same time, and which concerns itself with bespoke, often one-off, designs. It will concern itself more with design, development, integration, and proving, usually of several different projects in parallel but at different stages. Similarly, the management services division may produce nothing other than paper, and it will be organized to encourage and enable cerebral studies, analyses, simulations, and modeling.

## Tensions within Organizations

Note that each division in Figure 14.5 is set up to be an independently successful business in its own right. The whole is synthetic and, to a degree, organismic. Nonetheless, tensions may arise between the divisions; for example:

- Should the divisions cross charge each other when one works for another?
- Should the management services division propose solutions to studies that make use of products and software from other divisions within the group?
- What if two or more divisions find themselves competing for the same business?

Such tensions call for sound management and a long-term systems view. Consider the first bullet in the list above.

Prejudicing Social Capital

Cross charging between divisions would undoubtedly satisfy the relevant accountants, but then, as we have seen, accountants do not relate well to synergy. To understand the issues involved in something as apparently obvious as cross charging, consider a university department where the various lecturers cooperate to create interesting and varied courses. Each lecturer may present teaching modules for a variety of courses; all run within the same department. Suppose the department is now split, purely for reasons of accommodation, between two sites separated by a mile, and on different sides of a road that runs through the campus. Lecturers on the same course find themselves on different sides of the road.

A proposal is made that the department be formed into two groups, each with its local administrator, and that the two groups should be self-accounting. In particular, where lecturers from one side of the road give lectures to courses run by the group on the other side of the road, then there should be a charge for the service.

This proposal is going to cost money, to manage the cross accounting, and as a result, the cost of courses will rise. Additionally, as lecturers will now have to make returns on which courses they worked and when, there will be more bureaucracy. Inevitably, course managers in one group will be less inclined to ask for lecturers from the other group because of the higher cost, and the variety and quality of courses will fall as a result.

But what has really changed? The lecturers are still the same people in the same department; still able to cooperate. Only the situation has changed, and that change is psychological rather than substantive. Previously, there would have been a spirit of cooperation, reciprocity, and trust, so that a contribution from one lecturer on another's course would have been balanced by the second lecturer contributing to the first's course; in other words, reciprocal behavior.

Cross charging between groups, departments, and divisions of the same business or enterprise prejudices trust, cooperation, reciprocity, and synergy. More generally, practices that institutionalize boundaries may make such organizations "more manageable," but they also create bureaucracy, reduce social capital, and hence prejudice the cohesion and effectiveness of the organization.

Objectivity

The second bullet above concerns objectivity. If the management services division in Figure 14.5 is to undertake studies, and if the output from those studies is to be credible, then the work must be undertaken objectively. Proposing products of another division within the group, as part of a solution system, will be regarded with suspicion by customers and competitors alike. If, on the other hand, the proposal does not contain such products, the sister division will be unhappy, and the tension will be real.

In the long term, unless the management and services division is objective, and seen to be objective, then it will go out of business; customers will be few and far between. If, on the other hand, the management and services division justifies the emergent properties, capabilities, and behaviors of the solution system and its parts, and if the sister division is not able to compete with other suppliers to provide those properties, then this inability should serve as a wake-up call to the sister division. If it does, then interaction between the two divisions will be synergistic in the long run, but with integrity retained.

## Competition versus Cooperation

The third bullet concerns divisions competing for the same business. This situation, not uncommon, presents options to the executive: should only one division compete; should divisions be encourage to cooperate and create a single, group bid; or should the divisions simply bid independently and let the winner take all?

There is no simple answer. If only one division is allowed to bid, then the group as a whole may lose if that division is not successful, and the suppressed divisions will feel cheated. Allowing, say, two divisions to bid should increase the prospects of winning on a simple probability basis, but it divides the group's resources, and the cost to the group (bids may cost millions of dollars) will be twice what it need be. On the other hand, having the divisions mutually compete

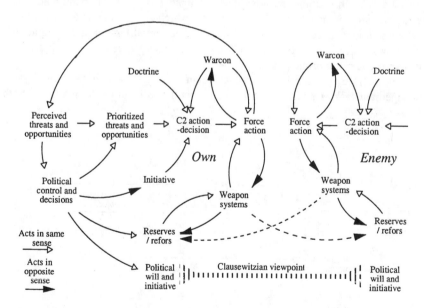

**Figure 14.6** *Command and control causal loop model. Refors: reinforcements. Warcon: war consumables (weapons, stores, fuels, and so on).*

may develop their edge and could lead to survival of the fittest—which may have longer term merit in business terms. Perhaps the best option, at least in principle, is to encourage a group bid, with each division cooperating and contributing. This should throw the might and expertise of the whole group behind the bid, and give the best prospect of winning for the least cost; that is, value for money. And the costs of the bid? Probably, these should lie where they fall, to use the quaint contracting term; meaning there should be no cross charging.

However, as those familiar with this kind of situation know only too well, there is so much more to it than objectivity, logic, synergy, and psychology. Customers have preferences, sometimes for particular suppliers, sometimes to ensure that no one supplier gets too much business—regardless of product suitability and quality. Bidding for new business can be a fascinating, as well as a highly creative activity, within the systems management arena.

## Organized Military C²

Military command and control is the management of warfare or conflict. Figure 14.6 shows a CLM for command and control (C²) with opposing blue and red systems. Each military system is shown as being under political control, in line with the Clausewitz [5] dictum that "war is an extension of politics."

The management function presents as one in which information is gathered, threats and opportunities are identified, decisions are made and forces are ordered to take action. In making decisions, the commander and the command team – the executive—are cognizant of military doctrine, which suggests courses of action in different circumstances based on political or experiential factors. In taking action, the forces utilize resources; without such resources, forces are ineffective, so resource centers are prime targets for enemy forces.

Figure 14.6 shows principal management factors starkly, but essentially the management task is no different for the executive of a large company. In both

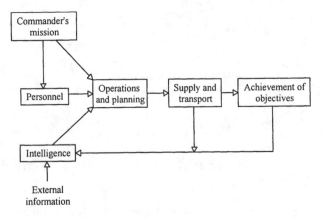

**Figure 14.7** *Functions of war (Peacock).*

instances, there is a competitive paradigm that would be familiar to the Chinese general Sun Tzu [6], who lived around 400 B.C., and whose teachings have been used extensively in Japanese business practices. The essence of the issue is that blue control of red, in any cybernetic sense, is impractical because red is continually influencing the situation as red, too, seeks to exercise control over blue.

While command [7] equates to executive in the nonmilitary world, control equates broadly to senior management. Looking at the CLM from a different viewpoint reveals the functions of war [8] (see Figure 14.7) from which may be generated the principles of war, which appear in a variety of forms. The following is according to [9] Peacock: maneuver, objective, offensive, surprise, economy of force, mass, unity of command, simplicity, and security. These principles, and their application, are at the heart of military doctrine, and are evident when U.N. peace enforcement operations take place, with a coalition of forces acting, at the time of writing, usually under U.S. operational control.

## Government C²

Governments have their equivalent of command and control systems with which they deploy national police forces, intelligence services, emergency services, and so on. They also have the ability to commandeer medical resources, power, shelter, food, water, and transport facilities to deal with natural disasters, civil unrest, and emergencies.

Internationally, intelligence channels are augmented by diplomatic facilities

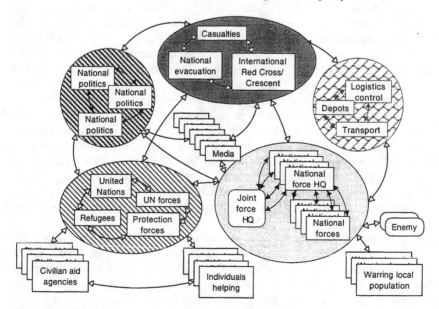

**Figure 14.8** *Interacting command and control in peace operations.*

that seek, essentially, to maintain and raise social capital between nations by overcoming social, political, and economic barriers and boundaries.

During U.N. peace operations, command and control can become complex, even confused; see Figure 14.8. The figure shows a typical arrangement, part of which is by design, but much of which is ad hoc. Military forces, the U.N., individual nations, civilian aid agencies, and so on, each have their information and management systems. All of these systems should be able to communicate. It is important, for instance, for the Red Cross/Crescent to coordinate its actions with military forces, so as to protect doctors, nurses, medical facilities, and so on. On the other hand, any suggestion that the Red Cross/Crescent is associated with the military may prevent them from doing their humanitarian work, due to mistrust of those in need of help.

As a result of the continually changing situation inherent in peace operations, the communications systems and the various centers represented in the figure are in a state of flux. It is not really possible to predesign such a fluid system: instead, the system works–or fails to work–because of continuing additions, subtractions, alterations, and amendments to the people, the communications, the information systems, and the architecture.


## SYSTEMS ORGANIZATION AT INDUSTRY LEVEL 4 [10]

Industry in the West is widely financed through the issuing of stocks and shares, which are dealt freely in open stock markets. Historically, it was, and is, different in Japan. At the end of the 19th century, large Japanese companies were funded through zaibatsu. These were family-owned holding companies that controlled industrial empires consisting of one major company in each of the major industrial sectors: steel, shipbuilding, insurance, finance, and construction. Each zaibatsu owned a bank; deposits in the bank were a major source of investment for companies in the zaibatsu. The zaibatsu were tightly controlled industrial groups and contributed to the phenomenal growth in industrial power in Japan.

After World War II, zaibatsu were eliminated by the U.S., and a new form of association emerged, the keiretsu. There was, and is, no holding company at the head of a keiretsu. Instead, a keiretsu consists of some 20 companies, one in each industrial sector. Again unlike the zaibatsu, the companies are not legally bound. Instead they are cross connected by owning each other's equity, and by reciprocal obligation; each company owns a portion of every other company's equity in a circular pattern. Also in each keiretsu are a bank, an insurance company, and a trading company.

The circular pattern of the keiretsu may be significant. There is a tradition that circular structures are particularly strong, offering no corners through which to be breached. Certainly, the keiretsu is strong and seemingly impregnable. Although each company in a keiretsu may have shares, few are likely to come up for sale on the stock exchange, since they will be owned–and retained–by other companies in the keiretsu. The notion of mounting a dawn raid, for instance, in which a predator buys up shares in an unsuspecting company with a view to gaining control, is unlikely to happen to a member of a keiretsu. Should any member company get into difficulty, the others would be obliged to help, if only to preserve the circle and, essentially, themselves.

The keiretsu, with its sense of reciprocal obligation between members is an example of high industrial social capital, and it evidently works well in a society where reciprocity and obligations are matters of honor. In the West, a zaibatsu might have been viewed as a virtual monopoly, particularly because it operated under a singular holding company. Whether a keiretsu would be viewed with favor in the West because there is no singular legal entity is not clear. The keiretsu is certainly an interesting approach to management at the industrial level.

## CREATING THE CULTURE

Creative organizations are, among other things, societies or groups of people working together, but also living together, eating together, and often socializing. Many people spend more time at work with colleagues than they do with their spouse and children. Work places develop, and exhibit culture. Such culture is tangible when first entering, say, an engineering factory, or a marketing consultancy. It is evident in the way people behave, talk, offer hospitality–or not, work in groups or individually, submit to dominance–or not, position desks across room corners or facing the door, and so on. In some cultures, humans are "plugged in" to machines as operators, while in others the humans gather around tables for discussions.

So, what should the culture of a creative organization be like, and how can it

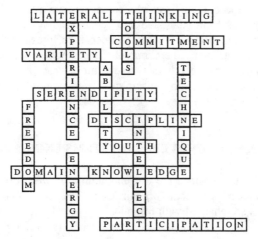

**Figure 14.9** *Aspects of creativity.[11]*

be achieved? The question presupposes both an answer and a method, but is a culture something that can be deliberately created, or does it, like Topsy, just grow?

We have already seen that control is inimical to creativity. Figure 14.9 shows some of the factors that seem common in creative environments, and it is evident that they would not survive the attentions of the control freak. Lateral thinking and freedom would be seen as threatening in a control environment, while variety would be suppressed to a level or degree approved by the controller.

The figure also presents some apparent paradoxes: youth is seen with experience, freedom coexists with discipline and commitment, and so on. The tension between these factors contributes to creative environments. If we take brainstorming as a typical example, then we see anarchic lateral thinking being encouraged within a disciplined framework that suppresses negative thoughts. So, if a young, inexperienced person suggests something that an older, more experienced person believes is infeasible, then the latter is not expected to criticize or belittle.

Eventually, after the session is over and the ideas that have been generated are being sifted through, it will be necessary to bring some experience to bear, but without the inexperience of youth it is unlikely that those "way out" ideas would surface, and every so often it is one of those ideas that enables a major jump forward. Some older people, of course, still have young, creative minds, but on average, youth will be more creative. Marrying youth with experience is the route to creative success.

The occurrence of a brainstorming session within an organization already suggests a system that is open to new ideas. Organizations that are open, that question what they do and why they do it–so-called thinking or learning organizations–are on the path to a creative culture.

Creativity is key to systems thinking and systems engineering. However, there is a time and place; the person who comes up with a much better design idea just as the last part is being integrated into the finished system solution at the end of a 15-year program might do better to keep it to himself! Within an overall systems organization, then, there may be areas where a creative culture is essential, but there may be others where a more prosaic approach is needed. Creative freedom and tension in a disciplined framework seem to be the order of the day.


## PROJECT MANAGEMENT VERSUS SYSTEMS ENGINEERING

As the title indicates, tensions are not uncommon between project management and systems engineering in some organizations. The reasons are not hard to see. Project managers are charged with bringing the project in on time, within budget, and to the requisite quality standard. Systems engineers working within the project seek to spend time at the start, eradicating errors in specifications, balancing, structuring, arranging, and so on, none of which "cuts metal or delivers code."

Project managers soon become impatient at the perceived lack of progress, and arguments that "time spent now is time saved later" tend not to appeal without proof. Similarly, when systems engineers indicate that something is not quite right and that rework will be necessary, the project manager, looking in vain on his program chart for anything called rework, will be tempted to forge ahead regardless, rather than eat into his contingency, which may not be in the plan and about which he may say nothing[3] to the systems engineers.

Project managers frequently employ tools that may be inadequate for the development of complex systems. A cornerstone of project management is the program evaluation and review technique, which is, as the name implies, a flexible approach to program management in which logical plans are drawn up showing sequential activities and their durations in a network. Using such networks, it is possible to calculate the overall time of the project by finding the route from start to finish that takes the longest time; the so-called critical path. Activities may be reordered or rearranged, such that the overall calculated duration is reduced. As progress is made, and as actual times supplant predicted times in the network, the technique encourages reevaluation and possible rearrangement. PERT charts rarely show rework in the sense of feedback and recycling. Instead, rework, if there were any, would appear as a timed activity inserted into the universal left-to-right flow.

We have seen earlier that this approach is reductionist, and overlooks the fact that the various activities represented in the network form a system with interactive parts (see Figure 12.16 and Figure 12.17 on page 368). When developing a PERT chart, it is necessary to know start times and durations for activities–factors that are, at best, educated guesses. It is not necessary, using PERT, to know how activities interact, however. (Interpretive structural modeling, on the other hand, produces activity networks based solely on the logic of their interactions and mutual contributions.)

When it comes to program reevaluation and review, not surprisingly perhaps, project management and systems engineering will see changes in allocated activity times and the rearrangement of activity sequences in a different light. Project management has a tool, the PERT chart, while systems engineering has no equivalent, and PERT rules, almost as a cult, in many projects. All of which leaves systems engineers feeling unhappy and undervalued.

While some tension is no doubt useful and creative, most is neither necessary nor productive. Both parties would contribute more effectively to the objectives of their business and their customer if:

---

[3] In some cultures, a project manager is allocated a contingency, so that he may–if necessary–go a little over time and/or budget. However, he may also be offered a bonus if he does not "eat into" his contingency; making him loathe to let project team members know of the contingency's existence.

- Systems architects, again using nonlinear dynamic simulations as a means of managing the complexity and anticipating the counterintuitive outcome, designed the whole program, from conception to replacement, including the process of creation;
- Project managers were also trained, practicing and experienced systems engineers; and
- PERT was replaced, or substantially brought up to date, so that the activity network was the output from a dynamic, nonlinear simulation.

A good project manager is worth his or her weight in gold. The ability to lead a disparate team in creating a good solution to a complex problem, within prescribed time, budget, and quality limits, is rare and highly valued. If systems thinkers, architects, and engineers were to present project managers with sound, realistic, and flexible program plans and effective tools to do the job, then the relationship between the two parties would be synergistic rather than antagonistic.

## TO PHASE OR NOT TO PHASE

### Stages and Phases

The process of creating a system falls naturally into stages, if only because the attitudes of mind, the skills, and the tools of practitioners change as the process unfolds. The conceptual stage can be quite abstracted, imaginative, and intensely creative, with mental openness a prerequisite. By the time proving ensues, the subjects are hard, tangible, and measurable; imagination is not high on the list of desirable practitioner attributes.

Because there are natural stages, it does not follow that the creative process need be broken up into phases, where a phase implies that something is completed, some kind of end is reached, and a new start is about to be made. Natural stages would, ideally, follow on from each other seamlessly, with no discernable slowing down, and certainly no stop. Breaks in the process cause a loss of momentum, divert the attention and minds of the practitioners, allow data that has already been acquired to fade; and prejudice the optimal conceptual systems solution, since the environment within which that concept was derived will be constantly changing.

Complex projects often develop strands of activity; for instance, one subsystem may be simple and quick to design, so that it may proceed into specification while another part is still in design. Each strand would then be proceeding seamlessly, whereas a phase suggests that all strands should be kept in synchronism, that is, all should be designed before any proceeds to specification.

Phasing slows the creative process. Seamless development is, in principle, faster and cheaper, simply because there are no gaps, momentum is maintained throughout the process, and there is less time for the problem and solution spaces to evolve and change.

## Phased Defense Procurement

Defense procurement has employed phases, such as: prefeasibility, feasibility, project definition part one (PD1), project definition part two (PD2), preproduction, and production. Prima facie, these have little to do with the stages that are evident in, say, a systems engineering waterfall process. Defense procurement phases are about managing the business of procurement to be fair, to be seen to be fair, and to obtain the best value for (public) money.

The desire to be seen to be fair, and to promote competition as a way of driving down price, resulted in defense procurement phases for a single project being competed, and for the winner of any one phase to be barred from bidding for any other. Gaps developed between phases to accommodate open competition. Winners of competitions started the next phase with no contemporary experience, and only the written word output from the previous phase as a guide. Winners of the preproduction phase may have differed from the full production phase winners, thereby introducing systems made by different manufacturers, with consequent support systems incompatibilities.

The overall procurement process, now hopefully and mercifully superseded, stands as a testament to bureaucratic folly. The process was reductionist, protracted, expensive, wasted the very taxpayers' money it was intended to save, and produced solutions to problems long since overtaken by events.

Within phases, such as PD1 or PD2, for example, defense procurers would also introduce milestones and reviews. Milestones, curiously, referred to times and not distances; a milestone would indicate a time by which certain activities should be completed. Typical of reviews were preliminary design review (PDR) and critical design review (CDR), the titles being largely self explanatory. A CDR might, for instance, constitute a milestone, meaning that the review should occur at a particular time. During creative phases, payments would be made to contractors for the successful achievement of milestones.[4] The set of activities with reviews, milestones, and payment milestones provided the procurer with the ability to micromanage projects; to criticize the substance of, and outcome from, activities; and to criticize any perceived lack of progress.

Achieving milestones became vital within supplier organizations: no milestone, no money. To ensure success, contractors would cease all creative activity several weeks (six to 12 weeks would not be uncommon), to bring all activities associated with a milestone or review up to a date, to create extensive review documents, presentations, simulations, and prototypes. These were showered on visiting procurement personnel, who would attempt to see the solid tree of progress through the forests of documentation.

---

[4] This is poor English, but it makes a kind of sense, and is typical of the contractual language in vogue.

There was, and is, undoubted advantage in this process of review. Bringing the various strands of a project together provides a comforting assurance of control, and creates a solid baseline for the following stages of work. It is, of course, an expression of the waterfall archetype, and as such is evidence of a careful, balanced approach. Unfortunately, it also consumed a large amount of effort and time, such that either there was less time left to undertake essential work, or projects overran, prejudicing quality or budget, respectively.

## ELICITATION AND REQUIREMENTS

A glance at systems engineering conferences programs and journals will show that there is activity in the areas of requirements elicitation and requirements engineering; both terms have been largely overlooked in this book so far.

In requirements elicitation, an elicitor gains information from a subject expert that serves, in principle, to act as the expert's vision of (part of) some problem space and/or solution system. An elicitor may sit down with, for example, a naval principal weapons officer (PWO) to elicit from him what he does, how he does it, what information he needs, in what form, and with what timeliness. The elicitor then uses the elicited information to develop a requirement for a system to aid, support, or perform some task on behalf of the PWO.

There are lacunas in such a process. The PWO may have been at sea for 15 to 20 years, working up through the ranks, gaining experience, becoming part of a team, attending staff college, and finally becoming the leader of the team. He will have forgotten more about "fighting the ship" and naval operations, than the systems analyst could ever aspire to know. The PWO will be so familiar with his subject that he will be unable to appreciate how little the analyst knows.

The elicitors will receive data, but will lack the context in which to understand it and turn it into information. If they are conscientious, they may take a trip to sea on an exercise to observe things in action, and may come to believe that they understand. However, unless they could somehow acquire the background and experience of the PWO, they will never understand the full extent of the PWO role; both when things are going right and when they are most definitely going wrong!

Every job has its wrinkles. Ask someone how he or she sweeps the floor–in detail: there is more to it than meets the eye.[5] If sweeping the floor has its wrinkles, then a PWO job has many more, and each PWO will develop his own style and way of working. Can, and should, the analyst elicit the wrinkles, too? If so, should it be from one PWO, or from a sample? Essentially, there appears to be a flaw in the process; perhaps the only person who could reasonably develop a requirement in support of a PWO is the PWO himself, and he is unlikely to have the inclination, experience, expertise, technological background, and so on, to

---

[5] How *do* you get rid of that last bit of dust at the very end?

undertake such a task. And the PWO was just an example; this situation applies generally to elicitation.

The act of stating a system requirement implies a major disjoint in a systems engineering process that would, in an ideal world, be entirely seamless. A customer who presents a statement of system requirement has (or should have) already undertaken a major, most creative, and challenging part of the systems engineering process. Before issuing a definitive statement of requirement, the customer has supposedly explored the problem space; circumscribed the solution space; developed a variety of solution concepts; tested and compared the options dynamically; selected the preferred solution concept(s); and developed the concept(s) into a full description, or descriptions, of the solution system(s) in the solution space.

Having, presumably, undertaken this essential foundation work in camera, the customer then (generally) makes neither the concept development process, nor the data (mathematical and simulation models, operational analysis, and so on) generated by it, available as part of the statement of system requirement.[6]

Fracturing the system engineering process, as a customer-stated system requirement does, appears also to be unscientific, in the sense that both continuity and currency of evidence are prejudiced. When a customer provides detailed requirements for a system solution, it runs counter to the philosophy of systems engineering, which concerns itself with synthesizing holistic solutions to problems that it has addressed, and to which it has found solutions, resolutions, or dissolutions [12].

Requirements specification is not a universal practice; much of the commercial world does not use it, at least not in the sense of a customer writing down in detail what he or she wants some manufacturer to make. You do not specify to a car manufacturer, a washing machine manufacturer, or a fashion designer what he or she should make. You go into the marketplace and shop for what you want. In each of these instances, the boot is on the other foot: the creator may research what you, his customer, needs or wants, and may provide it unbidden and in competition with other suppliers.

Japanese car manufacturers, for instance, send employees and their families to live in the U.S.A., using cars on a daily basis to go shopping, take children to school, go camping, and so on. From this, the manufacturers can see where shortfalls in their cars lie, and can introduce new features to make the cars more suited to their environment.

Even in the defense procurement arena, requirements specification has not always existed in its current form. In the 1950s and '60s, it was the practice for defense customers to approach systems houses with their problems, rather than with preconceived solutions. The typical systems house, which prided itself on its

---

[6] I much prefer "future vision" to "statement of requirement;" the term in general use. Future vision encompasses the idea of a solution system operating dynamically in its operational environment with other, interacting systems. By comparison, the term, "statement of requirement," is both imperious and opaque.

objectivity and independence, would address the problem, using modeling and simulation where appropriate, and would involve the customer and user in the development and assessment of possible solutions. Once a shared future vision of the solution system had been achieved, the systems house would then create, or manage the creation of, the solution in full, putting out subcontracts as needed for development.

The process worked well, but infuriated large manufacturers of defense equipment and platforms, which were interested less in objectivity and more in selling their particular equipment. Governments were influenced to starve the systems houses of funding, and neither the objective, independent, creative expertise of the systems houses, nor the process as it once existed, is available to the same degree and at the same level today.

Some defense customers have likewise recognized the problem, and have issued "cardinal point specifications." These are bare specifications identifying, as the title suggests, only the most fundamental features and objectives of a new system.

For third-world defense equipment customers, the opportunity to specify requirements may not exist. Instead, like commercial customers, they shop in the marketplace.

## Managing Customer Control

In spite of the foregoing, it may be possible for an organization to provide an effective, optimal solution to a customer's requirements, partly by "managing" the customer. In many requirements, customers (or their contracts staff) insert a clause to the effect that the contractor should propose alternative solutions to the requirement (they mean to the problem), if these can be shown to afford better, cheaper, simpler, and/or faster solutions. It would be a bold contractor who responded to that opportunity exclusively, but it may be possible to propose two solutions: one to the requirement as posed; and a second—a quite different and supposedly better alternative—to solve, resolve, or dissolve the problem.

Meeting the requirement and solving the problem can be quite different. A requirement arose in a wealthy Middle Eastern country equipped with squadrons of fighter aircraft in constant activity. The requirement was for a computer-generated electronic tote board that would show the status of each and every aircraft. Aircraft might be: serviceable on the ground, in flight, on turnaround, at various levels of servicing, or away having major repairs.

A systems designer visited, to understand the problem in more depth. It was essentially simple. He acquired some transparent plastic sheet and some grease pencils from the local market and drew up a table, with columns for aircraft designation and the various states in which each aircraft might be.

This state board was fitted to the wall in the operations control room, and kept up to date by a clerk using a grease pencil and a piece of cloth. This handcrafted state board solved the problem so well that the customer withdrew the requirement. The systems designer lost the job, but gained the customer's trust, which led to future business.

An alternative way to "manage the customer" is to find out what the problem really is, analyze it, conceive solutions in the proper systems engineering style, and then present the holistic, synthetic, organismic, and optimal solution concept to the customer with the evidence, for instance, of nonlinear dynamic simulation to support the case. Until the advent of affordable simulation tools and minimalist methods, this would have been expensive; now it is quite feasible. If the customer is not still convinced, then at least the contractor has a sound basis for judging whether or not to tackle the original customer's requirement.

**Specification Trees**

At systems engineering levels 1, 2, and 3, technological and sociotechnical systems are generally specified formally. Specifications are the outcome of design. There is an overall system design and specification, and there are also designs and specifications for subsystems that integrate and interact to create the whole system.

Solution specifications define in some detail what the systems engineering contractor is going to create. As such, they must meet not only the objectives of the customer, but also the objectives of the business. It may be necessary to modify the solution system in some way, so that it nearly meets the customer's requirements, or meets them in a way that the customer did not anticipate. Sometimes, this is the only approach that is viable in business terms.

This necessity can be achieved with integrity by developing a number of specifications:

- *The customer's requirement specification (CRS)*–This is derived from the customer's documentation, presentations, and so on.
- *The system requirement specification (SRS)*–This is the company's interpretation of the customer's requirement for the overall system solution, which the company can build within time and budget constraints, and so on. The SRS represents the solution that is acceptable to the company in business terms.
- *The performance and design requirement specification (PDR)*–This details the emergent properties, capabilities, and behaviors for each of the elaborated subsystems, together with design layouts, drawn from the SRS.

The SRS will undoubtedly differ from the CRS, so the customer should be presented with the SRS as an interpretation of what he, the customer, asked for, and as a statement of what the company or corporation intends to build. Once accepted, the SRS becomes the contractual basis for all further work.

This approach has two benefits:

1. It allows the systems engineering company to build something that is viable in business terms; that is, a system that can be built in time to a sensible budget and that will make a profit.
2. It also allows the systems engineering company to go back to the beginning of the overall systems engineering process, to define the problem space, conceive solutions within a circumscribed solution space, undertake full supporting analyses and simulations, and essentially conceive a solution in which the company has faith. It is this full-blooded, in-house solution concept that then constitutes the SRS, and it is legitimized only when the customer approves it as the solution to his problem/requirement.

The second item shows that full scientific systems engineering can still be undertaken on behalf of a customer seeking to impose inappropriate or inadequate requirements–to that customer's benefit, and with that customer's full knowledge, acquiescence, and approval.

### Looking the Gift-Requirement in the Teeth

What if the customer is adamant that he alone understands the problem and that his solution, in detail, is the only one that is acceptable? A systems engineering company then faces choices. It could:

- Build the solution required by the customer blindly, employing methods and technology that will ensure that the project comes in on time and budget, and gamble that the customer and the user are satisfied. This is the so-called "engineers' approach," in which the engineer supposedly concerns himself only with what he is building, but not what it may be used for (e.g., guns, nuclear, biological, and chemical weapons, etc.). This is unscientific and nonoptimal, but it might be good business in the short term. If the users are dissatisfied, then it may prove to be poor business in the longer term.
- Build the solution as in the first bullet, but clearly denying any responsibility over fitness for purpose. Even if the customer accepts the proviso, which is questionable, an unsuccessful solution system will still be blamed on the supplier, not on the procurer.
- Walk away from the business. This may be a drastic step. It is quite possible to go to the wall, clinging to your principles.

Systems engineering is a high-integrity practice. Choosing between the three bullets above will never be easy.

## PROJECT VERSUS FUNCTIONAL VERSUS MATRIX ORGANIZATION

### Organizational Paradigms

Systems companies, businesses, enterprises, and so on, face an internal organizational dilemma: should they organize themselves around projects, functions, or in some matrix fashion, where functions supply skills to projects as needed?

Project and functional organizations are shown in Figure 14.10. In the figure, projects run horizontally, while functions run vertically. In a company that is project oriented, each project "owns" its project personnel, so that they work exclusively on the one project. Functionally oriented companies are organized about centers of expertise. Systems designers, for instance, would be owned by a system design manager, and would work on several projects, perhaps in sequence, perhaps in parallel, according to need.

The practical differences between the two organizations are significant. In project organizations, the emphasis is on completing the overall project within time, budget, and quality constraints. Project managers may place a so-called Chinese wall around their projects, implying that "their" personnel are not allowed to leave, and even that new ideas and methods are unwelcome, since they, too, may prejudice progress. It is not unknown for engineers to spend 12 to15 years on one project, during which time they have worked diligently and effectively. At the end of the project, however, with no new ideas or methods in their repertoire, they

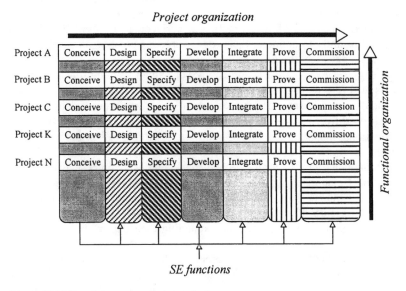

**Figure 14.10** *Function versus project organization.*

are effectively 12 to 15 years out of date, and may be unemployable.

On the other hand, functional organizations group together people with like disciplines and expertise. Because they are not dedicated to any one project, their allegiance is to their discipline and their discipline leader or manager. Over a number of years, such groups may become centers of excellence, building up their knowledge and expertise within the boundaries of their function and discipline. They may know little about project outcomes and overall corporate business success.

The third option, a matrix organization, may be a cross between project and functional organizations. It may be driven by a business imperative, in which a business manager is tasked with managing several projects within a business sector. To achieve this, he leases staff from functional managers who work on one or more projects for which he is responsible. The functional manager chooses who is to work on which project, which indicates that the individual so directed has responsibilities to the functional manager for his technical quality, and to the project manager for his performance to time and budget. Such arrangements can involve a degree of bureaucracy, to account for whom is supplied by which functional manager to work on which project, for how long and at what cost to the business sector manager.

Matrix management enthusiasts declare this to offer the best of all worlds, since the project organizational element makes the management of progress and costs direct and apparent, while the functional management element ensures that the staff members employed on projects are current, practiced, and expert in their discipline.

Practice is important. A systems practitioner may be experienced and knowledgeable, but unless he is in practice, his work will be slow and prone to errors. In other walks of life, practice is believed to be essential. Sportsmen practice constantly to maintain their edge. Surgeons practice on substitute bodies before tacking the real thing.

In engineering, practice is apparently not valued. While project management focuses effort on progress, with a view to speedy achievement of targets, functional management ensures that practitioners are practiced by continually working on different projects, using the latest tools and methods. This suggests that functional management should, in principle, offer speedier and more effective progress.

Accountants tend to prefer project management to functional management. With project management, it is straightforward to accumulate the costs incurred within the Chinese wall. Functional management can, from an accountant's viewpoint, obscure the full costs of individual projects, and can create internal empires within a company, intent on expanding their capability, tools, expertise, and so on.

Matrix management has its own problems. Staff members may not be sure for whom they are working, who writes their annual performance reports, and so on. Such confusion may prejudice morale, motivation, and dedication. In effect, the

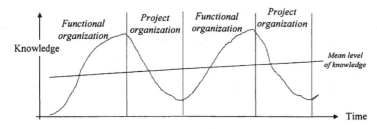

**Figure 14.11** *Switching between functional and project organization.*

matrix approach reduces social capital, hopefully supplanting it with micro-resource management.

Overall, there seems to be no right answer to the issue of management organization, and it may be significant that organizations switch periodically from one to another; see Organizational Systems on page 117.

## Dynamics of Organizational Change

Switching patterns are illustrated in Figure 14.11, which shows the notional variation of corporate knowledge being applied to activities under project and functional organizations, respectively. How might this periodic switching come about?

During periods of functional organization, the functional groups develop expertise, acquire new tools, and develop or apply new methods; these are their stock in trade. Such activity and investment cost money and time, and although work on projects may improve as a result, there is unlikely to be any noticeable and consequent reduction in project durations and costs. A function's principal motivation may be to undertake activities more effectively, more innovatively, and more thoroughly, rather than more quickly. In any event, project activity durations have a habit of expanding to fit the time available in the plan.

Accountants will notice if the company overheads rise, as they may under functional organization, and will push for a change to project organization, with its reductionist clarity of management control and accounting. During periods of project-oriented organization, functions may become vestigial, and there is relatively little learning and there are few new ideas. Corporate knowledge and expertise fall away compared with that of competitors.

To counteract this, lateral connections may emerge between projects, where workers on one project seek advice or ideas from workers on another project who are ahead of them in the project cycle, and who may have already met and solved problems. Ad hoc lateral connections give way over time to more formalized expertise groups, and the enterprise or business is on its way to switching back to functional organization.

Periodic switches are undoubtedly upheavals, but they are to be expected. In effect, the business is circulating around the life cycle map discussed earlier, but instead of collapsing completely, it switches rapidly from project to functional organization. Looking at the graph again, it can be seen that this periodic switching may be a naturally stable condition, in which the business grows and learns in spurts, but keeps itself lean and mean between spurts. Overall, corporate knowledge, expertise, and capability grow with time.

Advocates of matrix management propose that it obviates the need for such switching, and that corporate growth should be improved in consequence. The jury, as they say, is still out on that one.

## MAINTAINING THE EDGE

Organizations, businesses, enterprises, and so on, can become stale, even moribund, or "stuck in a rut." Such organizations may continue to do what they do well, and they may be successful. Yet the writing may also be on the wall, since such organizations will find themselves severely challenged by changes in their environment or market, to which they may be unable, or even unwilling, to respond.

It is very tempting, particularly during lean years, to close down the variety within an organization and to hide like a tortoise in its shell, waiting for the sun to come out. Systems organizations are not exempt; they too have been known to shed the bright, intelligent imaginative systems engineers during lean years, only to find that there is none available when recession lifts, and they desperately need to revitalize themselves. As the system life cycle map discussed earlier shows, the answer is to maintain and continually refresh complementary variety within the organization. This practice maintains stability, while avoiding tendencies toward becoming moribund. The map also shows that the danger is of dominance suppressing the variety.

What does this mean in practical terms? The dominance that suppresses variety in most organizations is the finance and accounting function, which often sees no measurable value in variety, synergy, and complementation. Holding on to skills that are not of immediate and direct use during a recession would be anathema to many accountants. Accountants may be the ultimate reductionists, of course, and many are uncomfortable even with the idea of "system." Paradoxically, financial directors may be innate systems thinkers, realizing – as many evidently do–that investment in the right place at the right time can engineer useful change and create valuable interchange.[7]

Maintaining the edge in systems thinking, engineering, and management implies having complementary groups of practitioners, with personnel being continually retrained, circulated, and even replaced with "new blood." Competition between such groups will sharpen performance, and practice is

---

[7] The process of change from being an accountant to becoming a financial director, or vice president, must be akin to the metamorphosis from caterpillar to butterfly, so different are their respective characteristics.

essential, too. Newcomers can be trained to the company's practices and ethics, and enabled to practice on nonthreatening problems before–and in between–tackling live systems.

One of the most effective ways of maintaining edge is to operate in competitive teams (see Figure 14.4). This practice may seem profligate in the use of resources, but can prove so valuable in terms of increased quality and reduced activity times, that the increased value of the resource to the organization more than outweighs increased setting-up costs.

## SUMMARY

A broad look at management first examined different concepts of management: by control, by incentive, and by competition. Likely features of creative organizations were highlighted, with suggestions for how they might be invoked – and inhibited.

An overview of organized management systems; that is, those that hold different organizations together, including industrial, military and governmental, culminated with a brief overview of Japanese zaibatsu and keiretsu. The relationships between project management and systems engineering were explored, followed by the pros and cons of project phasing.

The ubiquitous topics of elicitation and requirements followed, with the controversial conclusion that system requirements, in which customers specified in detail what they want, prejudiced scientific systems engineering. It was, however, possible for systems engineering organizations to overcome the difficulty by backtracking to the original problem, and tackling the creation of a solution from the start, with continuity. Customers could then be provided with the best solution.

Comparisons were drawn between internal company organizations by project, by function, and by matrix. No ideal solution was found. Instead, it was observed that the tendency for some organizations to switch periodically between functional and project management may be a means of advancing company knowledge and performance, while at the same time remaining relatively lean.

Strategies for maintaining the creative edge were outlined, based on life cycle theory, introduced earlier.

## ASSIGNMENTS

1. You work in an enterprise that is run along project lines. Business appears good from within each of the projects, in that they are working reasonably toward time and budget constraints, but, overall, the company is not doing too well, and there seems to be some difficulty in attracting new business. The last twenty or so bids for new projects have all failed,

at significant cost to the company. Develop a reasoned analysis of the situation, citing as the root cause of the issue that the company has fallen behind the competition in terms of capability, methods, and tools, and propose in broad terms a strategy for recovery.

2.  You are the CEO of a systems engineering corporation, and receive a routine requirement to build a new system to render navy destroyers and capital ships transparent: the "cloak" of science fiction. The requirement indicates broadly how the technology is intended to work, and states that the successful contractor will be supplied with sealed "black boxes," provided from government sources. These, when fitted into the overall system for which you are to tender, will be guaranteed to provide the nucleus of the capability. Your corporation will be required to underwrite the whole systems as "fit for purpose," and sanctions may be imposed if your overall solution is unsuitable. How do you respond, and why?

3.  Adapt/convert the principles of war so that they apply, in practical terms, to the business of manufacturing consumer durables.

4.  During a brainstorming session that you are leading, a graduate engineer suggests that a new main battle tank should consist of a manned "mother" vehicle accompanied by three to five robotic offspring tanks, each able to navigate rough terrain and engage targets allocated by the mother tank. A retired senior army officer, once a tank squadron commander and now an employee, ridicules the idea as farfetched and totally impractical. What action do you take at the time, and later? What outcome do you expect from your two actions?

5.  A project manager developing a one-off, en-route radar system for export decides that, to save money, he will dispense with systems engineering staff members once they have completed the design specifications. Predict the outcome, and, in particular, suggest whether money will, indeed be saved.

## REFERENCES

[1] van Creveld, M., *Command in War*, Cambridge, MA: Harvard University Press, 1985.

[2] Napoleon I, *Oxford Dictionary of Quotations*, Oxford, UK: Oxford University Press, 1981, p. 359.

[3] Hitchins, D. K., "Systems Creativity," *IEE Proceedings*, Vol. 135, Pt. A, No. 6, 1988, pp. 407-418.

[4] Miller, George A., "The Magical Number Seven, Problems of Perception," *The Psychological Series*, Harvard Vol. 63, No. 2, 1956.

[5] Clausewitz, Carl von, *On War*, Princeton, NJ: Princeton University Press, 1989.

[6] Wee, Chow How, K. S. Lea, and W. H. Bambang, *Sun Tzu War and Management*, Singapore: Addison Wesley, 1991.

[7] van Creveld, op. cit.

[8] Peacock, W. E., *Corporate Combat*, London, England: Maple Vail, 1984.

[9] Ibid.

[10] Womack, J. P, D. T. Jones, and D. Roos, *The Machine that Changed the World*, New York, NY: Rawson Associates, 1990.

[11] Hitchins, D. K., *Putting Systems to Work*, Chichester, England: Wiley, 1992.

[12] Ackoff, R. L., *Creating the Corporate Future*, New York, NY: Wiley, 1981.

# Chapter 15

## Societal Systems Evolution

*Piecemeal social engineering resembles physical engineering in regarding the ends as beyond the province of technology.*

*Sir Karl Popper*
The Poverty of Historicism, Volume 3, p.21

### OUTCOME AS THE MEASURE OF SYSTEMS

Earlier, the measurement of systems was associated with management by control, which is not universally successful. A significant difficulty arises with systems, in that it is not necessarily obvious what should be measured. Indeed, it seems as though the most important aspects of systems are often those least amenable to measurement.

To understand the issue, consider school examinations. It is widely accepted that the way to measure the progress of students is to examine them periodically, to score their performance, and to set the score against some standard gauge. This is useful, in that it tells teachers who is absorbing and regurgitating information to order.

However, we are all aware that retention of knowledge is important, together with the ability to make use of that knowledge when required. We also know that, as children are given a broad education, not all of it will be of direct use to every child. So, if we wish to establish the value of school education, perhaps we should be assessing pupils several years after they have left school, to see how their school education has helped them, and where it may have fallen short.

It is helpful to describe this process as measuring outcome, as opposed to measuring output, which, like school examinations, is much more immediate. In general, people avoid measuring outcome. It cannot be done until some time after the event, and, in consequence, any remedial action is going to take time; both for action, and to take effect. However, it is important.

Outcome is important in medicine. We all know of the apocryphal surgeon whose operation was a success, although the patient died. That attitude is characteristic of output measurement. If we were to measure the effectiveness of

431

some medical procedure, it might be in the degree to which quality of life was restored and for how long that restoration lasted; in other words, the outcome.

Dealing with complex systems successfully requires an outcome-oriented attitude. We may consider that the successful delivery of a product to a customer on time and within budget is the measure of success. However, when the product fails to please the user, we may find that, although the output was successful in its own terms, the outcome is unsatisfactory. In going from output to outcome, we move from execution of process to quality of product.

Outcome orientation is important for socioeconomic systems, too. Socioeconomic systems are often large and diffuse, with long time constants, nonlinear dynamic behavior, multiple points of stability, and counterintuitive response patterns. For these reasons, outcome may be difficult to predict; indeed, the outcome may be a permanently moving picture, rather than some straightforward state or condition.

- What will be the outcome of creating a green belt around a city within which no more building is allowed?
- What will be the outcome of a change in the law that forbids ownership of weapons?
- What will be the outcome of enabling people to vote using the Internet?
- What will be the outcome of banning smoking altogether?
- What will be the outcome of reducing the frequency of trains on a busy commuter line?

Each question suggests an event, followed by a period of turbulence before settling toward some new state, which may be chaotic, catastrophic, oscillatory, or steady. Each question opens up a variety of factors, and the prospect of complex interactions and reverberations. Will reducing the frequency of trains: cause road traffic jams, cause riots at railways stations, cause demonstrations, lose money for the railway, allow trains to run on time, make recruitment harder for city businesses, make mail deliveries late, encourage coach companies to start up new routes, and so on?

When addressing social issues, and when undertaking socioeconomic systems engineering, it is prudent to look to immediate output, but to be concerned with longer term outcome [1]. The following topics will examine social engineering, which does not appear to concern itself with outcome, and will pose some conceptual futuristic views of the ways in which societies and populations might evolve. These should be seen as possible goal states, some more desirable than others, toward which societies might aspire.


## SOCIAL ENGINEERING

It is the nature of the world that it is governed by change. The world has been in a state of continual change since it started; driven by energy from the sun, the rain, and the earth's rotation. Biological, or Darwinian, evolution is a powerful tool, shaping flora and fauna, converting energy into substance and back into dissipating energy, and shaping the land through the vegetation and animals.

Since the advent of man, social evolution has overtaken Darwinian evolution. The rate at which human societies have evolved is astonishing. We have already met the Old Kingdom of ancient Egypt, which flourished around 2500 B.C. on the banks of the Nile. Here was a feudal, agrarian society, with a vibrant economy before money was invented. The people, at once simple and sophisticated, were close to the land and the animals that lived around them. The people had a powerful faith in their god king, who was the intercessor between them and their pantheon of gods, some of whom controlled the annual inundation of the river, which brought fertility to the Nile valley. For perhaps the first time in history, food was so plentiful that life could be more than a continual search for sustenance.

The faith of these people brought them together to build the pyramids; tombs for their kings so that those kings could continue after life to care for their people from their place among the stars (a case of self-interest motivated by a powerful belief system). It is unlikely that we could build the Great Pyramid today. It would be too big an undertaking for our modern societies; we lack the all-consuming faith. Besides, risk analysis would be against it, as would cost-benefit analysis. The ancient Egyptians had not been exposed to ideas such as socialism, the withdrawal of labor, mass production and Taylorism, management, and so on. Indeed, they had met none of the ideas, great and not so great, that have shaped and fractured today's societies.

One of the differences between then and now is the rate of change. The ancient Egyptian nation lasted, with a few fits and starts, for over 3,000 years. Today, it is hard to find a nation that has remained unchanged for 1,000 years, and many show much more recent emergence. Compared with the ancient Egyptian religion, which remained substantially the same for 3,000 years, Christianity is less than 2,000 years old, and Islam is less than 1,400 years old.

In recent centuries, change has accelerated. Technology has enabled us to move faster; to reach farther. The globe has an increasing, aging population; presently over 6 billion and rising fast. Not surprisingly, this rise is coupled with dwindling resources such as fossil fuels and minerals, while renewable resources are *not being* renewed such as forests, soil, fresh water, fish, unpolluted air, bio-diversity, and so on. There is a continuing species extinction, presently occurring at a rate of some five to 15 species per day; faster, apparently, than that which wiped out the dinosaurs 64 million years ago!

With so many people on the planet, the predominance of systems, both in numbers and in size, appears to be of human systems. Everywhere people spread, they take their technology with them, be it the simple technologies of food preparation, or the more complex technologies of transportation and communication. But people systems predominate, as families, communities, societies, cultures, organizations, clubs, associations, corporations, industries, economies, armies, and so on.

It is not surprising, then, that some human groups seek to manipulate, control, influence, command, and control other human groups. For generations, politicians have indulged in social engineering, although many would be uncomfortable to have their efforts so categorized.

Social engineering by various governments of all political persuasions has led to large urban conurbations, supposedly easier to support and manage, but containing increasingly atomized and fragmented societies. In the Western world, female freedom and independence, traceable in part to the contraceptive pill, are leading to increasing female assertiveness, promiscuity, and aggression. These trends have been accompanied by the virtual demise in some cultures of, first, the extended family and, latterly, the nuclear family. Yet, the contraceptive pill was introduced to help married women to plan their families. Sex education in schools, intended to reduce unwanted teenage pregnancies, has been accompanied by an alarming increase in such pregnancies. Social engineering is generally piecemeal, with counterintuitive effects. Those indulging in social engineering are unable, or unwilling, to foresee outcomes.

In the West, we see dwindling faith in organized religions, coupled with a resurgence in alternative beliefs in paganism, crystals, alternative medicines, and conspiracy theories. Commercialism is emerging as a self-justifying dogma. And,

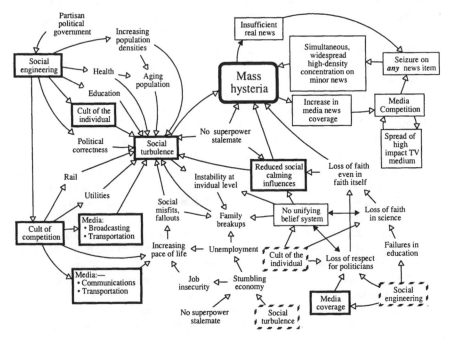

**Figure 15.1** *Social engineering and the effects of modern media on societal tranquility.*

to accompany the trends, we have coordinated mass hysteria, orchestrated by commercialized media; see Figure 15.1.

The figure notionally draws together some of the threads emanating from political social engineering in typical Western societies. Social engineering presides over changing, often growing, populations that have a consequent tendency to social turbulence. Different political dogmas address the turbulence in different ways: some by suppression; some by offering mass entertainment. In ancient Rome, the forum was the coliseum; today it is radio, television, newspapers, magazines, football, theater, cinema, clubs, fashion, and travel to exotic places.

In the West, political social engineering has extolled the cult of the individual and the cult of competition. The figure shows how these cults can and have led to uncontrolled media coverage. Freedom of the media (TV, newspapers, and so on) is vaunted as a bastion of democracy, and so it may be, but, as the figure shows, it is possible for the media to indulge itself in feeding frenzies that can stun societies.[1] The media will have this effect only on societies that are so predisposed, but such predisposition may also be the result of piecemeal social engineering.

Energized societies may be operating on the edge of chaos. Figure 15.2 shows a possible scenario, based on a simple model of chaotic interactions. An energized society sits on the edge of chaos (at left, on the graph), behaving regularly and with no excessive turbulence. An impulse, in the form of some unexpected event that disturbs society and triggers a media feeding frenzy, pushes the society into a short-lived state of chaos, from which it recovers over time. Before it can fully

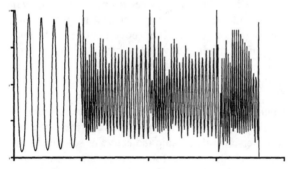

**Figure 15.2** *On the edge of chaos. From the left, behavior is regular until disturbed. The first disturbance pushes the system into chaos, from where it gradually returns, only to be pushed over the edge again and again. The last push, at the right, arrives too soon for adequate recovery, and the system becomes unstable.*

---

[1] A graphic example can be seen in the untimely death of Princess Diana of Wales, which brought large parts of the U.K. to a stop, with people who had never met her distraught, crying, depressed, and so on. The situation persisted for several days, sustained by nonstop blanket media coverage.

recover, a second impulse occurs, and then a third. The fourth impulse in the graph occurs too soon after its predecessor for sufficient recovery to have taken place, and the whole system becomes unstable as a result. The model suggests that "free" societies, some of which may indeed exist near the edge of chaos, are generally robust and can weather many storms, but that there may be conditions, or frequencies of occurrence, that push them over the edge into instability.

Is this how societies break down? It is unlikely to be quite so straightforward, but the model does point to a potential threat to future societies; one among many. Paradoxically, and again counterintuitively, the threat to order in this instance appears to be partly from the unfettered media that are seen as the bastion of democracy. As Karl Popper observed at the start of this chapter, piecemeal social engineering has little regard for the outcome of its machinations.

Living on the hyperactive social brink may be exciting and creative, but it may also prove exorbitantly expensive, as activity indicates disorder, work, motion, and so on. This is what we see in many cities where individuals live and work at a fast pace, consuming resources, expending energy, receiving high salaries, and spending at high rates, too. As well as individuals, groups, societies, and cultures find themselves in the same situation.

Social instability and a high tempo are indicators of "self-organized criticality [2]." The term describes a state or condition, that can be reached by systems comprised of many interacting parts; see Chaotic on page 40.

The archetypal example considers the formation of a pile of sand caused by the continual dropping of grains of sand on to a flat, circular disk. After a time, the sand forms a pile, which slowly rises to become a cone. Eventually, a point of quasistability is reached; at this point, there will be occasional avalanches of sand, which will cause the pile to fall below the critical height. Adding further grains will build the cone again, so that it rises above the critical height, before another avalanche brings grains tumbling down. The height of the pile may fall below or rise above critical height, but always returns to the critical height; hence, self-organized criticality.

Not surprisingly, there are more avalanches with just a few grains of sand, and fewer avalanches with many grains of sand. Unexpectedly, however, a clear mathematical relationship can be identified between these factors, of the general form $y = a.x^b$, a power-law relationship indicative of so-called "weak chaos." This general relationship is also to be found in the frequency of earthquakes of different severities, the frequencies of meteorites of particular sizes entering the upper atmosphere, the pattern of noise in electrical conductors, the distances between cars on a busy highway, the price movements on the stock exchange, and so on.

Self-organized criticality is a particularly interesting phenomenon because of its ubiquity and simplicity, and because it is a dynamic, whole system phenomenon; that is, not dependant on understanding interactions between the parts for explanation. It is also interesting because it helps us to understand and appreciate how systems work at the macro level, including large, people systems.

Societies that find themselves on the hyperactive edge of chaos may have brought themselves to that level (like some stock market crashes) by tighter coupling between the parts of an overall, nonlinear system. Since being in such a state makes a society susceptible to risks, it may be essential to loosen its own "coupling" to survive.

Note that the above topic has had to make recourse to nonlinear dynamic systems behavior even to describe, let alone think about, such issues; the linear-predictive world of engineering cannot address them. Perhaps a comprehensive systems approach can. Perhaps, 21$^{st}$ century systems engineering will, perforce, extend above its engineering management roots, and move its center of activity upwards through the levels to the top level: socioeconomic systems engineering.

## THE OUTLOOK

Civilizations come and go; that is the way of things. This time, however, things *are* different. Already there are too many people, and too few resources. So, we need highly efficient food, energy, and technology systems to sustain Western-style urban societies. Moreover, we have sufficient weapons arsenals to eradicate life on the planet several times over.

As this extended civilization in which we presently live decays, we can expect: a collapse of the efficient food chains that bring foods from around the globe; international and internecine squabbling over dwindling resources, particularly fossil fuels; and nuclear weapons ending up in the hands of "barons" and thugs; and so on.

Of course, it doesn't have to be that way, but the augurs are not favorable. Societies will need to become increasingly efficient; that is, they will need to live in balance with their environments. As populations age, societies must self sustain where the majority do not work, yet have the vote. Fundamentally, there appear to be some irreversible trends that we are going to have, literally, to live with.

### Increasing population

Some of the factors affecting population in particular are shown in Figure 15.3. The figure is based on the classic births-population-deaths cycle; see Figure 1.17 on page 31, around which are developed the effects of medical advances, increasing both population and average population age; a water shortage, affecting the population as potable water, and through farming shortages; and a land shortage, affecting biodiversity and food production. While the earth's human population has in the past fallen from time to time, usually because of pandemics such as the Black Death and influenza, there is an inexorable trend upwards. War is always an unwelcome possibility, and to be sure we have recently witnessed the unedifying spectacle of ethnic cleansing as a means of selectively reducing population.

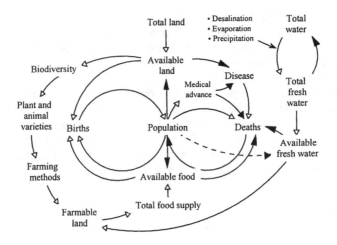

**Figure 15.3** *Increasing human population: CLM.*

Science fiction writers are fond of the idea that we can escape from an overcrowded earth to other planets in other solar systems. It seems unlikely to be practicable for the vast majority, if only because of the phenomenal energy that would be required to move whole populations to remote solar systems.

Could we, perhaps, control our population? As the tragedy of the commons[2] showed, self-control by one section is unlikely to be matched by self-control from another. Human nature would simply fill the space–again, and again.

## Spreading concrete

As humans sprawl across the planet, we lay concrete with houses, streets, highways, industries, malls, and so on. Every square foot of concrete covers and kills a part of the living, breathing soil. Towns that were once some distance apart are now adjacent suburbs, part of the urban sprawl emanating from some city. Eager to escape the city, workers the world over choose to live ever farther from their work, commuting daily over multilane highways, and polluting their world with noise and exhaust fumes.

While the spread of concrete may not be materially responsible for the rapid diminution of biodiversity, it is certainly not helping to slow it, nor is it assisting recovery.

---

[2] The tragedy of the commons refers to a situation that arose in medieval England, where commoners became entitled to graze livestock on common land. Livestock numbers grew and the common land became over-grazed. It was in nobody's short-term interest to stop, or even reduce, his flock's grazing, since, if he stopped, another commoner would seize the opening to graze his flock instead. The outcome was the effective loss of the common land to all, since it ceased to support grass and grazing. The example of the tragedy of the commons is an archetypal instance of the potential social effects of noncooperative social behavior.

**Atmospheric Pollution**

The litanies of global warming need no repetition. Apart from the supposed heating aspects, changes in the proportions of gases in our atmosphere may present other problems to flora and fauna, including us. Oxygen levels in the air have been much higher in the past than they are today; it has been suggested, for example, that many of the larger dinosaurs existed because of this, and that increasing $CO_2$ levels, caused by volcanic action, may have contributed to the demise of the dinosaurs 64 million years ago. Could further $CO_2$ rises affect earth's flora and fauna today? Some trees and planktons flourish in $CO_2$ rich atmospheres, while others do not.

The proportions of gases forming our atmosphere are the result of stability in a complex, nonlinear, dynamic system. There are many natural feedback mechanisms tending to restore oxygen and $CO_2$ levels in particular. One of the characteristics of nonlinear systems is that they exhibit multiple points of stability, and that, as conditions change, they may suddenly, and without warning, switch from one stable state to another. It may be that the linear change patterns that many pundits anticipate in our atmosphere may not happen at all. Instead, there could be a sudden and catastrophic change; it certainly happens in some models....

**Fresh Water Shortages**

In a world that is two-thirds covered in water, it seems strange that fresh water is in such short supply. Water cannot really be destroyed; after all, it circulates in the never-ending water cycle of evaporation and precipitation; and water in underground aquifers may remain there for thousands of years. However, in some parts of the world, water is being drunk and used for farming at greater rates than it is being replaced in aquifers, rivers, lakes, and wells. Some parts of the earth are becoming "hollowed out" by fresh water extraction, leaving an unstable crust.

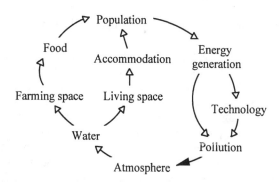

**Figure 15.4** *The population-pollution cycle.*

**Energy Shortages**

Ever-increasing populations and industrialization together indicate an exponential rise in energy demands. A simple population-pollution cycle, shown in Figure 15.4, suggests that there may be a natural population-regulating mechanism, with ever-increasing populations being inhibited by pollution from their energy and technology demands. The model suggests an unpleasant outlook, with the human race polluting not only itself, but also the planet's biosphere.

For political reasons, many governments propose to meet their future energy needs from renewable sources, that is, wind, wave, and solar power. They propose to reduce dependence on fossil fuels because these release $CO_2$ on burning. Nuclear energy is "green," but is eschewed by politicians as a dangerous energy source.

Depending on wind, wave, and solar power sounds sensible, although countries will need a lot of alternative systems to meet today's energy demands, and have little hope of meeting tomorrow's needs. Advocates of alternative energy tacitly assume that there will always be wind and sun. However, studies of global warming and experiences of shifting weather patterns suggest that their assumptions are suspect; they may have little sun and no wind for extended periods in locales that were previously ideal. When they eschew nuclear energy, today's politicians are gambling with tomorrow's societies and economies.

## SYSTEM THINKING: SOCIETAL EVOLUTION [3]

Is there a way out of this long-term dilemma? Well, yes, of course there is always a way out of such situations. The question is: will the escape route be acceptable?

Coming back to the present for a moment, many major cities face a perennial problem of overcrowded airports, and the consequent need to add new runways, and to build additional airports around the city. Local residents, who would thereby experience even greater noise pollution, invariably greet proposed additions with vociferous opposition. Wildlife groups oppose new airfields because of the environmental damage they might do. Yet, there is often a simple solution.

Many major cities exist around river estuaries, or near a coastline, where a floating airport could be constructed from interconnected concrete caissons. Locating floating airports offshore would prevent noise and air pollution on shore, would have no material effect on wildlife, and, since there is no land involved, would avoid interminable wrangles over land ownership and usage. Constructing floating airports would provide a stimulus for the local construction industries, and each airport would generate its own electricity from wind, wave, and tide; becoming a net provider to the local power grid. All around, this would be a win-win situation, you might think. Not so, judging by the number of floating airports around the world. Opposition to such ideas is based, not on technical issues, but on political and psychological issues. From this viewpoint, a floating airport would be: risky, a hazard to shipping, unstable, disorienting, a danger to seabed ecology, constantly on the move, and so on.

The floating airport is just one example. In essence, although there may be sound escape routes from a social or political issue, it is by no means certain that any of these will be even considered, let alone followed. Bearing that factor in mind, let us return to the longer term issues facing humanity on earth.

If we accept that the major, long-term trend of expanding human population will continue, then the other factors identified above seem set to follow: pollution, famine and drought, pandemics, biodiversity collapse, and so on. The most immediate issues concern energy, food, and water, and these problems are not worldwide–yet.

## Mountain Retreats

In principle, there is ample land for the current human population, but large tracts are uninhabitable, including: mountainous areas, deserts, tundra, and the Antarctic. We can, as a species, choose to open up these areas, using and developing our technologies to create new living spaces.

Most people would not welcome a return to primitive living in the hills, nor would such a retreat support large populations. Instead, we could create communities in remote areas, living at urban densities, and protected from environmental extremes. Such communities would be largely self supporting, but would also form part of the international community, making extensive use of

**Figure 15.5** *Conceptual future mountain community.*

communications. Such facilities would enable some in the community to work and play in international forums. Some might work within tiers of agile lean volume supply organizations, for instance.

Figure 15.5 shows an impression of such a future mountain community. The facility is large; supporting perhaps 30,000 to 50,000 people in a high-rise complex that contains all the features of a modern city: doctors, hospitals, schools, colleges, theaters, playgrounds, malls, and so on. There are residential areas, commercial areas, internal transport systems, and so on. The whole is enclosed in a transparent dome, to create a comfortable, temperate environment for the inhabitants, sheltering them from winds as well as from snow and ice–which provides the community with its ample fresh clean water supplies, and enables it to grow most of its own produce using hydroponics. Energy is largely solar, but wind energy is also plentiful, and is extracted from outside the dome. (One of the problems with wind energy is that it is most plentiful where people choose not to live, for example, rugged coastlines and mountains. That would not be the case here.)

While one such community would hardly dent the population pressure, there could be hundreds, or even thousands of such domed cities. Technologically, there are few barriers, except perhaps for the dome itself. To build such a large, transparent dome might be beyond our current technological capabilities–or would it? It could be built on geodesic principles, and it could be made of robust materials. Certainly, the domes would be a technological challenge.

Mountain communities might suit some, but others might be unhappy with the remoteness and the supposed isolation. Others would undoubtedly feel that they were "living in a goldfish bowl," although, from inside, the dome would be invisible. An alternative conceptual life style for the future might be for societies to form island communities.

## Island Communities

The term 'island communities' refers not to islands in the sea, but to islands on the land. Figure 15.6 shows how such future communities might look, conceptually. The figure shows several structures as high-rise, or 3-D communities, mutually separated by virgin territory.

In this concept, it is important to maintain biodiversity, and even to reverse its contemporary trend downwards. That objective has been addressed by confining human activity to inside the structures, and by building the structures both upwards and downwards, under the ground. Communications between communities is electronic, optical, or physical; by underground tunnels. The objective is to bar any human activity or interference in the biosphere surrounding the island communities, which should then return to their natural state.

As before, communities would be self-contained, housing perhaps 75,000 to 100,000 people. Each would have its own doctors, hospitals, entertainment, communications, and so on, and it would be simple for people to visit other

**Figure 15.6** *Conceptual future island communities.*

communities via the tunnels, preventing any atmospheric pollution. Parts of each structure would be set aside for living, for working, and for entertainment, and perhaps even as holiday homes for people from other island communities, or for people who just felt like a change.

The concept addresses the issues of population overcrowding and of biodiversity, but it falls short on other counts. Food production and fresh water supplies would still be issues, and water would be needed to establish and maintain the external land as it returned to a more natural state–always supposing that not to be desert.

The island community concept has few technological pitfalls; it does, however, present some psychological ones. Some people do not like living in high-rise structures, while others would be unable to resist invading the surrounding land. However, as population pressures increase, and as land resources dwindle, there may come a time where people no longer have the choice....

### Littoral Enclaves

A significant variation on the island community theme would be to construct littoral enclaves; see Figure 15.7. While these could be created on any shoreline, they would perhaps be at their most effective on shores near desert areas.

**Figure 15.7.** *Conceptual littoral enclaves.*

As the figure shows, the structures would exist partly on land and partly under the sea. Habitats would be interconnected either by submerged tubes, or by submersible vehicles. As before, each structure would be self-contained, although one social group could comprise several interconnected structures, each of which might be dedicated to different purposes: vacationing, industry, university, and so on. (Hotels and resorts[3] are being constructed already along these lines, with part of the hotel under the sea, and with guest rooms having windows under the ocean.)

The food supply would be largely hydroponics based, as before, supplemented by food from the sea, which would be farmed for vegetable as well as animal sustenance. Each of the habitats would have extensive desalination facilities, using tidal rise and fall as a means of driving osmotic filtration. Excess fresh water would be used to revitalize the adjoining desert areas, making them suitable for food production and the restoration of local biodiversity. A coastline of such developments would result in a strip of land perhaps 50 to 100 miles deep being restored, and perhaps being developed as a $CO_2$ "soak;" that is, a means of absorbing excess $CO_2$ from the atmosphere.

Because of the ready availability of fresh water and food, littoral enclaves could sustain significant populations, although to retain social coherence, sensible limits might constrain[4] populations to 100,000 to 150,000. As with the previous

---

[3] The Atlantis Resort in the Caribbean, for example.
[4] It is possible that there is an optimum size for a society or social group, that maximizes social capital. Certainly, social capital is high within extended family groups, and low within large, impersonal cities.

notions, the littoral enclaves would take part in international activities as part of agile lean volume supply systems. There seem to be few technological challenges in this concept. Psychological issues may emerge, however, with people happy to holiday in such enclaves, but less inclined to take up permanent residence. Since these conceptual solutions are envisaged as lifestyles of the future, it is reasonable to suppose that people will be more amenable to such living arrangements when the population rises from today's six billion to 12, 15, 30, or 100 billion.

**Submersible Cities**

In these conceptual routes for social evolution, we have so far avoided the obvious fact that the seas contain over 98% of the biosphere on planet earth. It seems inevitable that we will eventually have to spread into the sea, as pressure on the land mounts.

Figure 15.8 shows one concept; based on complementary systems principles. The figure shows four spheres; the observer is standing on a fifth, looking outwards through the transparent dome or globe. The five spheres each have different functions. At left, with fields and trees visible, is a farming habitat able to grow food, and support animal life: it is replete with microorganisms, worms, insects, and a diverse bio-structure. In the submerged part of the farming habitat–as in all the others–is a fresh water system, using tidal rise and fall to partly power desalination. The fresh water is used for farming, for maintaining appropriate humidity levels, and for creating clouds above the ground to protect vegetation. In practice, there will be little need for more fresh water once the habitat is operating, since the water in the globe will continually recycle. Instead of a biosphere on the outside of a globe, here is one on the inside of a globe.

The second sphere is for living, relaxing, and taking vacations. In such a society, where people would be expected to live for very long times, work would no longer be an obligation: people would work if they wanted to, but many people might be too old, mentally if not physically, for regular work, and will seek to learn more, pursue social interactions, and create new ideas, new things, and even new technologies.

To further this last end, the third sphere is an activity habitat, where activities may range from sporting to industrial, with laboratories, workshops, sports halls, and so on. The fourth sphere, at the right, is not transparent, and it contains industrial facilities for extracting minerals from seawater, for recycling waste, for creating and forming materials, and so on. These may be used by those in the third sphere to maintain a state of repair in the existing habitats, to construct new spheres, and to develop new technologies for the convenience and use of the people in the habitats, and for export to other habitats.

The final sphere, from which the observer looks out, is a habitat command sphere. It controls the environments in all the spheres, and is able to anticipate the arrival of inclement weather by submerging the habitat set, maintaining the habitat

**Figure 15.8** *Conceptual submersible habitats.*

spheres in close contact at all times. Each habitat can also be rotated, allowing sunlight to reach different parts.

Interchange between the spheres is usually by electrically driven submersible; these vessels are also used for exploration and for farming the seabed and the sea surface. The spheres can move slowly both on the surface and when underwater; the command sphere coordinates their passage. They can also drop anchor, and use the rise and fall of tides to generate power, which may be shared between spheres. Power may also be derived using temperature differences between different depths in the sea.

The command sphere also coordinates business activities with other habitats, and with shore-based organizations. Exports and imports will be active, although in the nature of things it seems likely that barter, rather than money, may be the exchange medium. Groups within the habitat may operate as tiers within one or more global lean volume supply systems, ensuring that those in need of a business challenge can always find it. Together the five spheres form a complementary set; an idealistic, holistic, synthetic, and organismic solution to a population and pollution problem.

Optimizing the solution system would be important, too. Since there could be many thousands, even millions, of such habitats, it would be vital that they lived in balance with their environments; renewing everything that they used, and polluting nothing. This would imply characteristics about the resources they harvested, their ability to recycle, how (if at all) they disposed of waste, what effect they had on surrounding seawater temperature and evaporation, and even the effect on the numbers of people that such habitats could accommodate. It would also be important that the economics of each habitat were in balance. In such circumstances, the idea of making a profit would give way to one of parity between inflows and outflows, so that each habitat received what it needed in exchange for giving other habitats what they needed.

So, what is wrong with the notion of societies living in submersible, complementary habitats in the way described? Well, not a lot, really. Apart from the transparent domes, the technology does not seem to be beyond our grasp. We already have submersible and semisubmersible devices (oil rigs and platforms) that routinely cross the Atlantic Ocean, normally on the surface, but "ducking under" to avoid storms. We also have enclosed bio-habitats,[5] with many varieties of flora and fauna interacting to create a stable whole. We have the technology to extract minerals from seawater, and we even have the so-called "station-keeping" facilities, which ships and aircraft use to maintain formation.

## PURSUING THE GOAL

It is one thing to dream up futuristic lifestyles and ways of managing increasing populations, but it is another for these things to come about. Yet, why not be goal oriented? Instead of starting from now and predicting outcomes, why not look toward favorable outcomes, and seek routes to their achievement?

In terms of difficulty, socioeconomic systems engineering will be akin to nailing jelly to a wall. The ideas expressed, and the methods used, in this book are system-type independent, and as such are as applicable to socioeconomic systems as to any other systems.

Unlike piecemeal social engineering, socioeconomic systems engineering has to be vitally concerned with outcome. We presently lack adequate models of societies and social behavior, however. Without these, it would be difficult-to-impossible to assess likely outcomes. Relatively simple societal models have been presented–see Figure 10.35 on page 305 as an example–and these can be useful, but there is some way to go before we can predict outcome with any confidence. At present, we can identify many of the factors, but the timeframes of interactions between those factors are difficult to gauge, and completeness is an issue too.

The challenges facing systems engineering in the 21st century are many and various at all five levels in the systems engineering hierarchy, but nowhere is the challenge greater than at the socioeconomic level.

---

[5] Consider, for example, the Eden project in Cornwall, England.

## SUMMARY

This chapter has looked at outcome as a reasonable measure of systems, and has concluded that, although more difficult, it is greatly to be preferred to measuring output. Social engineering was briefly examined, with its checkered past and questionable present. Trends in human societies and populations were explored, leading to some conclusions about the uncertain prospect facing future societies. Four views were presented of conceptual future environments designed to accommodate expanding human populations in different ways. Finally, the challenges facing systems engineering in the 21st century were highlighted.

## ASSIGNMENTS

1. Your organization manufactures a complete range of personal computers, from the small handheld, to the large professional desk machines. Statistics show that the mid-range model is making the most profit, so the organization decides to concentrate on that and discontinue the others. Predict the effect of this decision on the workforce and the business.
2. A university decides to raise entry standards and student charges. Predict the likely effects in the short term and in the longer term, say in 10 years.
3. The distances between vehicles on a busy highway obey the rule $y = a.x^b$, indicating that there are more vehicles that are close to each other, and fewer that are further apart. Explain, in qualitative terms, how this might come about, what effect it may have on the capacity of the highway, and what might be done to increase that capacity.
4. Conceive a future habitat in which people would live on the coastline of Antarctica. Identify how it may self sustain, and its advantages and disadvantages, and suggest the social and psychological effects that living in such an environment might have on people in a crowded future world.

## REFERENCES

[1] Warfield, J. N., *Societal Systems*, Salinas, CA: Intersystems Publications, 1989.

[2] Bak, P., and K. Chen, "Self-Organized Criticality," *Scientific American*, Vol. 264, No. 1, 1991.

[3] Hitchins, D. K., "Systems Engineering in the 21st Century," *Proc. 9th Annual Int. Symposium of the International Council on Systems Engineering*, Brighton, England, June 6-11, 1999.

# Appendix A

## Configuration Entropy
## as a Useful Measure of Systems

### MEASURING N² CHARTS

A useful way to represent entities and their relationships is the so-called $N^2$ chart [1]. This is a rectangular chart with N rows and N columns; hence $N^2$. Unlike a normal chart that has row and column headings, an $N^2$ chart locates the entities on the leading diagonal matrix boxes.

|   | A | B | C | D | E |
|---|---|---|---|---|---|
| 1 | ■ | 1 | 1 | 1 | 1 |
| 2 | 1 | ■ | 1 | 1 | 1 |
| 3 | 1 | 1 | ■ | 1 | 1 |
| 4 | 1 | 1 | 1 | ■ | 1 |
| 5 | 1 | 1 | 1 | 1 | ■ |

**Figure A.1** *Simple N² chart.*

Figure A.1 is an example of a simple $N^2$ chart. The black rectangles in the figure, A1 to E5, represent entities, while the other rectangles represent interfaces. In this instance, all the entities have interfaces, or relationships, with each other. A zero or a blank in any rectangle would indicate an absence of an interface or relationship.

## Scoring N² Charts

$N^2$ charts can be scored by multiplying the entities, taken two at a time, by the distance between them. In the figure, for instance, there are two "1s" between A1 and B2; that is, at B1 and at A2. Overall, the score for this $N^2$ chart is:

- 8 interfaces at distance 1 from the leading diagonal= 8;
- 6 at distance 2 = 12;
- 4 at distance 3 = 12; and
- 2 at distance 4 = 8.

This makes a total score of 40. Since each of the interfaces is filled, this is the maximum score for this binary $N^2$ chart.

In general, the maximum score for a binary $N^2$ chart is given by:

$$\text{Score} = 2* \sum^{(N-1)} k*( N-k) \tag{A.1}$$

## Minimizing N² Chart Entropy

Having all the interfaces filled, as above, represents a disordered condition. Were the entities individuals in a room, they would all be talking at once; each talking to every one of the others and each trying to hear every one of the others through the resulting cacophony. This represents, then, a high-energy, high-entropy state, from which little energy can escape as useful work. Were the $N^2$ chart representative of a committee, or a bureaucracy, everyone would be working, relating to each other, yet going round in circles. The net output for all that work would approach zero.

**Table A.1**
Combinations of N Items

| N | $^NC_1$ | $^NC_2$ | $^NC_3$ | $^NC_4$ | $^NC_5$ | $^NC_6$ | $^NC_7$ | $^NC_8$ | $^NC_9$ | $^NC_{10}$ | $\sum^N(C)$ | $N^2$ Score |
|---|---|---|---|---|---|---|---|---|---|---|---|---|
| 1 | 1 | | | | | | | | | | 1 | 0 |
| 2 | 2 | 1 | | | | | | | | | 3 | 2 |
| 3 | 3 | 3 | 1 | | | | | | | | 7 | 8 |
| 4 | 4 | 6 | 4 | 1 | | | | | | | 15 | 20 |
| 5 | 5 | 10 | 10 | 5 | 1 | | | | | | 31 | 40 |
| 6 | 6 | 15 | 20 | 15 | 6 | 1 | | | | | 63 | 70 |
| 7 | 7 | 21 | 35 | 35 | 21 | 7 | 1 | | | | 127 | 112 |
| 8 | 8 | 28 | 56 | 70 | 56 | 28 | 7 | 1 | | | 255 | 168 |
| 9 | 9 | 36 | 84 | 126 | 126 | 84 | 36 | 9 | 1 | | 511 | 240 |
| 10 | 10 | 45 | 120 | 210 | 252 | 210 | 120 | 45 | 10 | 1 | 1023 | 330 |

Using Combination Mathematics

Entropy is concerned with the number of different ways in which entities can be arranged. Combination mathematics indicates as follows:

The maximum number of ways of combining N entities:

$$\sum^N \text{(combinations)} = 2^N - 1 \qquad (A.2)$$

Table A.1 shows how this formula is derived, by adding the combination of N items taken one at a time, two at a time, three at a time, and so on. By inspection of the table, it can also be seen that that the score for an $N^2$ chart $= 2\{^N C_2 + {}^N C_3\}$.

Mapping between $N^2$ Score and Entropy

Entropy $= k * ln$ (number of ways) $= k * ln (2^N - 1)$ \qquad (A.3)

where k is Boltzmann's constant and
  *ln* is the natural logarithm

From Table A.1, it can be seen that the $N^2$ chart maximum score increases and decreases consistently as the function $(2^N - 1)$ increases and decreases. In other words, as N varies, there is a conformal mapping between the $N^2$ chart maximum score and the number of different ways that N items can be arranged.

It follows that, for an arbitrary system, one in which not all interfaces are filled, rearranging an $N^2$ chart to minimize the $N^2$ chart score also minimizes configuration entropy for that system.

$N^2$ Chart Entropy Index

Since the maximum $N^2$ chart score corresponds with maximum entropy (and hence with the minimum energy escape from within the system), reducing the $N^2$ chart score corresponds to increasing the energy escape in the form of useful work.

Simply by rearranging a set of systems to reduce their configuration entropy/$N^2$ chart score, it may be possible for the resulting grouping to do more work and to retain less energy.

A simple index of entropy can be used to gauge any $N^2$ chart as follows:

Entropy index $=$ (actual $N^2$ score) / (maximum $N^2$ score) \qquad (A.4)

Rearranging an $N^2$ chart to minimize the entropy index, also minimizes the configuration entropy.

## PRACTICAL EXAMPLE OF ENTROPY MEASUREMENT AND REDUCTION

A major airline's base storage depot has 12 groups of spares and consumables arranged within a large rectangular hangar. Items are "picked" from groups as needed. Individual "picks"[1] visit one, two, or more groups in sequence, according to the "pick list."

The pattern of "pick movements" around the hangar is uneven, with some categories occurring more frequently than others. It is becoming apparent that a considerable amount of unnecessary traveling is occurring around the hangar, taking time and effort, and potentially causing a safety hazard. The rectangular room is the only suitable space available. Can anything be done to improve the situation?

(This is a typical problem facing storage depots which are used not only by major transport industries, but by logistics depots, supermarket depots, wholesalers, and many, many more.)

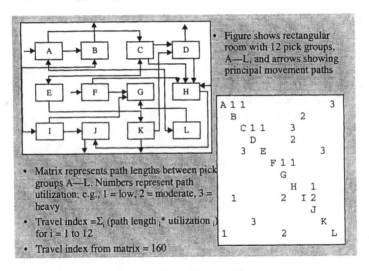

**Figure A.2** *The pick-list problem.*

Figure A.2 shows a diagram, top left, of the pick areas in the airline's base storage depot together with the paths taken by operatives picking typical loads from various stocks. Some "picks" occur more frequently than others, and these have been ranked in the $N^2$ chart, at bottom right. A travel index has been formed, bottom left, showing the overall travel between pick areas being taken. The travel index is 160, and provides a simple approximation of the total distance being traveled.

---

[1] The term "pick" is widely used as a noun as well as a verb in the logistics industry.

Figure A.3 shows the results after rearranging the $N^2$ chart to reduce configuration entropy, and relocating the pick groups accordingly. The travel index was reduced by a factor of 3.2 from 160 to 56, indicating a major reduction in the overall work of traveling. The rearrangement actually increased the separation between some pick piles, but the sum of all the link lengths was reduced from 79 to 36; that is, by 54%. By including information about the number of times individual links were used, the clustering algorithm that rearranged the $N^2$ chart was able to minimize busier routes preferentially: see that interfaces marked "3," for the busiest, have been placed closest to the leading diagonal.

This example shows the practical advantages of using entropy and entropy reduction as a measure to guide system improvements. Being extensive, entropy applies to the whole system, not just each part, and so overall optimization is feasible. Curiously, the whole system has been improved in terms of efficiency and effectiveness, without the nature of the work being altered in any way. Major savings have been made simply by rearrangement or reconfiguration.

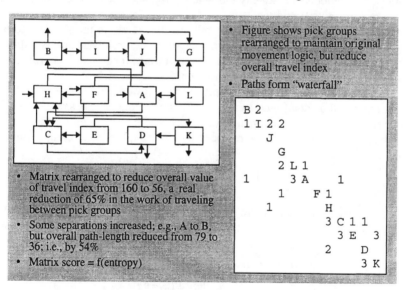

**Figure A.3** *Reducing entropy to reduce work.*

## MEASURING PROCESS ENTROPY

As processes are systems too, it should be possible to measure the configuration entropy of a process or group of processes and use the measure to direct reductions in entropy leading to greater efficiency and effectiveness.

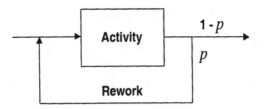

**Figure A.4** *Rework cycle.*

One view of an ideal business, production or commercial organization sees it as a laminar flow of substance/information through sequential processes—the so-called value-added chain. Ideally, a straight line could represent this flow pattern, indicating minimum configuration entropy; that is, without any disorder in the pattern of workflow. Any deviations from the laminar, straight flow would equate to turbulence and increased entropy. Turbulence is caused typically by inadequate training/practice/tools, a hierarchy introducing decision delays, late supplies from subcontractors, rework, and so on.

Consider rework in particular. Rework may be viewed in the simple figure as a form of feedback in which some of the output, being unsatisfactory, is fed back, to be reworked. If there is a probability, *p*, of rework, then there is a probability, 1- *p*, of not feeding back, that is, of moving on. See Figure A.4.

If activity takes time, T, and the probability of rework $= p$ then, on average, a proportion *p* will be done again, and will take further time *p*.T. A small proportion of *this* rework will need done again, taking $p.p.T = p^2T$.

Overall, the time T can be replaced by $T/(1+p+p^2+p^3+...)$ or by using the Binomial Theorem:

$$T_{total} = T/(1-p) \tag{A.5}$$

In Figure A.5, *p* is a measure of the turbulence/disorder in the flow caused by rework. The figure shows three sequential processes: $P_1$, $P_2$, and $P_3$. The panel at

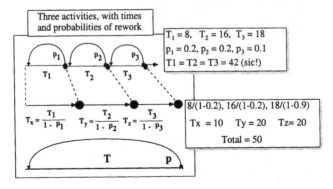

**Figure A.5** *Process sums.*

the right shows the relevant details for each process: duration, and probability of rework. The sum of the times for each process is 42 units.

The lower panel at right shows the calculations for the equivalent times of each activity, now including rework: $T_x$, $T_y$, and $T_z$. These times are marginally longer than the corresponding raw times, T1, T2, and T3, summing to 50 units.

So, the network of three processes has been simplified to a single process, duration T, with feedback probability, $p$. This overall probability $p$ is an indicator of the likelihood of feedback, or turbulence, or entropy.

Instead of tracking each activity, estimating each T and each $p$, and then combining them, it is easier to rearrange equation A.5:

Total time = Direct path time / (1 - $p$)

$p$ = 1 - (direct path time/total time)                                    (A.6)

where:  $p$ is the probability of feedback/rework;
        Direct path time is the time the whole project would take with no rework, no deviations, and no decision delays; and
        Total time is the time for the whole project, including all the rework deviations and decision delays.

Direct path time can be estimated for real projects from PERT planning and similar activities that conventionally ignore rework and presume ideal working.

Total time can be estimated from historical records. In the absence of sound records, the simplest approach would be to build a model to predict how long a project will take, given task duration and rework at typical levels for the organization. By comparing the time taken with no rework with that incorporating

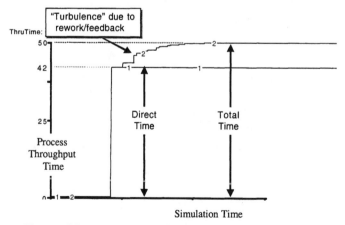

**Figure A.6** *Process time with, and without feedback, from simulation of Figure A.5.*

rework, a value for $p$ can be estimated. See Figure A.6, which shows the results from simulating the process model of Figure A.5, using the same figures in both cases. Note that the direct time is 42 units and the total time is 50 units, in both figure and graph. Additionally, the graph shows the turbulence due to a package of work being fed back for rework.

Shannon's [2] entropy theorem indicates that entropy of a probability distribution is the sum of the terms $p.\ln(p)$ for all discrete values of $p$ in the distribution. While not strictly applicable in basic form to our problem, we can nonetheless use[2] the Shannon entropy theorem to give us some idea of the entropy suggested by our model.

Figure A.7 shows the Shannon entropy relationship, where entropy corresponding to $0 < p < 0.37$ is scaled from $0 - 100\%$. The simulation above showed that the direct path time was 42 units, while the total time was 50 units, giving a value for $p = 1 - (42/50) = 0.16$. Using the Shannon entropy graph, this equates to about 80% of maximum disorder. The value in the example of 80% is high because the example used very high amounts of feedback. If a real project had values for activity rework of 0.2 (20%) there would, indeed, be major disturbance and turbulence.

Using the approach indicated above, it would be practicable and useful to gauge the overall entropy of projects, to identify and eradicate sources of disorder in process, and to assess the likely reduction in total time resulting from the change.

The value of this approach, as ever using entropy, is that it measures the whole, not just the parts, or just some aspect.

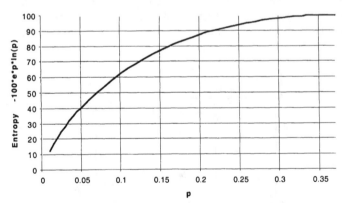

**Figure A.7** *Rework probability, p, versus entropy. The graph is derived from Shannon's entropy theorem.*

---

[2] Or, perhaps, misuse it? Using Shannon's entropy theorem in this manner might best be viewed as systems engineering convenience, rather than scientific insight.

## SUMMARY

Two practical and pragmatic methods of using entropy as measures of systems, and as a guide to the optimization of systems, have been presented. Although both are mathematically based, neither requires mathematics or significant mathematical knowledge in use.

---

## REFERENCES

[1] Hitchins, D. K., *Putting Systems to Work*, Chichester, England: Wiley, 1992, pp. 135 – 147.

[2] Shannon, C. E. "A Mathematical Theory of Communication," *Bell System Technical Journal*, Vol. 27, July and October 1948, pp. 379-423 and 623-656.

# Appendix B

## Set Theoretic Proof of Method: Rigorous Soft Method

The proof will be presented in two forms: diagrammatically, and using set theory notation.

### DIAGRAMMATIC REPRESENTATION

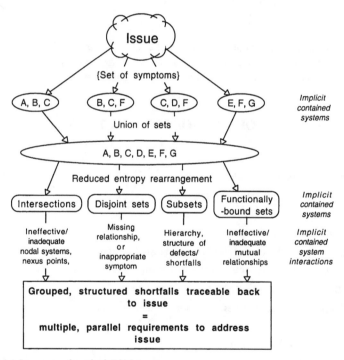

**Figure B.1.1** *Rigorous soft method (RSM) in sets.*

The rigorous soft method process is shown diagrammatically in Figure B.1, using set-theoretical terms:

- Symptoms imply sets of possible causes; implicit contained systems.

- These sets may be unified and rationalized to form a union of sets. (Although not shown in the figure, relationships between the implicit contained systems are retained throughout the process.)
- Implicit systems and their interactions form a pattern of interactions that is untangled to reduce its perceived entropy (using the $N^2$ chart), revealing implicit containing systems and nodes.
- These may occur in different form, according to the issue, and the effectiveness of the synthesis; RSM is largely self auditing.
- A failure to find sufficient symptoms or errors, such as collecting some symptoms that are irrelevant, or that refer to another issue, results in distinctive patterns.
- A disjoint set, for example, means that the $N^2$ chart has clustered into two or more unconnected parts; this may be caused by introducing an irrelevant symptom, or by making an omission or error in identifying causal relationships.

The output from the analysis is a set of requirements to address the original issue. Using the SEPP, these may be used to create and choose between optional conceptual design solutions to solve, resolve, or dissolve the issue.

## SET THEORY PROOF OF RIGOROUS SOFT METHOD

Issue $\Rightarrow$ {symptoms}

symptom $\Rightarrow$ {implicit contained systems}

(non-functional, one-to-many mapping)

$\therefore$ Issue $\Rightarrow$ {$ics$} $\cup$ {$ics$} $\cup$ {$ics$} = E

Within E there are intersections:

{$ics$} $\cap$ {$ics$} $\cap$ {$ics$} which constitute an issue nexus or node

Within E there may be subsets found by rearranging E:

$\alpha \subset E$, $\beta \subset E$ which constitute higher level systems

of problems, or problem themes: functional, many-to-one

mapping.

Within E there may be disjoint sets,

$\lambda \cap \mu = \varphi$, which result either from relationships not evident

in the analysis or owing to systems being identified at

different hierarchy levels or may arise from misidentification

of symptoms at the start.

where:

E is the issue space, the union of all implicit contained systems;
*ics* is an implicit contained system;
α, β, μ, and λ are subsets of E; and
φ is a null, or empty set.

# About the Author

**Professor Derek K. Hitchins** is now a part-time consultant, teacher, visiting professor, and international lecturer. Formerly, he held the British Aerospace (now BAE Systems) Chairs in Systems Science and in Command and Control, Cranfield University at RMCS Shrivenham. Prior to that he held the Chair in Engineering Management at City University, London.

Derek joined the Royal Air Force as an apprentice and retired as a wing commander after 22 years, to join industry. His first industry appointments were as the system design manager of the Tornado F3 Avionics, technical coordinator for UKAIR CCIS, and U.K. technical director for the NATO air command and control system (ACCS) project in Brussels. He subsequently held posts in two leading systems engineering companies as marketing director, business development director, and technical director before becoming an academic in 1988.

He was also the inaugural chairman of the Institution of Electrical Engineers (IEE) Professional Group on Systems Engineering, and the inaugural president of the U.K. chapter, International Council on Systems Engineering (INCOSE). INCOSE presented him with their prestigious Pioneer Award in 1998.

His current research is in system thinking, system requirements, social psychology and anthropology, Egyptology, command and control, system design and world-class systems engineering. He is a member of a select, international team developing advanced methods of modeling and predicting the behavior of large-scale systems, including socioeconomic systems, and conflicts. He published his first book, *Putting Systems to Work,** in 1992, and is presently writing a new book called *Exploding the Pyramid Myths.*

*Available as a free download from his website, www.hitchins.org/prof.

# Index

system
architecture, 24, 26, 29-30, 38, 92-93, 96, 176, 230-240, 253-254, 276, 412
belief, 68-70, 126, 170-171, 187, 189, 433
boundary, 12, 80, 125, 128, 257, 299, 316, 333, 383, 406, 408, 412, 424
closed, 4-5
complementary, 48, 119, 248, 265, 276, 280, 375, 445
complementary sets, 13, 113, 335, 446
control, 14, 35, 173, 350, 370-371, 393, 401, 411, 463
definition, 8, 26, 108, 159
directive control, 16
human activity, 4, 57, 376, 399
information management (IMS), 37-38
integration, 81-82
legacy, 81, 369-373
life cycle theory, 107, 115, 118-119, 120-121, 208, 427
operational maintenance, 378
organismic, 12, 58
people, 34, 138, 433, 436
principles, 26
process model, 16
system of systems, 80-81, 84-85, 87, 89, 334

system of systems, 84
systematic engineering, 311
systems engineering problem-solving paradigm (SEPP), 78-79, 130, 159, 315-316, 460
systolic cumulative selection, 280
tacit knowledge, 65, 72, 170-172, 263
teleology, 6
teleworking, 83
thermodynamics, second law, 5, 9, 41, 103-104
trade off, 291, 323, 355, 357
transport
general equations, 7
integrated, 81-83
model, 14, 20
truancy, 45, 142-143
U.N. peace operations, 214, 302-304, 308, 411-412
unit production cost, 23, 325, 383
United Nations (U.N.), 73, 86, 214, 301, 303, 411-412
value for money, 45, 47, 58-60, 78, 81-82, 281, 316, 410
von Bertalanffy, Ludvig, 4, 7, 42
von Clausewitz, Carl, 410, 428
weapon systems, 220, 270, 284, 337
weighting and scoring, 50, 60
world maps, 64-65
zaibatsu, 412-413, 427

# Recent Titles in the Artech House Technology Management and Professional Development Library

Bruce Elbert, Series Editor

*Advanced Systems Thinking, Engineering, and Management,* Derek K. Hitchins

*Critical Chain Project Management,* Lawrence P. Leach

*Decision Making for Technology Executives: Using Multiple Perspectives to Improve Performance,* Harold A. Linstone

*Designing the Networked Enterprise,* Igor Hawryszkiewycz

*Engineering and Technology Management Tools and Applications,* B. S. Dhillon

*The Entrepreneurial Engineer: Starting Your Own High-Tech Company,* R. Wayne Fields

*Evaluation of R&D Processes: Effectiveness Through Measurements,* Lynn W. Ellis

*From Engineer to Manager: Mastering the Transition,* B. Michael Aucoin

*Introduction to Information-Based High-Tech Services,* Eric Viardot

*Introduction to Innovation and Technology Transfer,* Ian Cooke and Paul Mayes

*ISO 9001:2000 Quality Management System Design,* Jay Schlickman

*Managing Complex Technical Projects: A Systems Engineering Approach,* R. Ian Faulconbridge and Michael J. Ryan

*Managing Engineers and Technical Employees: How to Attract, Motivate, and Retain Excellent People,* Douglas M. Soat

*Managing Successful High-Tech Product Introduction,* Brian P. Senese

For further information on these and other Artech House titles, including previously considered out-of-print books now available through our In-Print-Forever® (IPF®) program, contact:

Artech House
685 Canton Street
Norwood, MA 02062
Phone: 781-769-9750
Fax: 781-769-6334
e-mail: artech@artechhouse.com

Artech House
46 Gillingham Street
London SW1V 1AH UK
Phone: +44 (0)20 7596-8750
Fax: +44 (0)20 7630-0166
e-mail: artech-uk@artechhouse.com

Find us on the World Wide Web at:
www.artechhouse.com